Lecture Notes in Compu42

Commenced Publication in 1973
Founding and Former Series Editors:
Gerhard Goos, Juris Hartmanis, and Jan

Tansu Alpcan   Levente Buttyán
John S. Baras (Eds.)

Decision and Game Theory for Security

First International Conference, GameSec 2010
Berlin, Germany, November 22-23, 2010
Proceedings

 Springer

Volume Editors

Tansu Alpcan
Technical University
Berlin, Germany
E-mail: alpcan@sec.t-labs.tu-berlin.de

Levente Buttyán
Budapest University of Technology
and Economics
Budapest, Hungary
E-mail: buttyan@crysys.hu

John S. Baras
University of Maryland
College Park, MD, USA
E-mail: baras@umd.edu

Library of Congress Control Number: Applied for

CR Subject Classification (1998): E.3, C.2.0, D.4.6, K.4.4, K.6.5, H.2.0

LNCS Sublibrary: SL 4 – Security and Cryptology

ISSN	0302-9743
ISBN-10	3-642-17196-6 Springer Berlin Heidelberg New York
ISBN-13	978-3-642-17196-3 Springer Berlin Heidelberg New York

springer.com

© Springer-Verlag Berlin Heidelberg 2010
Printed in Germany

Typesetting: Camera-ready by author, data conversion by Scientific Publishing Services, Chennai, India
Printed on acid-free paper   06/3180

Preface

Securing complex and networked systems has become increasingly important as these systems play an indispensable role in modern life at the turn of the information age. Concurrently, security of ubiquitous communication, data, and computing poses novel research challenges. Security is a multi-faceted problem due to the complexity of underlying hardware, software, and network interdependencies as well as human and social factors. It involves decision making on multiple levels and multiple time scales, given the limited resources available to both malicious attackers and administrators defending networked systems. Decision and game theory provides a rich set of analytical methods and approaches to address various resource allocation and decision-making problems arising in security.

This edited volume contains the contributions presented at the inaugural Conference on Decision and Game Theory for Security - GameSec 2010. These 18 articles (12 full and 6 short papers) are thematically categorized into the following six sections:

- "Security investments and planning" contains two articles, which present optimization methods for (security) investments when facing adversaries.
- "Privacy and anonymity" has three articles discussing location privacy, online anonymity, and economic aspects of privacy.
- "Adversarial and robust control" contains three articles, which investigate security and robustness aspects of control in networks.
- "Network security and botnets" has four articles focusing on defensive strategies against botnets as well as detection of malicious adversaries in networks.
- "Authorization and authentication" has an article on password practices and another one presenting a game-theoretic authorization model.
- "Theory and algorithms for security" contains four articles on various theoretic and algorithmic aspects of security.

Considering that decision making for security is still a research topic in its infancy, we believe that this edited volume as well as the GameSec conference will be of interest to both researchers and students who work in this area and have diverse backgrounds.

November 2010

Tansu Alpcan
Levente Buttyán
John Baras

Organization

GameSec 2010, the inaugural Conference on Decision and Game Theory for Security, took place on the campus of Technical University of Berlin, Germany, during November 22–23, 2010. GameSec brings together researchers who aim to establish a theoretical foundation for making resource allocation decisions that balance available capabilities and perceived security risks in a principled manner. The conference focuses on analytical models based on game, information, communication, optimization, decision, and control theories that are applied to diverse security topics. At the same time, the connection between theoretical models and real-world security problems are emphasized to establish the important feedback loop between theory and practice. Given the scarcity of venues for researchers who try to develop a deeper theoretical understanding of the underlying incentive and resource allocation issues in security, GameSec aims to fill an important void and to serve as a distinguished forum.

Steering Committee

Tansu Alpcan	Technical University of Berlin & T-Labs., Germany
Nick Bambos	Stanford University, USA
Tamer Başar	University of Illinois at Urbana-Champaign, USA
Anthony Ephremides	University of Maryland, USA
Jean-Pierre Hubaux	EPFL, Switzerland

Program Committee

General Chair	*Tansu Alpcan*	Technical University of Berlin/ T-Labs
TPC Chairs	*John Baras*	University of Maryland
	Levente Buttyán	Budapest University of Technology and Economics
Publicity Chairs	*Albert Levi*	Sabanci University, Istanbul
	Zhu Han	University of Houston
Publication Chair	*Holger Boche*	Technology University of Berlin / HHI
Finance Chair	*Slawomir Stanczak*	Technology University of Berlin / HHI

| Local Chair | *Jean-Pierre Seifert* | Technology Universityof Berlin / T-Labs |
| Secretary | *Christine Kluge* | Technology University of Berlin / T-Labs |

Sponsoring Institutions

Industry Sponsors

Gold Sponsor: *Deutsche Telekom Laboratories* (T-Labs)
Silver Sponsor: *Fraunhofer Heinrich Hertz Institute* (HHI)

Technical Co-sponsors

IEEE Control System Society
International Society of Dynamic Games (ISDG)
In-cooperation with *ACM Special Interest Group on Security, Audit and Control* (SIGSAC)
Co-sponsored by the *IEEE Multimedia Communication Technical Committee*

Technical Program Committee

Imad Aad	Nokia Research, Switzerland
Eitan Altman	INRIA, France
Sonja Buchegger	KTH, Sweden
Mario Cagalj	University of Split, Croatia
Srdjan Capkun	ETH Zurich, Switzerland
Lin Chen	University of Paris-Sud 11, France
John Chuang	UC Berkeley, USA
Sajal K. Das	University of Texas at Austin, USA
Merouane Debbah	Supelec, France
Mark Felegyhazi	ICSI-Berkeley, USA
Jens Grossklags	Princeton University, USA
Are Hjorungnes	University of Oslo, Norway
Eduard A. Jorswieck	Technical University Dresden, Germany
Iordanis Koutsopoulos	University of Thessaly, Greece
Jean Leneutre	Telecom ParisTech, France
Xiang-Yang Li	Illinois Institute of Technology, USA
Li (Erran) Li	Bell Labs., USA
M. Hossein Manshaei	EPFL, Switzerland
Pietro Michiardi	EURECOM, France
John Mitchell	Stanford University, USA
Refik Molva	EURECOM, France
Pierre Moulin	University of Illinois at UC, USA
Ariel Orda	Technion, Israel
David C. Parkes	Harvard University, USA
George C. Polyzos	AUEB, Greece

Table of Contents

Authorization and Authentication

Theory and Algorithms for Security

Design of Network Topology in an Adversarial Environment

Assane Gueye, Jean C. Walrand, and Venkat Anantharam

University of California at Berkeley, EECS Department, Berkeley CA 94720, USA
{agueye,wlr,ananth}@eecs.berkeley.edu

Abstract. We study the *strategic* interaction between a network manager whose goal is to choose (as communication infrastructure) a spanning tree of a network given as an undirected graph, and an attacker who is capable of attacking a link in the network. We model their interaction as a zero-sum game and discuss a particular set of Nash equilibria. More specifically, we show that there always exists a Nash equilibrium under which the attacker targets a *critical* set of links. A set of links is called *critical* if it has maximum vulnerability, and the *vulnerability* of a set of links is defined as the minimum fraction of links the set has in common with a spanning tree. Using simple examples, we discuss the importance of critical subsets in the design of networks that are aimed to be robust against attackers. Finally, an algorithm is provided, to compute a critical subset of a given graph.

Keywords: Network Topology, Connectivity, Graph Vulnerability, Spanning Trees, Minimum Cut Set, Game Theory, Nash Equilibrium, Linear Programming, Blocking pairs of polyhedra, Polymatroid, Network Flow Algorithm.

1 Introduction

In this work, we aim to study the *strategic* interaction between a network manager whose goal is to choose a spanning tree of the network as communication infrastructure, and an attacker who tries to disrupt the communication tree by attacking one link in the network.

The network topology is given as a connected undirected graph. The formulation of the problem extends naturally to the manager choosing a k-connected component and the attacker selecting $k' \geq k$ links to attack. For example, if $k = 2$, this problem models the situation where the manager is choosing a primary communication tree and a backup tree in the presence of an attacker who can attack more than 2 links in the network. The discussion in the present paper, however, focuses attention only on the case $k = k' = 1$.

In general, each tree has a given cost which is the loss seen by the manager when one of the edges of that tree is attacked. This cost (or a function of it) goes to the attacker. Also, it is conceivable that the attacker incurs some cost by attacking a link. The goal of the network manager is to minimize the cost of attack while the attacker is trying to maximize the net attack reward.

T. Alpcan, L. Buttyan, and J. Baras (Eds.): GameSec 2010, LNCS 6442, pp. 1–20, 2010.

In a non-adversarial environment, choosing a minimum cost spanning tree (MST) of the graph would be optimal for the network manager. Algorithms for calculating the MST have been studied extensively since the work of Kruskal [15] and Prim [20].

In this paper, we will assume that every tree has equal cost. It is also assumed that the cost of attacking any given link is zero for the attacker. These assumptions will be relaxed in subsequent studies of the problem.

The communication networks community has spent a lot of effort studying the reliability/robustness of networks. The interested reader is referred to [21] and [12] and the references therein. Robustness has mostly been considered against *non-strategic* phenomena (e.g. random failures). However, network disruption can also be due to malicious attackers. The nature of the attack can be varied. In an *availability* attack, the attacker might be launching a denial of service (DoS) attack on some node/link, or simply jam a communication channel. In a *confidentiality* attack, the attacker could be choosing a link and observe/analyze the traffic that it carries. An *integrity* attack could also be launched, where the attacker will try to modify the traffic (or generate traffic) for a target link/node.

These problems have received a lot of attention specially in the area of mobile and *ad-hoc* networks [5], [1] and mostly in a non-strategic framework. In an environment where the adversary is cognitive, most of the results found in the literature do not apply any more. For example, in the graph connectivity problem considered here, when the attacker strategically chooses the edge to attack, it is no longer obvious how the network manager should choose a spanning tree. For example, if the network manager were to always choose a specific MST, the attacker could compute this MST and attack one of its links to disconnect the network.

To understand how the network manager should choose a spanning component as well as how an attacker could break the communication, we model their interaction as a game where:

- the manager's strategy set is the set of spanning trees of the graph;
- the attacker's strategy set is the set of links;
- the goal of the manager is to minimize the average cost of getting attacked while the goal of the attacker is to maximize that cost.

We assume that the network topology is known to both players. All trees have the same cost, and there is no cost of attack. We would like to understand the structure of the (or at least some) set of Nash equilibria of this one-shot, zero-sum game.

The organization of this paper is as follows. In the next section, we present the model of the game considered in this paper. The notion of critical subset is discussed in subsection 2.1, followed by illustrative examples in subsection 2.2. The main result of the paper (the critical subset attack theorem) is presented in subsection 2.3 and a brief discussion of this result is provided in subsection 2.4. A proof of the theorem is provided in section 3. This proof requires the notions of *blocking pairs of polyhedra* and a characterization of the *spanning tree polyhedra*. A tutorial presenting those notions is given in appendix A. Section 4 presents an algorithm to compute a critical subset of a graph. The algorithm is essentially based on the theory of *polymatroids* which we discuss in section B. Concluding remarks and directions for future work are given in section 5.

2 The Game

The network topology is given by a connected undirected graph $G = (\mathcal{V}, \mathcal{E})$ with $|\mathcal{E}| = m$ links and $|\mathcal{V}| = n$ nodes. Let $\mathcal{T}$ be the set of spanning trees, and let N denote $|\mathcal{T}|$.

We consider the 2-player, zero-sum game where player 1 (the network manager) chooses a spanning tree according to some distribution on $\mathcal{T}$ to minimize the probability (which, for equal unit cost of trees, corresponds to the cost) that the spanning tree is disrupted. Player 2 (the attacker) chooses a link to attack according to some distribution on $\mathcal{E}$ to maximize this probability. A tree gets "disconnected" if the attacked link belongs to it. We aim to analyze the set of Nash equilibria of this game.

More precisely, let $\mathcal{A} := \{\alpha \in \Re_+^N \mid \sum_{T \in \mathcal{T}} \alpha_T = 1\}$ be the set of mixed strategies of the network manager and $\mathcal{B} := \{\beta \in \Re_+^m \mid \sum_{e \in \mathcal{E}} \beta_e = 1\}$ the set of mixed strategies of the attacker.

The manager wants to minimize and the attacker wants to maximize $C(\alpha, \beta)$ where

$$C(\alpha, \beta) = \sum_{e \in E} \sum_{T \in \mathcal{T}} \alpha_T \beta_e 1\{e \in T\}. \tag{1}$$

2.1 Critical Set

We first characterize some subsets of edges as being most vulnerable to attack.

Definition 1 (Critical Set). *For any nonempty subset of edges $E \subseteq \mathcal{E}$, define*

$$\mathcal{M}(E) := \min_{T \in \mathcal{T}} |T \cap E| \ and \ \vartheta(E) := \frac{\mathcal{M}(E)}{|E|}. \tag{2}$$

We call $\vartheta(E)$ the vulnerability *of E. It is the minimum fraction of links the set E has in common with a spanning tree. A nonempty subset E of edges is said to be* critical *if*

$$\vartheta(E) = max_{E' \subseteq \mathcal{E}} \{\vartheta(E')\}. \tag{3}$$

In other words, a subset of links is critical *if it has maximum* vulnerability. *The vulnerability of a graph G is defined as the vulnerability of its critical subset(s), and is denoted $\vartheta(G)$.*

For each $E \subseteq \mathcal{E}$ we define $\mathcal{T}_E \subseteq \mathcal{T}$ by:

$$T \in \mathcal{T}_E \Longleftrightarrow |T \cap E| = \mathcal{M}(E). \tag{4}$$

We will call any $T \in \mathcal{T}_E$ an E-minimal spanning tree.

Our notion of graph vulnerability is related to a notion which has previously been proposed in the graph theory literature (see [13], [8], [3]). However, to the authors's knowledge it seems to have not received a lot of attention. We briefly discuss those references in section 2.4.

2.2   Examples of Critical Sets

Let us illustrate the definitions with some examples, shown in Fig.1. For the network in Fig.1(a), all spanning trees must go through the middle link (called a *bridge*), so that $\vartheta(E) = 1$ if E is the set with only that link. That set is critical and the attacker can attack it and achieves the maximum cost of one.

In general, an edge that must be part of every spanning tree is called a bridge. Also, it is not difficult to verify that the vulnerability of a subset E is equal to the maximum value of 1 if and only if E is only composed of bridges.

The graph in Fig.1(b) contains 8 nodes and 14 links. It has one minimum cut set composed of the links 6 and 8. If $E = \{6, 8\}$, then any spanning tree contains at least one link in E. Thus, $|T \cap E| \geq 1$ for any tree T. Furthermore, there exists T such that $T \cap E = \{6\}$. Thus, $\mathcal{M}(E) = 1$, giving a vulnerability of $\vartheta(E) = 1/2$. This is the maximum vulnerability of this graph (verification is left as an exercise for the interested reader), which implies that $E = \{6, 8\}$ is a critical subset. If we consider the set of all links $E = \mathcal{E}$, then $|T \cap E| = n - 1 = 7$ for any tree T because any spanning tree contains $n - 1$ links. This set is also critical because $\vartheta(E) = \frac{7}{14} = 1/2$.

In general, there might be many critical subsets for a given graph. For instance, in Fig.1(b), as shown above, $E = \{1, 2, 3, 4, 5, 6, 7, 8\}$ is another critical subset. If $E = \{1, 2, 4\}$, choosing $T = \{3, 6, 7, 8, 9, 13, 14\}$ gives $T \cap E = \emptyset$. Hence, $\mathcal{M}(E) = 0$.

The minimum cut set of a graph is not always critical. In Fig.1(c) if $E = \{6, 8\}$ then $\vartheta(E) = 1/2$. However choosing $E = \{6, 8, 9, 10, 11, 12, 13\}$ gives $\vartheta(E) = 4/7 > 1/2$. One can show that $E = \{6, 8, 9, 10, 11, 12, 13\}$ is critical but $E = \{6, 8\}$ is not.

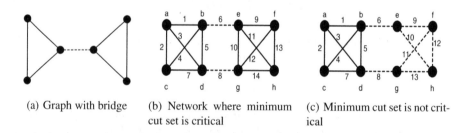

(a) Graph with bridge   (b) Network where minimum   (c) Minimum cut set is not crit-
                              cut set is critical   ical

Fig. 1. Illustrative network examples. Example 1(a) is a network that contains a bridge (dotted link). A bridge is always a critical set. The network in 1(b) is an example of graph where the minimum cut set (dashed links) corresponds to a critical subset. Example 1(c) shows a graph where the minimum cut set is not critical.

2.3   Critical Subset Attack

Next we give the structure of one particular class of Nash equilibria (NE) of the game defined above. First, we let

$$\alpha(e) := \sum_{T \in \mathcal{T}} \alpha_T 1\{e \in T\}, \text{ for } e \in \mathcal{E}. \tag{5}$$

Theorem 1 (Critical Subset Attack Theorem). *For each critical subset of edges, E, there exists a NE under which the attacker* uniformly and exclusively *targets the edges of the critical subset E and the network manager chooses only trees inside the set of E-minimal spanning trees. Specifically, the strategy of the attacker is*

$$\beta_e = \frac{\mathbf{1}_{e \in E}}{|E|}, \tag{6}$$

and the strategy of the manager is $\alpha \in \mathcal{A}$ *such that*

$$\begin{cases} \alpha_T \geq 0 \text{ if } T \in \mathcal{T}_E \\ \alpha_T = 0 \text{ otherwise} \end{cases} \tag{7}$$

$$\alpha(e) := \sum_{T \in \mathcal{T}} \alpha_T \mathbf{1}\{e \in T\} = \vartheta(E), \forall e \in E \tag{8}$$

$$\alpha(f) \leq \vartheta(E), \forall f \notin E. \tag{9}$$

The corresponding optimal payoff is equal to $\vartheta(E)$.

A proof of the theorem is provided in section 3.

2.4 Comments

A certain number of remarks are to be made about the previous result.

- The equilibrium strategy for the network α is such that each element of its support ($\mathcal{T}_E$) meets the critical set in the minimum number of links. Furthermore, the sum ($\alpha(e)$) of the probability assigned to the trees crossing each link $e \in E$ is the same for all links in the critical subset. This sum is equal to the vulnerability of the subset E.
- As we have seen in the examples of the previous section, a graph has in general many critical subsets. As a consequence, there might be many NE (each with a different α and β). There might even exist other Nash equilibria than the ones isolated above. However, because the game is zero-sum, all equilibria have the same payoff [24]. As a consequence, it is reasonable to use the terminology "vulnerability of a graph" for $\vartheta(G)$, defined earlier as the vulnerability of any critical subset of its links.
- Theorem 1 implies that every critical subset supports some Nash equilibrium (for instance the critical subset attack equilibrium).
- Knowing the critical subsets (the weakest points of the network) is important for the network manager. The example in Fig.2 is an illustration. Consider the network in Fig.2(a) whose vulnerability is equal to $\frac{3}{4}$. In all these figures, the critical subset is represented by the dashed edges. Suppose that the network manager has an extra link to add to this network and would like to know the optimal way to add this link. If the additional link is put in the position as in Fig.2(b), then the vulnerability of the graph becomes $\frac{3}{5} < \frac{3}{4}$ (the graph is always less vulnerable with an additional link). If instead the link is added as in Fig.2(c), the vulnerability of the graph is $\frac{2}{3} > \frac{3}{5}$ leading to a less robust network.

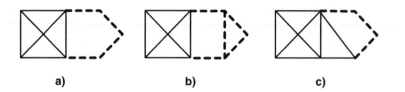

Fig. 2. Critical subset and topology design. Graphs (b) and (c) are two different ways of adding a link to graph (a) which have a vulnerability of 3/4. If it is added as in (b), then the vulnerability is $\frac{3}{5}$. If it is done as in (c), the vulnerability is $\frac{2}{3} > \frac{3}{5}$, which is leads to a less robust network.

– As was mentioned in section 2, the notion of graph vulnerability considered in this paper has been previously (with some differences) defined in a related but slightly different context. In [13], Gusfield discussed the consequences of Tutte [23] and Nash-Williams' theorem [19] and was particularly interested in the maximum number (M) of edge-disjoint spanning trees of a graph G. Two spanning trees of G are called disjoint if they have no edge in common.
Gusfield showed that

$$M = \min_{E \subseteq \mathcal{E}} \lfloor \frac{|E|}{Q(G_{\bar{E}}) - 1} \rfloor , \qquad (10)$$

where $G_{\bar{E}}$ is the graph resulting from deleting the edges in E from G, and $Q(G_{\bar{E}})$ is the number of connected components in $G_{\bar{E}}$. $\bar{E}$ denotes the complement of E in $\mathcal{E}$.

The quantity $\sigma(G) = \min_{E \subseteq \mathcal{E}} \left(\frac{|E|}{Q(G_{\bar{E}})-1} \right)$ was then used as a measure of the *invulnerability* of the graph, i.e. the smaller this is the more vulnerable the graph is, in the sense of Gusfield. In that paper, any minimizing set for this quantity was interpreted as *a set of edges whose removal from G maximizes the number of additional components created, per edge removed*. The main question that was asked in that paper was whether there exists a polynomial time algorithm to compute $\sigma(G)$.

Cunningham provided such an algorithm in [8]. Considering $\sigma(G)$ as the *strength* of G, he defined (in a non-game theoretic setting) an *optimal attack* problem as well as a *network reinforcement* problem. The optimal attack problem consists of computing the strength of G and determining a minimizing set. Cunningham considered edge-weighted graphs, with edge j having strength s_j; the strength of the graph is defined as $\sigma(G) = \min_{E \subseteq \mathcal{E}} \left(\frac{\sum_{j \in E} s_j}{Q(G_{\bar{E}})-1} \right)$, which corresponds to the invulnerability defined by Gusfield when $s_j = 1$ for all $j \in \mathcal{E}$. The network reinforcement problem of [8] is related to minimizing the cost of increasing the strengths of individual edges in order to achieve a target strength for the graph. For details, see [8].

Using *polymatroid theory* and *network flow analysis*, Cunningham provided polynomial time algorithmic solutions to both problems. In section 4, we discuss this algorithm in the context of the present paper.

A more recent paper by Catlin *et al.* [3] generalizes Gusfield's notion of invulnerability by imposing bounds on the number of connected components, $Q(G_{\bar{E}})$.

In the present paper, the critical subsets, in our sense, have been found to correspond to Nash equilibria of a zero-sum game. It is to be noticed that our definition of vulnerability verifies $\vartheta(G) = \sigma(G)^{-1}$. To see that, one needs to show that,

Lemma 1. *For any $E \subseteq \mathcal{E}$,*

$$\mathcal{M}(E) = Q(G_{\bar{E}}) - 1. \tag{11}$$

Proof. The ideas in the proof is as follows. Consider the different connected components of the graph when the edges in E are removed. Any spanning tree of the original graph has to connected those components, and this connection is done by only using edges in E. Since there are $Q(G_{\bar{E}})$ connected components, one needs *exactly* $Q(G_{\bar{E}})$-1 to connect them in a cycle-free way. A complete proof is given in [11].

It is interesting to note that, despite the fact that this metric ($\sigma(G)$) is more refined than the *edge connectivity* (i.e. size of minimum cut set), it has largely not been used in the graph theory community. One reason suggested by Gusfield is the complexity of its computation. As was stated earlier, Cunningham [8] has subsequently provided a polynomial time algorithm to compute $\sigma(G)$ as well as a minimizing subset.

Our result shows that, in a environment where the adversary is cognitive, $\vartheta(G)$ is indeed an appropriate metric of graph vulnerability.

From the discussion above, we can, by using Cunningham's algorithm, compute a critical set of a given graph. We present the details of the algorithm in section 4.

3   Proof of the Critical Subset Attack Theorem

In this section we present a proof of the critical subset attack theorem. The proof is done in two parts. In the first part (section 3.1), we show that the strategy pair given in Theorem 1 forms a pair of best responses to each other. In the second part of the proof (section 3.2), we show that for any critical subset, there indeed exists a probability distribution α that satisfies conditions (7-9).

3.1   Best Responses

Let (α, β) be a strategy pair. Observe that the attack cost is given by

$$C(\alpha, \beta) = \sum_{e \in E} \sum_{T \in \mathcal{T}} \alpha_T \beta_e 1\{e \in T\} = \sum_{e \in E} \beta_e \alpha(e). \tag{12}$$

Let $E \subseteq \mathcal{E}$ be a critical subset and assume that α satisfies the conditions (7-9). Then, any distribution β concentrated on E achieves the cost $\vartheta(E)$. This is the maximum possible cost achievable for the attacker. To see this, observe that for any β,

$$C(\alpha, \beta) = \sum_f \beta_f \alpha(f) \leq \sum_f \beta_f \vartheta(E) \leq \vartheta(E). \tag{13}$$

Now assume that β is uniform on a critical set E. Then the distribution α achieves the cost $\vartheta(E)$. This is the minimum possible cost. To see this, note that, for any α,

$$C(\alpha,\beta) = \frac{1}{|E|} \sum_{e \in E} \sum_T \alpha_T 1\{e \in T\} = \frac{1}{|E|} \sum_T \alpha_T |T \cap E| \geq \frac{1}{|E|} \sum_T \alpha_T \mathcal{M}(E) = \vartheta(E),$$

$$(14)$$

where the next-to-last inequality uses the fact that $|T \cap E| \geq \mathcal{M}(E)$ for all T.

3.2   Existence of the Equilibrium Distribution

The claim is that one can find $\alpha \in \mathcal{A}$ that satisfies (7-9). To prove that fact, we formulate an optimization problem and we show in Theorem 2 that the solution is the desired α.

Let A be the edge-tree incidence matrix with $A(f,T) = 1\{f \in T\}$ for $f \in \mathcal{E}$ and $T \in \mathcal{T}$. The *spanning tree polyhedron* $\mathcal{P}$ is defined as the vector sum of the convex hull of the columns of A and the nonnegative orthant $\mathcal{R}_+^m$ (see appendix A.2, and references [10], [7]). It is known [4] that

$$\mathcal{P} = \{\mathbf{x} \in R_+^m \mid \mathbf{x}(\mathcal{E}(P)) \geq |P| - 1, \text{ for all feasible partitions } P = \{V_1, V_2, \ldots, V_{|P|}\}\}.$$

$$(15)$$

In (15), $P = \{V_1, V_2, \ldots, V_{|P|}\}$ is *feasible* if each V_i induces a connected subgraph $G(V_i)$ of G (see appendix A). $|P|$ is the size of the partition. The notation $\mathbf{x}(\mathcal{E}(P))$ is defined as $\mathbf{x}(\mathcal{E}(P)) := \sum_{i \in \mathcal{E}(P)} x_i$, where $\mathcal{E}(P)$ is the set of all edges of G having endpoints in different members of the partition.

Theorem 2. *Let E be a critical subset of edges. Let $\mathbf{x}^* \in R^N$ be the solution of the following problem:*

$$\text{Maximize } \mathbf{1}'\mathbf{x}$$
$$\text{subject to } A\mathbf{x} \leq \vartheta(E)\mathbf{1}, \mathbf{x} \geq \mathbf{0}. \tag{16}$$

Then
   a) $\mathbf{1}'\mathbf{x}^* \leq 1$;
   b) $\mathbf{1}'\mathbf{x}^* \geq 1$;
   c) $A\mathbf{x}^*(e) = \vartheta(E), \forall e \in E$.
As a consequence, $\alpha = \mathbf{x}^*$ satisfies (7)-(9).

Proof. **a)** Let $\mathbf{w}(f) = 1\{f \in E\}$ for $f \in \mathcal{E}$. Note that $A'\mathbf{w} \geq \mathcal{M}(E)\mathbf{1}$, by definition of $\mathcal{M}(E)$. Hence, for all $\mathbf{x} \in R^N$ satisfying (16),

$$\mathbf{1}'\mathbf{x} \leq \mathcal{M}(E)^{-1}\mathbf{w}'A\mathbf{x} \leq \mathcal{M}(E)^{-1}\mathbf{w}'\vartheta(E)\mathbf{1} = 1, \tag{17}$$

since $\mathbf{w}'\mathbf{1} = |E|$.
b) The dual of the program (see [2]) is

$$\text{Minimize } \vartheta(E)\mathbf{y}'\mathbf{1}$$
$$\text{subject to } A'\mathbf{y} \geq \mathbf{1}, \mathbf{y} \geq \mathbf{0}.$$

The constraints of the dual program define the following polyhedron

$$\hat{\mathcal{P}} = \left\{ \mathbf{y} \in R_+^m, \ \text{s.t } A'\mathbf{y} \geq 1 \right\} . \tag{18}$$

By results of linear programming (strong duality [2]), the value of the dual program is identical to that of the original program. Now we would like to show that the value of the dual program is at least 1, i.e. $\vartheta(E)\mathbf{y}'\mathbf{1} \geq 1$ for all $\mathbf{y} \in \hat{\mathcal{P}}$.

An equivalent way of saying this is that $\boldsymbol{\gamma} := \vartheta(E)\mathbf{1}$ belongs to the set

$$b(\hat{\mathcal{P}}) = \left\{ \mathbf{z} \in R_+^m, \ \text{s.t } \mathbf{z} \cdot \hat{\mathcal{P}} \geq 1 \right\} , \tag{19}$$

where $\mathbf{z} \cdot \hat{\mathcal{P}}$ defines the inner product of $\mathbf{z}$ with any vector in $\hat{\mathcal{P}}$.

According to standard terminology (see Fulkerson [10, pg. 171] or Chopra [4]), this set is called the *blocker* of the polyhedron $\hat{\mathcal{P}}$. Since A is defined as (the transpose of) the incidence matrix of the spanning trees, $\hat{\mathcal{P}}$ in (18) is also the blocker of the spanning tree polyhedron $\mathcal{P}$ [4]. From the theory of blocking pairs of polyhedra (see appendix A), we have: if $\mathcal{B}$ is a polyhedron and $b(\mathcal{B})$ its blocker, then $b\,(b(\mathcal{B})) = \mathcal{B}$. ($\mathcal{B}$ and $b(\mathcal{B})$ are said to form a blocking pair of polyhedra.)

Thus, since $\hat{\mathcal{P}}$ is the blocker of $\mathcal{P}$, $b(\hat{\mathcal{P}}) = \mathcal{P}$. Now, $\mathbf{y}'\boldsymbol{\gamma} \geq 1$ for all $\mathbf{y} \in \hat{\mathcal{P}}$ is equivalent to saying that $\boldsymbol{\gamma} \in b(\hat{\mathcal{P}}) = \mathcal{P}$. From (15), this means

$$\gamma\left(\mathcal{E}(P)\right) \geq |P| - 1 \tag{20}$$

for all feasible partitions P, $\mathcal{E}(P) \subseteq \mathcal{E}$.

Now assume that this is not the case, i.e $\gamma\left(\mathcal{E}(P)\right) < |P| - 1$ for some P. Then

$$\sum_{i \in \mathcal{E}(P)} \gamma_i = \frac{\mathcal{M}(E)}{|E|} \sum_{i \in \mathcal{E}(P)} 1 = \frac{\mathcal{M}(E)}{|E|}|\mathcal{E}(P)| < |P| - 1, \tag{21}$$

which implies that

$$\frac{\mathcal{M}(E)}{|E|} < \frac{P - 1}{|\mathcal{E}(P)|}. \tag{22}$$

This means that $\mathcal{E}(P)$ is more vulnerable than E. Indeed, $|P| - 1$ is the minimum number of edges in common with $\mathcal{E}(P)$ that a spanning tree of G has.

Now, since the value of the dual program is at least 1, and the value of the primal program is at most 1, we can conclude that the value of the primal problem is one.

c) Note that, $\mathbf{1}'\mathbf{x}^* = 1$ and (17) imply

$$\mathbf{w}'A\mathbf{x}^* = \mathcal{M}(E). \tag{23}$$

By (16), we have $A\mathbf{x}^*(e) \leq \vartheta(E)$ for all $e \in \mathcal{E}$. Thus $\mathbf{w}'A\mathbf{x}^* \leq \mathcal{M}(E)$, but then by (23) we also have $A\mathbf{x}^*(e) = \vartheta(E)$ for all $e \in E$. Finally, we see that if $x^*(T) > 0$ for any $T \notin \mathcal{T}_E$, we would have $\mathbf{w}'A\mathbf{x}^* > \mathcal{M}(E)$, contradicting (23).

4   An Algorithm to Compute a Critical Subset

In this section, we present an algorithm to compute the vulnerability of a graph $\vartheta(G) = max_{E \subseteq \mathcal{E}} \{\vartheta(E)\}$, as well as a maximizing subset (i.e. a critical subset). The algorithm was first presented by Cunningham in [8]. For the sake of completeness, we discuss it here, and adapt it to the context of this paper. A summary of the steps of the algorithm is presented in section 4.1, and its details are discussed in section 4.2.

The discussion of section 4.2 needs the notions of *matroid* and *polymatroid*, which we present in appendix B.1.

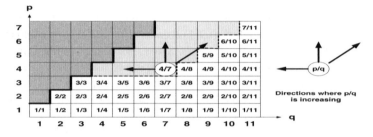

Fig. 3. An illustration of the 2-dimensional search algorithm to find the vulnerability of a graph. The dark (blue) region consists of p and q verifying $p/q > 1$. Since $\vartheta(G) \leq 1$, those values do not need to be tested. The light (blue) consist of values of p and q such that $\frac{p}{q} > \frac{p_0}{q_0}$ (here $\frac{p_0}{q_0} = \frac{4}{7}$). If $\vartheta(G) < \frac{p_0}{q_0}$, then, those values can be discarded from the test. The remaining (uncolored) values are the only ones that need to be tested.

4.1   Summary of the Algorithm

Observing that $\frac{\mathcal{M}(E)}{|E|}$ takes values in a finite set ($0 \leq \mathcal{M}(E) \leq |\mathcal{V} - 1|$ and $1 \leq |E| \leq |\mathcal{E}|$), we propose a *binary search* algorithm to find $\vartheta(G)$ *(BinarySearch2D)*. This requires an *oracle* to perform the test $\vartheta(G) \leq \frac{p}{q}$ for some p and q, $1 \leq p \leq |\mathcal{V} - 1|$ and $1 \leq q \leq |\mathcal{E}|$.

We show that such an oracle is equivalent to one that solves a minimization problem on the subsets of $\mathcal{E}$.

Solving this optimization problem will be further shown to be the same as finding a P-*basis* of some properly defined polymatroid P (those notions are presented in appendix B.1). A simple, greedy algorithm *(CunninghamMin)* will be used to find such P-basis.

The greedy algorithm will successively visit the edges of the graph, and for each edge, solve a minimization problem that is related to the first one. This last minimization can be solved by running a network flow algorithm on an appropriately defined graph.

4.2   Details of the Algorithm

In the process of computing the quantity $\vartheta(G) = max_{E \subseteq \mathcal{E}} \left(\frac{\mathcal{M}(E)}{|E|} \right)$, we first notice that, if there exists an oracle to test whether $\vartheta(G) \leq \frac{p}{q}$, then one will be able to compute $\vartheta(G)$ using an efficient search algorithm. Indeed, the values of p and q for which one needs to test are in a finite range. We illustrate this 2-dimensional search in Figure 3. Details of the algorithm will be discussed later.

Related to the test $\vartheta(G) \leq \frac{p}{q}$, we define the following problem (that Cunningham calls the *optimal attack* problem)

$$\text{minimize} \left(\frac{p}{q}|E| - \mathcal{M}(E) \right) , \tag{24}$$

where the minimization is carried out over all subsets of edges $E \subseteq \mathcal{E}$, and p and q are given numbers. The next lemma shows an equivalence between testing $\vartheta(G) \leq \frac{p}{q}$ and verifying whether the minimum in (24) is greater than or equal to zero.

Lemma 2. *For fixed values of p and q (define $\rho := \frac{p}{q}$), we have*

$$\vartheta(G) \leq \rho \;\Leftrightarrow\; 0 \leq \min_{E \subseteq \mathcal{E}} \left(\rho|E| - \mathcal{M}(E) \right) . \tag{25}$$

Proof. The proof of the lemma is as follows:

$$\vartheta(G) \leq \rho \Leftrightarrow \max_{E \subseteq \mathcal{E}} \left(\frac{\mathcal{M}(E)}{|E|} \right) \leq \rho \Leftrightarrow \frac{\mathcal{M}(E)}{|E|} \leq \rho, \; \forall \; E \subseteq \mathcal{E}$$

$$\Leftrightarrow 0 \leq \rho|E| - \mathcal{M}(E), \; \forall \; E \subseteq \mathcal{E}$$

$$\Leftrightarrow 0 \leq \min_{E \subseteq \mathcal{E}} \left(\rho|E| - \mathcal{M}(E) \right).$$

Now, we show how, by Lemma 1, we can rewrite the minimization using a function on subsets of the edges of the graph G. More precisely, we define $f(\cdot)$ such that $f(E) = |\mathcal{V}| - Q(G_E)$, where $Q(G_E)$ is the number connected components of the subgraph $G_E = (\mathcal{V}, E)$, that only contains the edges in E (in the terminology of Appendix B, $f(\cdot)$ is the rank function of the graphic matroid associated with G).

By definition of $f(\cdot)$, $f(\bar{E}) = |\mathcal{V}| - Q(G_{\bar{E}})$, where $\bar{E}$, denotes the complement of set the E. Using Lemma 1, we can write $\mathcal{M}(E) = |\mathcal{V}| - 1 - f(\bar{E})$.

The minimization in (24) can now be written as

$$\text{minimize}_{E \subseteq \mathcal{E}} \left(\left(\rho|E| + f(\bar{E}) \right) - \left(|\mathcal{V}| - 1 \right) \right) . \tag{26}$$

Thus, we can conclude that testing whether $\vartheta(G) \leq \rho$ is equivalent to testing

$$|\mathcal{V}| - 1 \leq \min_{E \subseteq \mathcal{E}} \left(\rho|E| + f(\bar{E}) \right) . \tag{27}$$

Since $f(\cdot)$ is the rank function of a matroid, it satisfies the hypothesis of Theorem 3 of appendix B. Using that theorem, the minimum in the RHS is achieved at an $P(f)$-basis of the vector $\rho\mathbf{1} \in R_+^{|\mathcal{E}|}$, where $P(f)$ is the *polymatroid* associated with $f(\cdot)$ (see appendix B.1). Thus, any oracle that computes a $P(f)$-basis for the polymatroid will suffice to compute a minimizer of (27) (and the minimum). Using such an oracle, we can now implement the following search algorithm that computes $\vartheta(G)$, as well as a critical set which is the minimizer provided by the algorithm when it terminates.

The search algorithm (summarized in Table 1) keeps a set of candidate values Pr for p, and for each $p \in Pr$, a range $\{q_{min}(p), \ldots, |\mathcal{E}|\}$ of values of q for which the test in (27) will be carried out.

Table 1. Left: Pseudocode of the *BinarySearch2D* algorithm to compute the vulnerability $\vartheta(G)$ of a graph and a critical subset. The algorithm *CunninghamMin* is discussed in Appendix B. The *update* method is presented in the right Table. **Right**: Pseudocode of the *Update* method.

BinarySearch2D

Input: connected graph $G = (\mathcal{V}, \mathcal{E}), \mathcal{V} = n, \mathcal{E} = m$

Output: $\vartheta(G)$ of G, $E \subseteq \mathcal{E}$ critical

```
 1 begin
 2    Pr = {1,2,...,n-1}
 3    qmin = {1,2,...,n-1}
 4    while |Pr|>0
 5       p <-- random(Pr)
 6       for q=m downto qmin(p)
 7          (E,minpq) = CunninghamMin((p/q)*1,G)
 8          if n-1 <= minpq then
 9             (Pr,qmin) = update(Pr,p,q)
10             goto 4
11          end    //if
12       end       //for
13       Pr = Pr-p
14    end           //while
15    return E, minpq
16 end               // begin
```

Update

Input: $Pr, p \in Pr, q \in \{q_{min}, |\mathcal{E}|\}$

Output: new Pr, q_{min}

```
 1    begin
 2       qmin(p) = q+1
 3       for j=p+1 to |n|-1
 4          qmin(j) = qmin(j-1)+1
 5          if qmin(j)>m
 6             Pr = Pr - j
 7          end    //if
 8       end       //for
 9       return Pr, qmin
10    end                //begin
```

At each iteration, for some $p \in Pr$ and $q \in \{q_{min}(p), \ldots, |\mathcal{E}|\}$, a call is made to the oracle; then Pr and q_{min} are updated. Pr is defined as $P_r = \{1, \ldots, |\mathcal{V}| - 1\}$ at initial time, and maintained as follows.

Since the vulnerability of a graph is always less than or equal to 1, the values of p and q for which $p/q > 1$ can be ignored from the test. These values correspond to the "dark" (blue) region above the first diagonal of Figure 3 (if the graph does not contain a bridge, one can eliminate the values in the first diagonal as well). This implies that for each p, there is a minimum value for q, call it $q_{min}(p)$; i.e. when p is considered in a given iteration, only values of q in the range $\{q_{min}(p), \ldots, |\mathcal{E}|\}$ need to be used for testing.

Also, if $\vartheta \leq \frac{p_0}{q_0}$ for some fixed (p_0, q_0), then $\vartheta \leq \frac{p}{q}$ for all $\frac{p}{q} > \frac{p_0}{q_0}$. As such, those values can be safely discarded from the set of values to be tested. In Figure 3, that set is represented by the "light" (blue) region for $p_0 = 4$ and $q_0 = 7$. It is the set of numbers that are located in the 135 degrees range, from the first diagonal to the horizontal axis (traveling counterclockwise). After removing this set, the values of $q_{min}(p)$ need to be updated for all $p \geq p_0$. If q_0 is the first value of q (starting from $|\mathcal{E}|$ going down) for which the test succeeds (i.e. $\vartheta(G) \leq \frac{p_0}{q_0}$), then $q_{min}(p_0) = q_0 + 1$, and for $p \in \{p_0 + 1, \ldots, \mathcal{V} - 1\}$, $q_{min}(p)$ is obtained by adding 1 to $q_{min}(p - 1)$. If $q_{min}(p) > |\mathcal{E}|$, then p can be removed from the set Pr of candidate values for p. If for some p, the test fails for all $q \in \{q_{min}, \ldots, |\mathcal{E}|\}$, then p can also be discarded from Pr. The algorithm stops when the test succeeds and $|Pr| = 1$.

For each value of p, the algorithm makes less than $|\mathcal{E}|$ calls to the oracle, and there are at most $|\mathcal{V}|$ possible values for p (this is the worst case). Thus, computing a critical subset will take a polynomial time provided that Cunningham's algorithm is polynomial. We will see that it is indeed the case.

5 Conclusion and Future Work

The paper studies a *1-connection* game where a network manager is choosing a spanning tree of a graph as communication infrastructure, and an attacker is trying to disrupt the communication tree by attacking one link of the graph. We discovered that for every critical subset of edges (a subset of edges of maximum vulnerability) there is a Nash equilibrium such that the attacker attacks uniformly at random over this subset of edges. The vulnerability of a subset of links E is defined as the minimum fraction of links it has in common with any spanning tree. More precisely, we show that there always exists a NE under which an attacker targets uniformly and exclusively a critical subset of links. The network manager chooses spanning trees that cross the critical set in the minimum number of edges and such that the sum of the probabilities of all trees going through any link in the critical set is the same. Since there exist, in general, multiple critical subsets, the NE of this game is typically not unique. We show, using a simple example, the importance of the critical subsets in the design of a robust network.

A polynomial time algorithm is presented, to compute the vulnerability of a graph as well as a critical set. The algorithm was previously presented in the literature. We discuss it and adapt it to the context of this paper.

A certain number of future directions are being explored by the authors. In the present paper, results have been obtained by assuming zero-attack cost for the attacker and an equal cost for all spanning trees in the network. Further investigations have shown that the notion of criticality of a set generalizes to the case where the attacker pays a certain cost to attack an edge. In this case, the definition of vulnerability needs a slight change to reflect the cost of attack.

Finally, in this paper, we only discuss the *1-connection* game in a graph. The case where the network chooses a *k-connected* component (for $k \geq 2$) and the attacker simultaneously attacks k or more links will be the subject of subsequent publications.

Acknowledgments

The authors would like to thank members of the Berkeley MURI and Netecon groups for their valuable input. Our special thanks go to Prof. Dorit Hochbaum for suggesting a set of very related papers. The work of the authors was supported by the ARO MURI grant W911NF-08-1-0233 and by the NSF grants CNS-0627161 and its continuation, CNS-0910702.

References

1. Awerbuch, B., Richa, A., Scheideler, C.: A Jamming-Resistant MAC Protocol for Single-Hop Wireless Networks. In: PODC 2008: Proceedings of the 27th ACM Symposium on Principles of Distributed Computing, pp. 45–54 (2008)

2. Boyd, S., Vandenberghe, L.: Convex Optimization. Cambridge University Press, Cambridge (March 2004)
3. Catlin, P.A., Lai, H.-J., Shao, Y.: Edge-Connectivity and Edge-Disjoint Spanning Trees. Discrete Mathematics 309(5), 1033–1040 (2009)
4. Chopra, S.: On the Spanning Tree Polyhedron. Operations Research Letters 8(1), 25–29 (1989)
5. Commander, C.W., Pardalos, P.M., Ryabchenko, V., Uryasev, S., Zrazhevsky, G.: The Wireless Network Jamming Problem. J. Comb. Optim. 14(4), 481–498 (2007)
6. Cordovil, R., Fukuda, K., Moreira, M.L.: Clutters and Matroids. Discrete Math. 89(2), 161–171 (1991)
7. Cornuejols, G.: Combinatorial Optimization: Packing and Covering. Society for Industrial and Applied Mathematics, Philadelphia, PA, USA (2001)
8. Cunningham, W.H.: Optimal Attack and Reinforcement of a Network. J. ACM 32(3), 549–561 (1985)
9. Edmonds, J., Fulkerson, D.R.: Bottleneck Extrema. Journal of Combinatorial Theory (8), 299–306 (1970)
10. Fulkerson, D.R.: Blocking and Anti-Blocking Pairs of Polyhedra. Math. Programming (1), 168–194 (1971)
11. Gueye, A., Walrand, J.C., Anantharam, V.: Understanding the Design of Network Topology in Adversarial Environment (2010),
    http://www.eecs.berkeley.edu/~agueye/index.html
12. Gupta, A., Kumar, A., Pal, M., Roughgarden, T.: Approximation Via Cost-Sharing: A Simple Approximation Algorithm for the Multicommodity Rent-or-Buy Problem. J. ACM 54(3), 11 (2007)
13. Gusfield, D.: Connectivity and Edge-Disjoint Spanning Trees. Information Processing Letters (16), 87–89 (1983)
14. Karger, D.R., Stein, C.: An o(n2) Algorithm for Minimum Cuts, New York, NY, USA, pp. 757–765 (1993)
15. Kruskal, J., Joseph, B.: On the Shortest Spanning Subtree of a Graph and the Traveling Salesman Problem. Proceedings of the American Mathematical Society 7(1), 48–50 (1956)
16. Fulkerson, D., Ford, L.: Flows in Networks, pp. 453–460. Princeton Univ. Press, Princeton (1962)
17. Matveev, A.O.: Maps on Posets, and Blockers. ArXiv Mathematics e-prints (November 2004)
18. Matveev, A.O.: On Blockers in Bounded Posets. Int. J. Math. Math. Sci. (26), 581–588
19. Nash-Williams, J.A.: Edge-Disjoint Spanning Trees of Finite Graphs. Journal London Math. Soc. (36), 445–450 (1961)
20. Prim, R.C.: Shortest Connection Networks and some Generalizations. Bell System Technology Journal 36, 1389–1401 (1957)
21. Robert, V.: Linear Programming: Foundations and Extensions. Springer, Heidelberg (2001)
22. Stoer, M., Wagner, F.: A simple min-cut algorithm. J. ACM 44(4), 585–591 (1997)
23. Tutte, W.T.: On the Problem of Decomposing a Graph into N Connected Factors. Journal of the London Mathematical Society (36), 221–230 (1961)
24. Neumann, J.v.: Zur Theorie der Gesellschaftsspiele. Mathematische Annalen 100, 295–320 (1928)
25. Welsh, D.: Matroid Theory. Academic Press, New York (1976)

A   Blocking Pairs of Polyhedra and the Spanning Tree Polyhedron

A.1   Blockers

Let N be a nonempty set that we will call the *ground set*, and let $\mathcal{J} = \{J_1, \ldots, J_p\}$ be a family of nonempty subsets of N. A subset J_0 of N is said to be a *blocking* set for $\mathcal{J}$ if $|J_0 \cap J_k| > 0$ for all $k \in \{1, \ldots, p\}$. The *blocker* of $\mathcal{J}$ is the family of all inclusion-wise minimal blocking sets of $\mathcal{J}$. As an example consider the graph $G = (\mathcal{V}, \mathcal{E})$ and let $N = \mathcal{E}$ the set of edges of G. Then the set $\mathcal{T}$ of spanning trees of G forms a family of subsets of $\mathcal{E}$. Any edge-cutset of the graph is blocking $\mathcal{T}$. The blocker of $\mathcal{T}$ is the set of all minimal cutsets of G.

In [6], [17], [18], the concept of blocker is defined as a mapping on families of subsets. More precisely:

Definition 2. *Given a ground set N, the* blocker *map $b(\cdot)$ is a function from the class $\mathcal{J}_N$ of all families of subsets on N to itself which associates to each family $\mathcal{J}$, its blocker*

$$b(\mathcal{J}) = \min\{J' : \ J' \subseteq N, \ J' \cap J \neq \emptyset, \ \forall J \in \mathcal{J}\}. \tag{28}$$

It has been shown [9] that if $\mathcal{J}$ is a family such that each element is not contained in another element (e.g. family of spanning trees), then the blocker map satisfies $b(b(\mathcal{J})) = \mathcal{J}$. As a consequence (since $b(b(\mathcal{J}))$ uniquely defines $\mathcal{J}$); $\mathcal{J}$ and $b(\mathcal{J})$ are said to form a *blocking pair*.

It is easy to see that if $\mathcal{J} = \{J_1, J_2, \ldots, J_k\}$ is the blocker of $\mathcal{J}' = \{J_1', J_2', \ldots, J_p'\}$, then $|J_i \cap J_j'| > 0 \ \forall \ J_j' \in \mathcal{J}'$ and for all $i = 1, \ldots, k$. Thus, any $J_i \in \mathcal{J}$ is actually blocking the family $\mathcal{J}'$.

A.2   Characterization of the Spanning Tree Family

We have seen above that the blocker of the set $\mathcal{T}$ of spanning trees is the set of minimum cuts of the graph. Let $\mathcal{M}$ be the *tree-link* incidence matrix of $\mathcal{T}$. It characterizes the *spanning tree polyhedron* $\mathcal{P}$ which is defined as the vector sum of the convex hull of the rows of $\mathcal{M}$ and the nonnegative orthant:

$$\mathcal{P} = conv\{x \mid x \ \text{is a row of} \ \mathcal{M}\} + R_+^m \tag{29}$$

where $m = |\mathcal{E}|$.

Next we give another characterization of $\mathcal{P}$. Recall that for a connected graph $G = (\mathcal{V}, \mathcal{E})$, a minimum cut partitions the node set $\mathcal{V}$ into two subsets $\mathcal{V}_1$ and $\mathcal{V}_2$, and includes all the edges having one end point in $\mathcal{V}_1$ and the other one in $\mathcal{V}_2$. Furthermore, the subgraphs, $G_i = (\mathcal{V}_i, E(\mathcal{V}_i))$, $i = 1, 2$ are connected. This notion can be generalized. Consider a partition $P = (\mathcal{V}_1, \ldots, \mathcal{V}_{k_P})$ of the nodes of G such that each subgraph $G_i = (\mathcal{V}_i, E(\mathcal{V}_i))$, $i = 1, \ldots, k_P$ is connected. Such partition is said to be *feasible*.

The spanning tree polyhedron of the graph G is characterized by the following proposition [4].

Proposition 1. *The spanning tree polyhedron of the graph G corresponds to the set*

$$\mathcal{P} = \left\{ x \in R^m_+ \mid \sum_{e \in \mathcal{E}(P)} x_e \geq k_p - 1, \ \forall \ P \ feasible \ partition \right\} ,$$

where $\mathcal{E}(P)$ denotes the subset of edges that go between vertices in distinct elements of the partition P.

The *blocking polyhedron* of $\mathcal{P}$ (corresponding to the minimal cuts) is given by (see [10],[4], [7])

$$\hat{\mathcal{P}} = \left\{ y \in R^m_+ \mid y \cdot \mathcal{P} \geq 1 \right\} .$$

In other words, $\hat{\mathcal{P}}$ consists of all nonnegative m-vectors y such that $y \cdot x \geq 1$ for all $x \in \mathcal{P}$.

Let $\hat{\mathcal{M}}$ be the $K \times m$ matrix whose rows correspond to the extreme points of $\mathcal{P}$.

Proposition 2. *The polyhedron $\hat{\mathcal{P}}$ is given by*

$$\hat{\mathcal{P}} = \left\{ y \in R^m_+ \mid \hat{\mathcal{M}}y \geq 1 \right\} .$$

B   Matroids, Polymatroids, and Network Flow

B.1   Matroids and Polymatroids

Let N be a finite set, and let $r : 2^N \to \mathbb{N}$ be a function from the family of subsets of N to the set of non-negative integers $\mathbb{N}$.

Definition 3. $M = (N, r)$ *is called a matroid if it satisfies the following properties:*

r.0: For all $J \subseteq N$, $r(J) \leq |J|$,
r.1: If $J' \subseteq J \subseteq N$, then $r(J') \leq r(J)$,
r.2: If $J, J' \subseteq N$, then $r(J \cup J') + r(J \cap J') \leq r(J) + r(J')$ (i.e. $r(\cdot)$ is submodular).

The subsets $I \subseteq N$ that verify $r(I) = |I|$ are called the independent sets of the matroid. Let $\mathcal{I}$ be the family of all independent sets. Sometime, the matroid is referred to by using the notation $M = (N, \mathcal{I})$.

An example of a matroid is the collection of cycle-free subsets of edges of a graph $G = (\mathcal{V}, \mathcal{E})$ on the ground set $\mathcal{E}$. It is called the *graphic matroid* of the graph. Its rank function is given by letting $r(E)$ be defined as the maximum size of a subset of edges in E that does not contain a loop. It is known to be equal to $r(E) = |\mathcal{V}| - Q(G_E)$, where $Q(G_E)$ is the number of connected components of the subgraph $G_E = (\mathcal{V}, E)$. The graphic matroid and its rank function will be very useful in the rest of this appendix.

More details about matroids can be found in [25].

In section 4, we have seen that, to compute the vulnerability of a graph, the search algorithm needs an oracle that solves

$$\min_{E \subseteq \mathcal{E}} \left(\mathbf{y}_0(E) + f(\bar{E}) \right) , \tag{30}$$

where $\mathbf{y}_0 = \frac{p}{q}\mathbf{1}$ for p and q given by the search algorithm. Notice that $\mathbf{y}_0(E) = \frac{p}{q}|E|$ for any subset of edges $E \subseteq \mathcal{E}$ of the graph. In this section of the appendix, we discuss how such an oracle can be built. We start by defining the notion of a *polymatroid*.

Definition 4. *A real-valued function $f(\cdot)$, defined on subsets of N, is called a* **polymatroid function** *if it verifies*

P.0: $f(\emptyset) = 0$,
P.1: If $J \subseteq J' \subseteq N$, then $f(J) \leq f(J')$ (i.e. $f(\cdot)$ is non-decreasing),
P.2: If $J, J' \subseteq N$, then $f(J \cup J') + f(J \cap J') \leq f(J) + f(J')$ (i.e. $f(\cdot)$ is submodular).

Given a polymatroid function $f(\cdot)$, the following polyhedron is called the **polymatroid** *associated to f:*

$$P(f) = \left\{ \mathbf{x} \in R_+^{|N|}, \; \mathbf{x}(J) \leq f(J), \; \forall J \subseteq N \right\}. \tag{31}$$

For any $\mathbf{y} \in R_+^{|N|}$, $\mathbf{x} \in P(f)$ is called a $P(f)$-basis of $\mathbf{y}$ if $\mathbf{x}$ is a componentwise maximal vector of the set $\{\mathbf{x}, \; \mathbf{x} \in P \text{ and } \mathbf{x} \leq \mathbf{y}\}$.

The matroid rank function defined above is an example of polymatroid function.

The following (max-min) theorem relates the minimizing subsets of (30) to the $P(f)$-basis of $\mathbf{y}_0$. The proof of the theorem can be found in [8].

Theorem 3. *Let $f(\cdot)$ be a polymatroid function on subsets of N. Then, for any $\mathbf{y} \in R_+^{|N|}$ and any $P(f)$-basis $\mathbf{x}$ of $\mathbf{y}$, we have*

$$\mathbf{x}(N) = \min\left(\mathbf{y}(J) + f(\bar{J}), \; J \subseteq N\right). \tag{32}$$

From this theorem, we see that an oracle that computes a $P(f)$-basis of $\mathbf{y}_0$ suffices for the minimization in (30). Let's see how such an oracle can be built.

The definition of $P(f)$-basis implies a very simple method for finding a $P(f)$-basis of any $\mathbf{y} \in R_+^{|N|}$. Namely,

start with $\mathbf{x} = 0$ and successively increase each component of $\mathbf{x}$ as much as possible while still satisfying $\mathbf{x} \leq \mathbf{y}$, and $\mathbf{x} \in P(f)$.

Implementing this simple and greedy algorithm might, however, not be so simple. In fact, it requires one to be able to compute, for a given $\mathbf{x} \in P(f)$ and any $j \in N$, the quantity

$$\epsilon_{max}(j) = \max(\epsilon : \mathbf{x} + \epsilon \mathbf{1}_j \in P(f)), \tag{33}$$

where $\mathbf{1}_j$ is the incidence vector of subset $\{j\}$. $\epsilon_{max}(j)$ is the maximum amount by which component j of $\mathbf{x}$ can be increased while keeping $\mathbf{x}$ in $P(f)$.

Verifying that a vector $\mathbf{x}$ belongs to the polymatroid can be done using the following idea: if $\mathbf{x} \notin P(f)$, then one can find a subset J for which $\mathbf{x}(J) \leq f(J)$ is violated. If $\mathbf{x} \in P(f)$ and $j \in N$, then any ϵ such that $\epsilon > \min_{J \subseteq N}(f(J) - \mathbf{x}(J), \; j \in J)$ will send $\mathbf{x} + \epsilon \mathbf{1}_j$ out of $P(f)$.

Also, if $\mathbf{x}$ is a $P(f)$-basis of $\mathbf{y}$, then for any $j \in N$, either $\mathbf{x}(j) = \mathbf{y}(j)$ or $\mathbf{x}(J) = f(J)$ for some subset J containing j. In fact, for all $j \in N$

$$\epsilon_{max}(j) = \min\left\{\mathbf{y}(j) - \mathbf{x}(j), \min_J(f(J) - \mathbf{x}(J), \; j \in J \subseteq N)\right\}. \tag{34}$$

If the minimum is achieved at $\mathbf{y}(j) - \mathbf{x}(j)$, then $\mathbf{x} \leftarrow \mathbf{x} + \epsilon_{max}(j)\mathbf{1}_j$ will satisfy $\mathbf{x}(j) = \mathbf{y}(j)$. Otherwise, there exists some $J_j \ni j$, such that $\mathbf{x}(J_j) = f(J_j)$ (J_j is said to be *tight*). Letting $\bar{J} = \bigcup_j J_j$, and $\mathbf{x}$ being the $P(f)$-basis obtained after running the greedy algorithm, it can be shown (see [8]) that $f(\bar{J}) = \mathbf{x}(\bar{J})$ (union of tight set is tight). For such $\bar{J}$, we have that

$$\mathbf{x}(N) = \mathbf{x}(J) + \mathbf{x}(\bar{J}) = \mathbf{y}(J) + f(\bar{J}) . \tag{35}$$

This is because $\mathbf{x}(\bar{J}) = f(\bar{J})$ and if $j \notin \bar{J}, \mathbf{x}(j) = \mathbf{y}(j)$.

Based on these observations, Cunningham [8] proposed a modified version of the greedy algorithm to compute a $P(f)$-basis, as well as a minimizing subset for the minimization in (32). The algorithm is presented in Table 2.

It starts with $\mathbf{x} = 0$ and $\bar{J} = \emptyset$. For each $j \in N$, the component $\mathbf{x}(j)$ is increased as much as possible: $\mathbf{x} \leftarrow \mathbf{x} + \epsilon_{max}(j)\mathbf{1}_j$. If the minimum in (34) is achieved at $\min_{J'} (f(J') - \mathbf{x}(J'), j \in J')$, then update $\bar{J} \leftarrow \bar{J} \cup J'$ where J' is a minimizer. At the end of the algorithm, $\bar{J}$ is a tight set and $\mathbf{x}$ is maximal. Also, it satisfies $\mathbf{x} \in P(f)$ and $\mathbf{x} \leq \mathbf{y}$, with $\mathbf{x}(N) = \mathbf{y}(J) + f(\bar{J})$.

To find a $P(f)$-basis, Cunningham's algorithm performs $|\mathcal{E}|$ computations of the the minimization below:

$$\min_{J} (f(J) - \mathbf{x}(J), j \in J \subseteq N) . \tag{36}$$

Now, all that remains is to find an algorithm that computes the minimization in polynomial time. This is the subject of the next section.

Table 2. Pseudocode of the oracle *CunninghamMin* that solves the minimization (36)

```
Cunningham
Input: Polymatroid function f, y ∈ R_+^{|N|}
Output: minimum eps, minimizer T

 1 begin
 2    x = 0
 3    J := {}
 4    for j in N
 5       eps := min(f(J')-x(J'): j in J')
 6       J'(j) := a minimizer
 7       if eps <= y(j)-x then J:=J U J'(j)
 8       else eps:= y(j)-x(j)
 9       end        //if
10       x= x+eps*1(j)
11    end    //for
12 end       //begin
```

B.2   Network Flow

In the notation of the last two sections, $\mathcal{E}$ below will be a ground set (N above), and subsets of $\mathcal{E}$ will be referred to using E (J and I above).

Let $G = (\mathcal{V}, \mathcal{E})$ be a connected graph and let $f(\cdot)$ the rank function of the graphic matroid that is associated to G. We have seen above that $f(E) = |\mathcal{V}| - Q(G_E)$. Let $P(f)$ be the polymatroid associated with $f(\cdot)$. An equivalent description of $P(f)$ is given as follows (see [8]):

$$P(f) = \left\{ \mathbf{x} \in R_+^{|\mathcal{E}|}, \ \mathbf{x}(\gamma(B)) \leq |B| - 1 \text{ for all } B, \ \emptyset \neq B \subseteq \mathcal{V} \right\}, \qquad (37)$$

where $\gamma(B)$ denotes the set of edges with both ends in B.

Recall that our goal is, for a given j, to find a subset E, $j \in E \subseteq \mathcal{E}$ that minimizes $f(E) - \mathbf{x}(E)$. This is equivalent to finding B that minimizes $|B| - 1 - \mathbf{x}(\gamma(B))$, with $j \in \gamma(B)$.

To find the minimizing subset of nodes, B, we define the following graph G' for a given polymatroid function $f(\cdot)$, $\mathbf{x} \in P(f)$, and edge $j \in \mathcal{E}$. The vertices of G' are $\mathcal{V} \cup (r, s)$ for new vertices r and s. Each $e \in \mathcal{E}$ is an edge of G', having the same ends and having capacity $\frac{1}{2} x_e$. There is an edge joining v to s for each $v \in \mathcal{V}$, it has capacity 1. There is an edge joining r to v for each $v \in \mathcal{V}$. It has capacity ∞ if v is an end of j, and otherwise it has capacity $\mathbf{x}(\delta(v))$. (Here $\delta(B) = \{e \in \mathcal{E}, \ e \text{ has exactly one end in } B \subseteq \mathcal{V}\}$, $\delta(v)$ is shorthand for $\delta(\{v\})$). This construction is illustrated in Figure 4(a). Its motivation is to ensure that $j \in \gamma(B)$ as can be seen next.

Now consider a cut in G' induced by the set $B \cup \{r\}$, where $j \in B \subseteq \mathcal{V}$. It is the set of links that have one end in $B \cup \{r\}$ and the other end in the complement of $B \cup \{r\}$. The capacity of such cut is (see an illustration in Figures 4(b))

$$|B| + \frac{1}{2}\mathbf{x}\left(\delta(B)\right) + \mathbf{x}\left(\gamma(\bar{B})\right) + \frac{1}{2}\mathbf{x}\left(\delta(B)\right) = |B| + \mathbf{x}(\mathcal{E}) - \mathbf{x}(\gamma(B)) \qquad (38)$$

$$= |B| - 1 - \mathbf{x}(\gamma(B)) + (\mathbf{x}(\mathcal{E}) + 1). \qquad (39)$$

The first term in the LHS of equation (38) corresponds to edges going from nodes in B to the sink s. There are $|B|$ of them, each having capacity 1. The next term corresponds to edges going from a node in B to a node in $\bar{B}$. The last two terms correspond to edges going from the root r to nodes in $\bar{B}$. For each such edge (r, u), the capacity is defined as $\frac{1}{2}\delta(\{u\})$. Let $e = (u, v) \in \delta(\{u\})$. Then, if $v \in B$ (i.e. $e \in \delta(B)$), then $\mathbf{x}(e)$ appears only in the capacity of (r, u); implying the term $\frac{1}{2}\mathbf{x}\left(\delta(B)\right)$. If, on the other hand, $v \notin B$ (i.e. $e \in \gamma(\bar{B})$), then $\mathbf{x}(e)$ appears both in the capacity of (r, u), and in that of (r, v), thus the term $\mathbf{x}\left(\gamma(\bar{B})\right)$.

Now, since a cut induced by a subset of edges B will have infinite capacity if $j \notin \gamma(B)$, a minimum cut in G' will indeed have the form $B \cup \{r\}$ with $j \in B$, hence, minimizing $|B| - 1 - \mathbf{x}(\gamma(B))$. As a consequence, any network flow algorithm can serve as an oracle for Cunningham's algorithm. Many polynomial implementations of network flow algorithms ([22], [14]) have been proposed since the proof of the Max-Flow Min-Cut theorem by Ford and Fulkerson [16] in 1962.

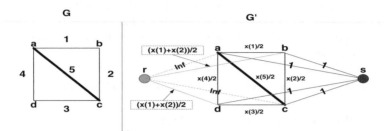

(a) Constructing the graph G' from G for the network flow algorithm.

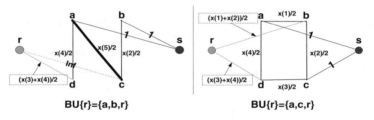

(b) Illustrating the cut induced by $B \cup \{r\}$

Fig. 4. Constructing the graph G' for the network flow algorithm. Figure 4(a) shows the construction of G' from G. The edge under consideration in this example is $j = 5$. Examples in Figures 4(b) show the cut induced by $B \cup \{r\}$ for $B \subseteq \mathcal{V}$. In the left figure, $B = \{a, b\}$ does not contain $j = 5$. The capacity of this cut is equal to infinity. In the right figure, $B = \{a, c\}$ which contains edge $j = 5$ (the only edge). As can be seen in the figure, the capacity of the cut induced by this choice of B is $2 + \mathbf{x}(1) + \mathbf{x}(2) + \mathbf{x}(3) + \mathbf{x}(4)$ which is finite.

Optimal Information Security Investment with Penetration Testing

Rainer Böhme and Márk Félegyházi

International Computer Science Institute, Berkeley, California
{rainer.boehme,mark}@icsi.berkeley.edu

Abstract. Penetration testing, the deliberate search for potential vulnerabilities in a system by using attack techniques, is a relevant tool of information security practitioners. This paper adds penetration testing to the realm of information security investment. Penetration testing is modeled as an information gathering option to reduce uncertainty in a discrete time, finite horizon, player-versus-nature, weakest-link security game. We prove that once started, it is optimal to continue penetration testing until a secure state is reached. Further analysis using a new metric for the return on penetration testing suggests that penetration testing almost always increases the per-dollar efficiency of security investment.

1 Introduction

Information security investment decisions have recently attracted the attention of researchers from computer science, economics, management science, and related disciplines. The emerging topic of the economics of information security aims at formalizing these decisions, but there is still a gap between the formal models and experiences in practice [1]. In particular, information gathering options of defenders of computer systems differ from other scenarios. Penetration testing (short: pentesing), the focus of our paper, is an example of proactive information gathering options specific to computer systems. Penetration testing is widely used in practice, but its effects have not been reflected in the information security investment literature.

In this paper, we build a model on a simplified version of the *iterated weakest link* (IWL) model of dynamic security investment [2,3] which emphasizes the role of uncertainty in security decision making. The original IWL model explains why a defender facing uncertainty about which threats are most likely to realize might defer security investment and learn from observed attacks where the investment is most needed. The benefits of more targeted investment may outweigh the losses suffered through non-catastrophic attacks, thereby increasing the *return on security investment* (ROSI). We extend the IWL model by an option to commission pentests as a means to reduce uncertainty. Indeed, waiting for actual attacks need not be the only way of gathering information to guide security investment. Uncertainty can also be reduced by observing pre-cursors of attacks or near misses [4], information sharing [5,6], or investment in information gathering. Penetration testing can be seen as information gathering prior to investing into protection against so-identified threats.

T. Alpcan, L. Buttyan, and J. Baras (Eds.): GameSec 2010, LNCS 6442, pp. 21–37, 2010.
© Springer-Verlag Berlin Heidelberg 2010

Penetration testing is also referred to as "ethical hacking" because the commissioned penetration testers investigate the target system from an attacker's point of view, reporting weaknesses rather than exploiting them. The aim of this work is to study the added benefits and costs of penetration testing to the entire system defense. The similarity between pentesting and attacks leads to the intuition that information revealed by pentests should be modeled in exactly the same way as information revealed by attacks. Yet there exist differences on the cost side: pentests cause calculable up-front costs, whereas costs associated with successful attacks are typically more volatile, much higher, and borne ex post. For all other modeling decisions, we stay close to the original IWL model, and we refer the reader to [2] for a more detailed discussion of its features.

This paper makes the following contributions:

- it provides a first attempt to study information gathering options by pentesting in the framework of the economics of security investments;
- it contains a proof that in this model, pentesting should be done consistently once started;
- it defines a metric for *return on penetration testing* (ROPT);
- and it demonstrates that pentesting not only increases total profit for the defender, but also increases (most of the cases) the per dollar efficiency of security investments.

The remainder of this paper is organized as follows. After recalling the context of related work in Section 2, we describe in Section 3 our approach to include penetration testing as information gathering step into an established model of security investment. Section 4 presents solutions of the model. Section 5 defines the ROPT metric and demonstrates how the model can be applied in investment decision making. The final Section 6 concludes with discussion and outlook.

2   Related Work

Information security investment have been studied from the economics perspective. Gordon and Loeb [7] formulate a basic economic model. They argue that taking both the risk profiles of vulnerabilities and the cost to protect them into account, the best investment strategy for a defender is to protect the mid-range of vulnerabilities.

Intrusion detection systems (IDS) build a solid line of defense against most outside attackers, but the systems are notoriously difficult to configure. Cavusoglu et al. [8] study the value of intrusion detection systems and argue that the main benefit of IDSs is not the increased detection rate, but the deterrence of the system and the increased availability of information for forensics. Using their analytical model, they found that an IDS is only valueable if the detection rate is high enough. The authors show that the threshold for an IDS to be valuable is determined by the attacker's benefit. The attacker's benefit is difficult to assess in practice [9], that makes the model difficult to apply in practice. In a subsequent paper, Ogut et al. [10] study intrusion detection policies using a decision-theoretic framework. They describe a scenario where defenders wait and gather more information about potentially malicious users instead of acting on IDS signals immediately. They propose an optimal waiting strategy as well as a myopic heuristic that relies on less parameters; hence it is easier to apply in practice.

Penetration testing is an important method to assess the vulnerability of a computer system before it is deployed. Geer and Harthorne [11] argue that penetration testing requires special skills because attacks are unknowable and hence innumerable in advance. They informally discuss the value of penetration testing and connect it to the formulation of the return on security investments (ROSI). The authors advocate the evaluation of penetration test results in the light of a risk assessment. Arkin et al. [12] provide an insight into software penetration testing practices. They mostly argue for better integration of testing during the development cycle of software systems. They agree with [11] that penetration test results should not be considered as a final checklist, but rather as a sample from the potential problems. They emphasize that decision makers often stop penetration testing after an initial round, because "having found the issues" gives them a false sense of security.

A leading survey of industry participants ([13], Fig. 20) reveals that the majority of responding firms performs penetration testing in practice. Yet, we are unaware of prior theoretical work that formalizes penetration testing as a specific tool available to the information security manager.

3   Model

Our model extends [2]. The *defender* operates a system that represents an asset of value a yielding a return r per period. The defender protects this system against a dispersed set of attackers. We do not distinguish between different attackers, rather we consider the group of attackers as a *single attacker* entity with enhanced capabilities. There exist n possible components of the system that are threatened by an attack.[1] Each threat can be prevented by investing into the protection of the specific component and we assume that a protection is always effective. The defender orders the threats according to their *expected* cost $\bar{x}_1 \leq \bar{x}_i \leq \bar{x}_n$, but the *true* costs to attack x_i is hidden from the defender. This reflects the opinion of many practitioners who remain skeptical about the quantifiability of attack probabilities but reckon it is possible to order threats by severity. There are no restrictions about the source of prior beliefs about this order. It can result from individual judgement, semi-formal aggregation of expert opinions, or formal calculations of threat prioritization using system models[14].

We model the true costs to attack as

$$x_i = \sup(0, \bar{x}_i + \chi_i) \quad \text{with} \quad \chi_i \sim \mathcal{N}/(\Delta x)^2 \quad \text{and} \quad \bar{x}_i = \bar{x}_1 + (i-1) \cdot \Delta x, \quad (1)$$

where $\mathcal{N}$ is a mean-free Gaussian random source with standard deviation $\sigma \geq 0$. This parameter controls the degree of uncertainty and it is key to analyze the usefulness of penetration testing as uncertainty varies. Figure 1 visualizes the

---

[1] Alternatively, the notion of 'components' can be substituted by 'attack vectors'.

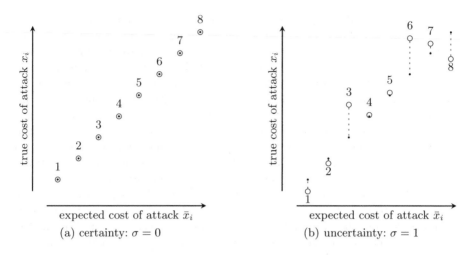

Fig. 1. The defender forms expectations about the order of attack costs for different threats (here: $i = 1, \ldots, 8$) but remains ex ante uncertain about the true costs

influence of σ in introducing noise, i.e., adding unknown offsets between actual and expected costs. Higher values of parameter σ indicate that the defender's order of threats differs more from the true order of costs to attack.

The order is relevant because in each round, the attacker exploits the *weakest link*. That is, she attacks the unprotected component with the least *true cost*[2] and loots a fraction of z from asset a. The attacker is opportunistic in that she attacks only if the benefits (i.e., the defender's losses) exceed the cost of attack. Hence, a secure state can be reached when all vulnerable components that can be attacked below the reservation cost of attack are protected.

We model the interaction between the defender and the attacker as a dynamic, discrete time, finite horizon, player-versus-nature game. In the initial round ($t = 0$), the defender chooses her *defense configuration* to protect against the k most possible threats $1, \ldots, k$. A unit cost of 1 is incurred per protection and round. As the realizations of $(x_1, \ldots, x_n)$ are unknown, any other initial configuration would lead to inferior outcomes on average.

In the following reactive rounds ($t = 1, \ldots, t_{max}$), four steps are iterated:

1. The defender chooses whether or not to commission a pentest at cost $c > 0$.
2. If a pentest has been executed, it succeeds with probability $p > 0$ and reveals the next weakest link i with true attack cost $i = \arg\min_j x_j$ over all unprotected links j. The defender protects i. This increases her defense cost by 1 in the current and all subsequent rounds.[3]

---

[2] The intuition is that our attacker model represents the ensemble of individual attackers that are likely to discover the weakest link.

[3] As defense costs are constant for each threat, the decision to defend upon revelation is cogent. Otherwise it is always better not to commission the pentest in step 1.

3. An attack occurs if at least one $x_i \le z \cdot a$. If so, the defender learns which link i was the weakest and incurs a loss of $z \cdot a$. Otherwise the defender learns that the system has reached a secure state.
4. The defender chooses whether to upgrade the defense configuration and protect against the threat revealed in the last attack. This increases her defense cost by 1 for all subsequent rounds.

Observe that steps 3 and 4 exactly correspond to the original model in [2], steps 1 and 2 are new to introduce pentests as means of information gathering. For simplicity, we do not consider sunk costs or interdependent defenses here, i.e., $\lambda = 0$ and $\rho = 0$ in the notation of [2]. Like the original model, the defender is risk neutral.

4   Analysis

Although the model is simple, its solution is not trivial. Figure 2 depicts an excerpt of the defender's optimization problem in extensive form. Observe the pairwise alternation of moves by player and nature and the repetition of steps 1 to 4 in each round.

The defender starts at node S and chooses the initial level of defense k. Nodes annotated with T are terminal nodes:

T_0: this singular case corresponds to knowingly indefensible situations, i.e., if $r \ge z$ and $a \cdot (r - z) < E\left[\frac{1}{n}\sum_i x_i\right]$. In this case, the defender refrains from investing in security and rather accepts the losses due to attacks in each round. The value of this node is

$$T_0 = t_{\max} \cdot a \cdot (r - z). \tag{2}$$

T_1: this case corresponds to the arrival at a secure state. The defender's goal is to reach a node of type T_1 as soon as possible. The deterministic value of these nodes is a function of t, k, and the number of successful pentests $|\mathcal{M}^+|$,

$$T_1(t, k, |\mathcal{M}^+|) = (t_{\max} - t + 1) \cdot (r \cdot a - t - k - |\mathcal{M}^+| + 1). \tag{3}$$

T_2: this is the case when the system is found indefensible only after revelation of realizations of nature. In an indefensible situation, the defender would always prefer T_0 over any T_2 (where costs for ineffective defenses are unrecoverably sunk). The deterministic value of these nodes is a function of t,

$$T_2(t) = (t_{\max} - t + 1) \cdot a \cdot (r - z). \tag{4}$$

Since the influence of nodes T_2 is negligible for the parameter settings used throughout this paper, we do not consider them in the analysis for brevity.

The dashed branches leading to an asterisk node are decisions not to pentest even though at least one pentest has been commissioned in an earlier round. Theorem 1 states that these paths are strictly dominated by the alternative choice and can indeed be eliminated to simplify the extensive form representation.

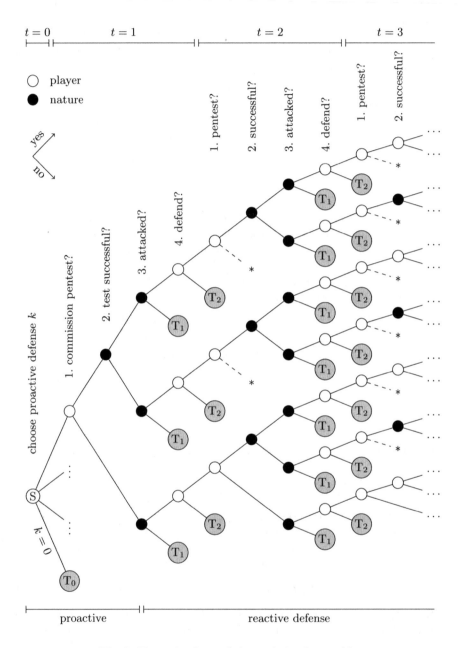

Fig. 2. Extensive form of the optimization problem

Theorem 1. *Once the defender starts pentesting, she will keep doing it until a secure state is reached.*

Proof. We use Lemma 1 proven in the appendix. It gives us the following expression for the total profit of a defender as a function of the initial defense k, the set of rounds in which pentests are commissioned $\mathcal{M} = \{m_1, \dots\}$, and a fixed number of unprotected components K:

$$G = \sum_{t=1}^{K}(a(r-z)-(k+t-1)) + \sum_{i=1}^{|\mathcal{M}|}(az-c-(K-i-m_i+2)) + \sum_{t=K+1}^{t_{max}}(ar-(k+K)). \quad (5)$$

The contribution of each pentest i to the total profit depends on the round m_i when the pentest is commissioned. The second sum of Eq. (5) shows that the marginal benefit of penetration testing increases with the number of rounds in the game. Thus, if a defender decides to commission pentests in round t, then she will keep doing it in each round $u > t$ until all weak links are discovered. $\square$

Pentests are successful with probability p. For $p < 1$, $\mathcal{M}$ can be partitioned ex post into two disjoint subsets $\mathcal{M} = \mathcal{M}^+ \cup \mathcal{M}^-$ of rounds with successful, respectively unsuccessful penetration tests. Since the pentests are independent events, we can simply multiply their contribution with their respective probability. Hence the expected value of (5) becomes:

$$E[G] = \sum_{t=1}^{K}(a(r-z)-(k+t-1)) + p\sum_{i=1}^{|\mathcal{M}|}(az-c-(K-i-m_i+2)) + \sum_{t=K+1}^{t_{max}}(ar-(k+K)).$$
$$(6)$$

An intuitive way to analyze the composition of the revenue is a graphical representation, as depicted in Fig. 3. The figures show a schematic representation with infinitesimally small rounds. The costs are proportional to the shaded areas defined by the asset value a, the loss due to attacks z, the total number of weak links K, the number of proactive defenses k, the cost of a pentest c and the probability of a successful pentest p.

The first figure shows the case when no pentests are commissioned and the weak links are discovered one-by-one until all K are protected. During this time, the attacker loots $(K-k) \cdot a \cdot z$ profit from the asset. While protecting the asset, the defender spends the proactive protection cost $(K-k) \cdot k$ and the reactive protection cost $\frac{K-k}{2}$. In round $K-k$, all weak links are protected and the defender maintains the defense cost K for all subsequent rounds. That prevents the attacker from looting the asset.

The second figure shows[4] that pentesting with $p = 1$ introduces two additional costs: the cost of pentests c and the cost of the resulting protection; both costs are shown as dark areas in Fig. 3. Pentesting has a benefit (the light grey areas in Fig. 3) of discovering weak links earlier than without pentests and this reduces

---

[4] We show the most likely case where the pentesting starts from the first round, but the figures can easily be adapted to the case when pentesting starts at a later round.

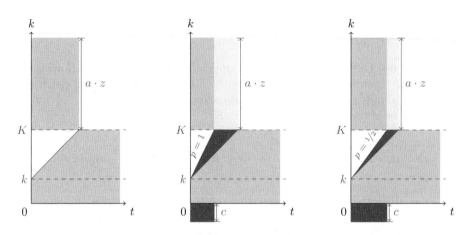

Fig. 3. Comparison of costs in scenario without (left) and with pentesting ($p = 1$, center) for infinitesimally many rounds; costs are proportional to areas; the success probability p defines the slope of the gradient towards reaching the secure level K (center versus right); note that K is a random variable unknown to the decision maker ex ante; pentesting is worthwhile if the expected value of the light area (savings) exceeds the expected value of the dark area (direct and indirect cost of pentesting)

the looting cost from the attacker. Having $p = 1$ doubles the speed of discovering weak links (the slope of protection costs is two) and halves the total looting cost. The defender chooses to perform pentests as long as the benefit due to prevented attacks is higher than the pentesting costs. The third figure shows a case when pentests are less efficient and hence their effect to reduce cost due to attacks decreases. Nonetheless, the defender has to pay the cost of pentesting for each try.

From Theorem 1, we know that the defender performs pentests from m_1 until the attacks stop. Then, we can write the expected number of pentest as:

$$E[|\mathcal{M}|] = \left\lceil \frac{K + 1 - m_1}{1 + p} \right\rceil \qquad (7)$$

The number of pentest depends on whether the last weakest link is protected following a pentest or an attack. Let ϵ be an indicator variable showing if the last weak link is fixed after a pentest or an attack. If the last weak link is discovered by a pentest, then $\epsilon = 1$, otherwise $\epsilon = 0$.

From (7) and using ϵ, we can derive the optimal number of pentests and the optimal time to start pentesting. The derivation is in the appendix. Substituting the optimal number of pentests into (6), we obtain an expression for the expected total profit with optimal number of pentests for a fixed K. The direct application of this expression, however, is impeded by the fact that the overall profit is largely determined by a discontinuous boundary condition and the randomness of K.

While k and m_1 are choice variables, K is a discrete random variable with known distribution but a priori unknown realization. Similar to [2], the optimal

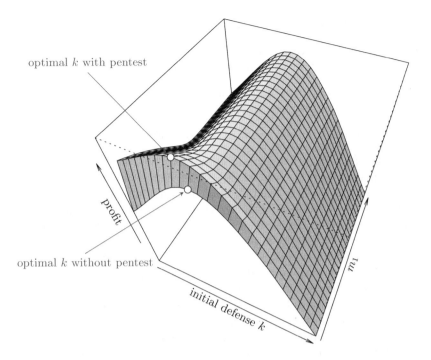

Fig. 4. Expected total return surface as a function of choice variables (k, m_1) for the following parameters: asset value $a = 1000$, return $r = 5\%$, loss given attack $z = 2.5\%$, profile of expected attack costs $(x_1, \Delta x) = (15, 1)$, uncertainty $\sigma = 4$, pentest cost $c = 0.5$, pentest success $p = 100\%$, $n = t_{\max} = 25$; $m_1 = 0$ means no pentest at all

defense strategy can also be found by numerically summing up the expected total profit over the domain of K and finding the maximum of a grid search for the tuple of choice variables (k, m_1). A result of a numerical maximization with a selected set of parameters is shown in Fig. 4, indicating the optimal strategies with and without access to pentests. For our example set of parameters, the total profit of the defender is optimal if she starts pentesting in the first round. We also observe that the initial investment in defenses k is lower in the case of pentesting as opposed to the case where no pentests are commissioned. The reason is that some resources spent on proactive protection are now reallocated to a more efficient discovery of weak links using pentests.

5   Return on Penetration Testing (ROPT)

Several definitions exist to measure the return on security investment (ROSI) [15]. We follow the approach in [2] and choose an indicator normalized by the average security investment per period [16]. Without pentesting we have,

$$\text{ROSI}_{\text{NPT}} = \frac{\text{ALE}_0 - \text{ALE}_{\text{NPT}} - \text{avg. security investment}}{\text{avg. security investment}}, \qquad (8)$$

where ALE is the *annual* (i.e., per period) *loss expectation* for two cases:

- ALE_0: a baseline case where no security investment is made,
- ALE_{NPT}: *with* security investment but *without* pentesting.

Higher values of $ROSI_{NPT}$ denote more efficient security investment. A natural extension to penetration testing is to define:

- ALE_{PT}: loss expectation *with* security investment *and* pentesting.

However, there is no straightforward way to measure the specific return on penetration testing by plugging both ALE_{NPT} and ALE_{PT} in the numerator of Eq. (8). The reason is that the fraction of security investment related to penetration testing is difficult to identify since it consists of direct costs c and indirect costs from defenses set up earlier than without pentesting (cf. Fig. 3). Yet another source of *indirect benefits* is not visible in Fig. 3. The possibility to do pentests can lead to a lower optimal initial defense k (cf. Fig. 4). If these benefits match or exceed the direct and indirect costs of pentesting, then the defender can face situations where she invests equal or less and still achieves higher security than without pentesting. In this case, funds are shifted from investment in protective measures towards spending on information gathering. This seemingly odd result once again demonstrates the special role of pentesting and the need to appropriately reflect it in security investment models. For such special cases, the normalizing term based on the simple difference becomes zero or negative. This would lead to undefined values for ROSI.

To fully characterize the effects of pentesting, we propose the following metric called *return on penetration testing (ROPT)*:

$$ROPT = ROSI_{PT} - ROSI_{NPT}, \qquad (9)$$

where ROSI is calculated according to Eq. (8) with ALE_{PT} and ALE_{NPT}, respectively. Consistent with the interpretation of ROSI as the dollar amount of prevented losses per dollar of security spending, ROPT can be understood as the dollar amount of *additionally prevented* losses per dollar if security investment is optimized *with* penetration testing.

Fig. 5 shows numerical values of ROPT as a function of uncertainty for the example set of parameters used in Fig. 4 and the pentest cost c. Observe for each curve that ROPT first increases with σ, then decreases until it approaches a constant value. The first increase in the ROPT values can be explained by the benefit of pentesting to gather additional information. A defender who commissions pentests invests less into proactive defenses and more into pentests and this benefit increases as uncertainty about true attack costs increases. ROPT starts to decrease when proactive defenses with pentest reach zero (e.g., $k_{PT} = 0$) meaning that a defender using pentests relies exclusively on reactive defenses, whereas a defender who does not commission pentests still invests in proactive

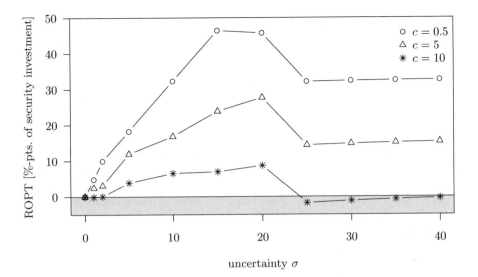

Fig. 5. Profile of return on penetration testing (ROPT) as uncertainty increases for varying cost of pentesting ($c = 0.5$ is half of the cost per protection measure and round)

defenses ($k_{NPT} > 0$). ROPT becomes constant when uncertainty is so high that both defense strategies avoid the proactive defense period ($k_{PT} = k_{NPT} = 0$).

The ROPT value is mostly positive (can be as high as 50%), meaning that pentesting brings a significant per dollar efficiency to security investments. However, if the pentesing cost c becomes relatively high ($c = 10$ is one order of magnitude higher than the defense cost of a weak link) then ROPT might turn negative, indicating that pentesting is a more costly security investment alternative than no pentesting. We emphasize that even in this case, the total profit of the defender increases with pentesting until the defense cost including pentesting reaches the looting cost. Thus we conclude that pentesting is a beneficial defense option for a wide range of parameters.

6   Discussion and Conclusion

In this paper, we leveraged the iterated weakest link model of [2] to propose a framework that accounts for penetration testing, an important information gathering option when making security investment decisions. To the best of our knowledge, this is the first paper that explicitly models penetration testing and shows its potentially catalyzing effect on the efficiency of security spending. We are also the first to propose ROPT, a metric to account for the efficiency of penetration testing.

Our model formalizes much of the informal discussion in other papers about security investments and pentesting. Ogut et al. [10] study intrusion detection policies and propose a model for optimal waiting time to act on intrusion signals. Our model recovers the same mentality by allowing the defender to invest less in proactive security and fix the weak links reactively after an attack occurs. Our model also formalizes the arguments of Geer and Harthorne [11] who emphasize that the results of penetration testing should be considered in the light of risk assessment rather then perceived as a security todo list. Arkin et al. [12] iterate on this view by stating that security decision makers should follow-up on the insights uncovered by pentesting. We proved that in the IWL framework, commissioning pentests is the best strategy for the defender until all weak links with feasible attack costs are protected.

We conjecture that our model captures the basic mechanisms in security investments with pentesting. Nonetheless, as any formal model, it has its shortcomings. We model the security investment process as a finite-horizon game between the defender and the nature player. A natural extension of this paper to consider the attackers as rational players. We acknowledge this future direction, but point to the fact that the profit functions of the attackers are relatively difficult to model [9]. There is some initial work to understand the profits of the attackers in real life [17], but we are lacking of deeper understanding to properly model attackers in security games. Our model assumes that the set of weak links does not change within the finite horizon of the game. There are two improvements one can consider regarding this assumption. First, the game is typically a dynamic game that can be considered as an infinite game with discounting. Second, the dynamics of changing weak links are worth exploring as well. Yet another direction involves further refinement of the model to capture even more specific details of security investment, such as the difference between black-box and white-box testing. This choice defines the distribution of information and should be modeled to affect the heuristic potential of pentesting to discover weak links similar to a real attacker.

Our paper provides a theoretical framework for penetration testing. While this exposition focused on the defender's decision, we note that this kind of model can also be solved for the cost of penetration testing to inform providers of pentest services and guide their price setting. One major question is how this model and its conclusions fit to real data from industry sources. Obtaining such a confirmation is a potential future work. Finally, we will extend the IWL framework considering other options for uncertainty reduction beyond penetration testing.

Acknowledgements

The first author received a Postdoctoral Fellowship by the German Academic Exchange Service (DAAD) to support his visit at the International Computer Science Institute.

References

1. Su, X.: An overview of economic approaches to information security management. Technical Report TR-CTIT-06-30, University of Twente (2006)
2. Böhme, R., Moore, T.W.: The iterated weakest link: A model of adaptive security investment. In: Workshop on the Economics of Information Security (WEIS), University College, London, UK (2009)
3. Böhme, R., Moore, T.W.: The iterated weakest link. IEEE Security & Privacy 8(1), 53–55 (2010)
4. Panjwani, S., Tan, S., Jarrin, K.M., Cukier, M.: An experimental evaluation to determine if port scans are precursors to an attack. In: Proc. of Int'l. Conf. on Dependable Systems and Networks (DSN 2005), Yokkohama, Japan (2005)
5. Gordon, L.A., Loeb, M.P., Lucysshyn, W.: Sharing information on computer systems security: An economic analysis. Journal of Accounting and Public Policy 22(6) (2003)
6. Gal-Or, E., Ghose, A.: The economic incentives for sharing security information. Information Systems Research 16(2), 186–208 (2005)
7. Gordon, L.A., Loeb, M.P.: The economics of information security investment. ACM Transactions on Information and System Security 5(4), 438–457 (2002)
8. Cavusoglu, H., Mishra, B., Raghunathan, S.: The value of intrusion detection systems in information technology security architecture. Information Systems Research 16(1), 28–46 (2005)
9. Barth, A., Rubinstein, B., Sundararajan, M., Mitchell, J., Song, D., Bartlett, P.L.: A learning-based approach to reactive security. In: Radu, S. (ed.) FC 2010. LNCS, vol. 6052, pp. 192–206. Springer, Heidelberg (2010)
10. Ogut, H., Cavusoglu, H., Raghunathan, S.: Intrusion detection policies for it security breaches. INFORMS Journal on Computing 20(1), 112–123 (2008)
11. Geer, D., Harthorne, J.: Penetration testing: A duet. In: Proc. of the 18th Annual Computer Security Applications Conference (ACSAC), Las Vegas, NV, USA (2002)
12. Arkin, B., Stender, S., McGraw, G.: Software penetration testing. IEEE Security & Privacy 3(1), 84–87 (2005)
13. Richardson, R.: CSI Computer Crime and Security Survey. Computer Security Institute (2007)
14. Miura-Ko, R.A., Bambos, N.: SecureRank: A risk-based vulnerability management scheme for computing infrastructures. In: IEEE International Conference on Communications (Proc. of ICC), pp. 1455–1460 (2007)
15. Böhme, R., Nowey, T.: Economic security metrics. In: Eusgeld, I., Freiling, F.C., Reussner, R. (eds.) Dependability Metrics. LNCS, vol. 4909, pp. 176–187. Springer, Heidelberg (2008)
16. Purser, S.A.: Improving the ROI of the security management process. Computers & Security 23, 542–546 (2004)
17. Kanich, C., Kreibich, C., Levchenko, K., Enright, B., Voelker, G., Paxson, V., Savage, S.: Spamalytics: An empirical analysis of spam marketing conversion. In: Conference on Computer and Communications Security (Proc. of ACM CCS), Alexandria, Virginia, pp. 3–14 (2008)

A Appendix

A.1 Lemma: Independence of Pentests

Let us write the total profit for the defender as

$$F = \sum_{t=1}^{t_{\max}} f(k, t). \tag{10}$$

In (10), $f(k, t)$ is the profit per round and can be written as follows:

$$f(k, t) = a(r - q \cdot z) - c_t, \tag{11}$$

where q is an indicator variable that takes the value of 1 if an attack is successful and 0 otherwise; and c_t is the cost in round t.

The defender chooses the initial defense k and fixes one defense at the time as discovered by the attacker. Let us now assume that without penetration testing, the number of rounds with successful attacks is K (where $k + K$ threats are warded off). In this case, we can write the total profit for the defender F as

$$F = \sum_{t=1}^{K} f(k, t) + \sum_{t=K+1}^{t_{\max}} f(k, t) = \sum_{t=1}^{K} (a(r - z) - (k + t - 1)) + \sum_{t=K+1}^{t_{\max}} (ar - (k + K)). \tag{12}$$

Let G be the total profit for the defender when commissioning pentests. We can show that penetration test independently contribute to the total profit of the defender.

Lemma 1. *For every fixed proactive defense k, the contributions to the expected total profit from individual reactive defenses to attacks and individual penetration tests are independent and additive.*

Proof. Let $\mathcal{M} = \{m_1, m_2 \ldots, m_{|\mathcal{M}|}\}$ be the ordered set of rounds where the defender commissions a pentest. Assuming $p = 1$ for now, we obtain

$$G = \sum_{t=1}^{K-|\mathcal{M}|} (a(r - z) - c_t) + \sum_{t=K-|\mathcal{M}|+1}^{K} (ar - c_t) + \sum_{t=K+1}^{t_{\max}} (ar - c_t),$$

where c_t is the cost at round t.

Let us now separate the saved losses due to pentesting in the second sum,

$$G = \sum_{t=1}^{K-|\mathcal{M}|} (a(r - z) - c_t) + \sum_{t=K-|\mathcal{M}|+1}^{K} (a(r - z) - c_t) + \sum_{t=K-|\mathcal{M}|+1}^{K} (az) + \sum_{t=K+1}^{t_{\max}} (ar - c_t)$$

$$= \sum_{t=1}^{K} (a(r - z)) + \sum_{t=K-|\mathcal{M}|+1}^{K} (az) - \sum_{t=1}^{K-|\mathcal{M}|} (c_t) - \sum_{t=K-|\mathcal{M}|+1}^{K} (c_t) + \sum_{t=K+1}^{t_{\max}} (ar - c_t). \tag{13}$$

We now develop the costs c_t for each period of the game. In the period of attacks, each pentest contributes one more to the total number of protected threats. In addition, each pentest costs c to perform. After the attacks stop, all links are protected and the defense cost remains $k + K$ for the rest of the game:

$$
G = \sum_{t=1}^{K}(a(r-z)) + |\mathcal{M}| \cdot (az) - |\mathcal{M}| \cdot c -
$$
$$
- \sum_{t=1}^{m_1-1}(k+t-1) - \sum_{i=1}^{|\mathcal{M}|-1} \sum_{t=m_i}^{m_{i+1}-1}(k+t+i-1) - \sum_{t=m_{|\mathcal{M}|}}^{K-|\mathcal{M}|}(k+t+|\mathcal{M}|-1) -
$$
$$
- \sum_{t=K-|\mathcal{M}|+1}^{K}(k+K) + \sum_{t=K+1}^{t_{max}}(ar - (k+K)).
$$

Note that if the last weak link is fixed by a pentest, then $m_{|\mathcal{M}|} = K - |\mathcal{M}| + 1$ and the 5th sum does not exist.

Now splitting the last but one sum results in

$$
G = \sum_{t=1}^{K}(a(r-z)) + |\mathcal{M}| \cdot (az) - |\mathcal{M}| \cdot c -
$$
$$
- \sum_{t=1}^{m_1-1}(k+t-1) - \sum_{i=1}^{|\mathcal{M}|-1} \sum_{t=m_i}^{m_{i+1}-1}(k+t+i-1) - \sum_{t=m_{|\mathcal{M}|}}^{K-|\mathcal{M}|}(k+t+|\mathcal{M}|-1) -
$$
$$
- \sum_{t=K-|\mathcal{M}|+1}^{K}(k+t-1) - \sum_{t=K-|\mathcal{M}|+1}^{K}(K-t+1) + \sum_{t=K+1}^{t_{max}}(ar - (k+K)).
$$

This algebraic manipulation allows us to separate the contribution of attacks and pentest to the total profit,

$$
G = \sum_{t=1}^{K}(a(r-z) - (k+t-1)) + |\mathcal{M}| \cdot (az-c) -
$$
$$
- \sum_{i=1}^{|\mathcal{M}|-1} \sum_{t=m_i}^{m_{i+1}-1}(i) - \sum_{t=m_{|\mathcal{M}|}}^{K-|\mathcal{M}|}(|\mathcal{M}|) -
$$
$$
- \sum_{t=K-|\mathcal{M}|+1}^{K}(K-t+1) + \sum_{t=K+1}^{t_{max}}(ar - (k+K)).
$$

Instead of writing the costs of pentesting per round, we rewrite them as a sum of costs per pentest,

$$
G = \sum_{t=1}^{K}(a(r-z) - (k+t-1)) + |\mathcal{M}| \cdot (az-c) - \sum_{i=1}^{|\mathcal{M}|} \sum_{t=m_i}^{K-i+1}(1) + \sum_{t=K+1}^{t_{max}}(ar - (k+K)).
$$

Finally, we can write the expression for the total profit as

$$G = \sum_{t=1}^{K}(a(r-z)-(k+t-1)) + \sum_{i=1}^{|\mathcal{M}|}(az-c-(K-i-m_i+2)) + \sum_{t=K+1}^{t_{max}}(ar-(k+K)).$$
(14)

The first sum is the contribution of attacks to the profit, the second sum shows the individual contributions of pentests and the last sum is the profit after the original attacks would have stopped without pentests.

A.2   Optimal Number of Pentests

Now we show a detailed derivation for the optimal number of pentests. Pentests are successful with probability p. Let $E[g]$ be the contribution of pentesting to the expected total profit (i.e., the second sum in (6)) and let us have a closer look at it. Clearly, penetration testing has to contribute a positive profit to be worth performing, i.e.,

$$E[g] = p\sum_{i=1}^{|\mathcal{M}|}(az - c - (K - i - m_i + 2)) > 0.$$
(15)

Now we use Theorem 1 and replace m_i by $m_1 + i - 1$,

$$E[g] = p\sum_{i=1}^{|\mathcal{M}|}(az - c - (K - m_1 - 2i + 3))$$

$$= p|\mathcal{M}| \cdot (az - c - K + m_1 - 3) + 2 \cdot \sum_{i=1}^{|\mathcal{M}|}(i)$$

$$= p|\mathcal{M}| \cdot (az - c - K + m_1 - 3) + |\mathcal{M}| \cdot (|\mathcal{M}| + 1)$$

$$= p|\mathcal{M}| \cdot (az - c - K + m_1 + |\mathcal{M}| - 2).$$
(16)

From (7) and using ϵ, we have:

$$K - m_1 = (1 + p)|\mathcal{M}| - 1 - \epsilon.$$
(17)

Hence, we can rewrite (16) as

$$E[g] = p|\mathcal{M}| \cdot (az - c - (1 + p)|\mathcal{M}| + 1 + \epsilon + |\mathcal{M}| - 2)$$

$$= p|\mathcal{M}| \cdot (az - c - p|\mathcal{M}| - 1 + \epsilon).$$
(18)

The series of pentests is worth performing if the expected profit $E[g]$ is positive, meaning that

$$E[g] = p|\mathcal{M}| \cdot (az - c - p|\mathcal{M}| - 1 + \epsilon) > 0.$$
(19)

Since $0 < m_1 \leq K$, we have $|\mathcal{M}| > 0$ from (7) and we can write the condition for pentesting:

$$az - c - p|\mathcal{M}| - 1 + \epsilon > 0$$

$$az - c - p\left\lceil\frac{K + 1 - m_1}{1 + p}\right\rceil - 1 + \epsilon > 0.$$

We can obtain the optimal number of pentests $|\mathcal{M}|^*$ as the value that maximizes (18),

$$|\mathcal{M}|^* = \max_{|\mathcal{M}|} p|\mathcal{M}| \cdot (az - c - p|\mathcal{M}| - 1 + \epsilon) = -p^2|\mathcal{M}|^2 + p|\mathcal{M}| \cdot (az - c - 1 + \epsilon).$$

Derivation gives us the maximum value as follows:

$$|\mathcal{M}|^* = \frac{az - c - 1 + \epsilon}{2p}, \tag{20}$$

where $0 \le |\mathcal{M}|^* \le \left\lceil \frac{K+\epsilon}{1+p} \right\rceil$. Note that the expression in (20) returns a real number that is optimal only asymptotically. The decision criterion can be discretized rounding off to the nearest integer or applying a randomized strategy.

A.3 Optimal Time to Start Pentesting

The substitution of (20) into (17) also gives us the optimal time to start pentesting m_1^*,

$$m_1^* = K - \frac{1+p}{2p}(az - c) + \frac{1 + 3p - \epsilon(1-p)}{2p}, \tag{21}$$

where $0 < m_1^* \le \left\lceil \frac{K+\epsilon}{1+p} \right\rceil + \epsilon$ holds.

A.4 Expected Total Profit of Pentesting

Substituting the optimal number of pentests into (6), we obtain an expression for the expected total profit with optimal number of pentests,

$$
\begin{aligned}
E[G] &= \sum_{t=1}^{K}(a(r-z) - (k+t-1)) + p\sum_{i=1}^{|\mathcal{M}|}(az - c - (K - i - m_i + 2)) + \\
&\quad + \sum_{t=K+1}^{t_{\max}}(ar - (k+K)) \\
&= K(a(r-z) - (k-1)) - \frac{K(K+1)}{2} + \\
&\quad + p|\mathcal{M}|^*(az - c - p|\mathcal{M}|^* - 1 + \epsilon)) + (t_{\max} - K)(ar - (k+K)) \\
&= K(a(r-z) - (k-1)) - \frac{K(K+1)}{2} + E[g] + (t_{\max} - K)(ar - (k+K)),
\end{aligned}
\tag{22}
$$

where $E[g]$ takes the values depending on the conditions in (20) as

$$
E[g] = \begin{cases}
0, & \text{if } |\mathcal{M}|^* = 0; \\
\left(\frac{az-c-1+\epsilon}{2}\right)^2, & \text{if } 0 < |\mathcal{M}|^* < \left\lceil \frac{K+\epsilon}{1+p} \right\rceil; \\
\frac{p}{1+p}\left\lceil \frac{K+\epsilon}{1+p} \right\rceil \left(az - c - K - 1 + \frac{p}{1+p}\left\lceil \frac{K+\epsilon}{1+p} \right\rceil\right), & \text{if } |\mathcal{M}|^* = \left\lceil \frac{K+\epsilon}{1+p} \right\rceil.
\end{cases}
\tag{23}
$$

Tracking Games in Mobile Networks

Mathias Humbert, Mohammad Hossein Manshaei,
Julien Freudiger, and Jean-Pierre Hubaux

LCA1, EPFL, Switzerland
{mathias.humbert,hossein.manshaei}@epfl.ch,
{julien.freudiger,jean-pierre.hubaux}@epfl.ch

Abstract. Users of mobile networks can change their identifiers in regions called mix zones in order to defeat the attempt of third parties to track their location. Mix zones must be deployed carefully in the network to reduce the cost they induce on mobile users and to provide high location privacy. Unlike most previous works that assume a global adversary, we consider a local adversary equipped with multiple eavesdropping stations. We study the interaction between the local adversary deploying eavesdropping stations to track mobile users and mobile users deploying mix zones to protect their location privacy. We use a game-theoretic model to predict the strategies of both players. We derive the strategies at equilibrium in complete and incomplete information scenarios and propose an algorithm to compute the equilibrium in a large network. Finally, based on real road-traffic information, we numerically quantify the effect of complete and incomplete information on the strategy selection of mobile users and of the adversary. Our results enable system designers to predict the best response of mobile users with respect to a local adversary strategy, and thus to select the best deployment of countermeasures.

Keywords: Location Privacy, Game Theory, Mobile Networks, Mix Zone.

1 Introduction

The advanced communication capabilities of mobile devices (e.g., WiFi or Bluetooth) enable the use of a new breed of mobile applications: mobile devices can directly communicate in a peer-to-peer wireless fashion and exchange *contextual* information, for example, about road-traffic conditions [16] or social presence [2, 25]. In such applications, mobile devices must unveil their identifiers (e.g., pseudonyms or cryptographic credentials) to authenticate and identify each other.

Yet, an adversary eavesdropping on such peer-to-peer wireless communications can, based on their identifiers, track mobile users. In order to protect their location privacy, mobile nodes can use multiple pseudonyms that they change over time. This approach has been adopted in cellular networks to achieve location privacy with respect to external eavesdroppers: cellular operators identify

T. Alpcan, L. Buttyan, and J. Baras (Eds.): GameSec 2010, LNCS 6442, pp. 38–57, 2010.
© Springer-Verlag Berlin Heidelberg 2010

their subscribers with a "Temporary Mobile Subscriber Identity" (TMSI). Every time a subscriber moves to a new geographical area, the cellular operator issues a new TMSI. The use of multiple pseudonyms has also been investigated to protect location privacy in mobile ad hoc networks [4, 10, 23]: in order to impede the linkability of old and new pseudonyms by using spatial and temporal correlation, pseudonym changes are coordinated in regions called *mix zones* [4]. In a mix zone, mobile users alter their spatial correlations by changing their pseudonyms, and their temporal correlations by: (i) remaining silent for a short period [17, 23], (ii) encrypting their communications [9], or (iii) using a mobile proxy [30]. We call these regions *active* mix zones. Mix zones must be carefully deployed in the network to reduce the cost they induce on users and to provide high location privacy. Indeed, the placement of mix zones affects their performance [18] and traversing mix zones incurs a communication overhead [31].

In contrast with most previous works on location privacy [3, 8, 10, 13, 23], we do not restrict our model to a global adversary. The cost might be prohibitive for an adversary to build and maintain a global eavesdropping system and to sort and process all the received information. Instead, we consider a *local adversary* with a limited budget and that eavesdrops on communications in only certain regions of the network. In the worst case, a local adversary has an unlimited budget and becomes global. The local adversary has to strategically deploy its eavesdropping stations to gather information from the network. Mobile users can take advantage of the presence of a local adversary and change pseudonyms in regions where the adversary has no coverage [6]. We call these regions *passive* mix zones.

In this paper, we investigate the strategic behavior of mobile users deploying active and passive mix zones to protect their location privacy and the behavior of a local adversary deploying eavesdropping stations to track mobile users. To do so, we develop a game-theoretic framework to predict the strategies of the adversary and of mobile users. We refer to these games as *tracking games*. We first analyze the interaction between users and the adversary in a single road intersection with *complete information*: the adversary and mobile users know each others' strategies and payoffs. We obtain one pure-strategy Nash equilibrium and one mixed-strategy Nash equilibrium [26]. We generalize the results to a network of intersections using the notion of supergames [11]. Then, we relax the complete information assumption because mobile users may not know the position of eavesdropping stations, and we study the *incomplete information* scenario. We prove the existence of one pure-strategy Bayesian Nash equilibrium [15] in the single road intersection game and extend the result to a network of intersections. Finally, we test our model using real road traffic statistics from Lausanne, Switzerland, and obtain two important results. First, in complete information scenarios, mobile users and the adversary tend to adopt complementary strategies: users place mix zones where there is no eavesdropping station, and the adversary deploys eavesdropping stations where there is no mix zone. Second, in incomplete information scenarios, the location privacy level achieved by mobile users depends on their level of uncertainty about the strategy of the adversary.

To the best of our knowledge, this paper is the first investigation of the strategic aspects of tracking games in mobile networks. Previous works aim at optimizing privacy-preserving mechanisms with respect to a worst case adversary [3, 10, 13]. In contrast, game theory allows us to further analyze the interactions between privacy-conscious nodes and the adversary in order to predict their strategies. In this direction, previous works investigate pursuit-evasion games (e.g., [20]) in which several users cooperate to locate one target user. Tracking games complement this existing work by considering a new type of game in which several users collaborate to protect their location privacy against a rational adversary equipped with local eavesdropping devices. Our results allow system designers to predict the strategies of a local adversary and mobile users with a limited budget. This paper is part of the trend of blending game theory with security to predict the strategies of the rational parties involved [1, 5, 7, 8, 12, 14, 21, 29, 33].

This paper is organized as follows. In Section 2, we present the system and threat models, and describe how mix zones provide location privacy. We introduce the game-theoretic framework in Section 3 and analyze it in Section 4. We provide the main numerical results based on real-traffic data in Section 5 and conclude in Section 6.

2   Preliminaries

In this section, we present the assumptions made throughout the paper. We also introduce mix zones and define a metric to measure location privacy.

2.1   Mobile Network Model

We study a system composed of mobile nodes moving in a road network of K intersections. Nodes are equipped with peer-to-peer wireless communication technologies (e.g., WiFi) and can communicate with other nodes in transmission range. Mobile devices identify each other using pseudonyms [27]. In order to prevent tracking by third parties, we assume that mobile nodes use multiple pseudonyms that they change over time. An offline Certification Authority (CA) run by an independent trusted third party provides mobile users, prior to entering the network, with a set of pseudonyms, such as public/private key pairs.

For each intersection, we assume the knowledge of accurate statistics: the parties know the number of vehicles per hour driving through any specific path, i.e., for each entering and exiting road pair. In practice, such information can be provided by city authorities in charge of road traffic optimization. Based on these statistics, we express the traffic intensity for each specific path in each intersection. The traffic intensity is defined in a normalized form as:

$$\lambda_i = \frac{n_i}{\mu_{\max}} \tag{1}$$

where n_i is the number of nodes going through intersection or road i per unit of time and $\mu_{\max}$ is the maximum number of nodes driving through any intersection

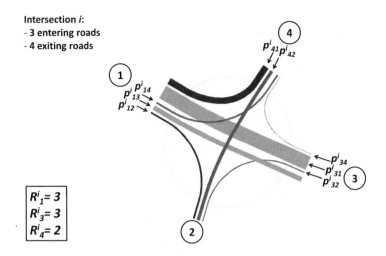

Fig. 1. Intersection i. Road 2 is one-way: no vehicle can enter the intersection from there. The width of each flow is proportional to the traffic intensity. Vehicles entering by roads 1 and 3 have 3 possible exits on each of them, whereas vehicles entering by road 4 have 2 possible exits.

of the network per unit of time. Figure 1 shows an example of one particular intersection i.

2.2   Threat Model

We consider a *local* adversary $\mathcal{A}$ that aims at tracking nodes: $\mathcal{A}$ has a limited number of eavesdropping stations to deploy in the network. As road intersections are strategic points of the network (through which all mobile nodes pass), we assume that the local adversary deploys its eavesdropping stations only at these places. Eavesdropping stations have a coverage area large enough to detect mobile nodes entering and exiting the intersection.

We assume a *passive* adversary: $\mathcal{A}$ cannot inject or modify messages [7]. $\mathcal{A}$ collects pseudonyms sniffed at every intersection where it has an eavesdropping station. Based on the collected information, it attempts to track the location of mobile nodes. Hence, the adversary threatens the location privacy of nodes [4].

2.3   Location Privacy Model

In order to defeat the tracking by an adversary, nodes can use multiple pseudonyms that they change over time. Nodes must coordinate pseudonym changes in regions called mix zones in order to prevent the spatial and temporal correlation of their location. We can distinguish between two types of mix zones: first, those that, besides the pseudonym change, request user action, such as turning their transceivers off [17, 23] or using a mobile proxy[30]; second, mix zones where the nodes merely take advantage of the adversary's lack of coverage to change pseudonyms without any other action [6]. In this paper, we refer to the former as *active mix zones* and the latter as *passive mix zones*. In the following, we consider active mix zones created using silent periods.

We now quantify the location privacy provided by active mix zones in the presence of an attacker that eavesdrops on communications. As proposed in a previous work [3], we measure the uncertainty of $\mathcal{A}$ in matching mobile nodes that enter and exit an active mix zone. The uncertainty of the adversary is measured with an information-theoretic metric, the entropy [32]. To generalize this measure to an entire intersection, we compute the normalized entropy for each incoming road k and sum over all possible incoming roads with a weighted factor based on traffic intensity λ_k. We then divide the result by λ_i to get a normalized entropy H_i at intersection i:

$$H_i = \frac{1}{\lambda_i} \sum_{\forall k} \lambda_k \frac{-\sum_{\forall j} p_{kj}^i \log_2 p_{kj}^i}{\log_2 R_k^i} \tag{2}$$

where R_k^i is the total number of possible outgoing roads when entering at road k in intersection i, and p_{kj}^i is the probability that a node coming in intersection i via road k leaves via road j. The normalized entropy H_i captures the uncertainty of the adversary about the direction of nodes exiting an intersection.

Assuming that the monitoring and correlation processes become more difficult for the adversary with a higher number of nodes within the intersection, the uncertainty increases with the number of nodes entering the mix zone. Thus, we assume that the mixing effectiveness at intersection i is $m_i = \lambda_i H_i$, where λ_i is the total traffic intensity at intersection i.

In passive mix zones, mobile nodes can change pseudonyms in regions where the adversary has no coverage while continuing to communicate. However, if nodes change pseudonyms in a region where the adversary eavesdrops, the mixing effectiveness becomes equal to zero because the adversary can easily link nodes before and after a pseudonym change. If there is no eavesdropping station, we have $m_i = 1$. Note that we assume that at least two nodes traverse a passive mix zone and change pseudonyms.

3   A Game-Theoretic Approach to Location Privacy

In order to model the interaction between a local adversary and mobile nodes wanting to protect their location privacy, we define a static game $G=(\mathcal{P}, \mathcal{S}, \mathcal{U})$. $\mathcal{P} = \{\mathcal{N}, \mathcal{A}\}$ is the players' set, where $\mathcal{N}$ corresponds to the aggregation of mobile nodes and $\mathcal{A}$ represents the adversary. $\mathcal{S}$ is the strategies' set. At any given intersection i, nodes can either *abstain* (A), deploy an *active mix zone* (M) or a *passive mix zone* (P), whereas the adversary can either *abstain* (A) or *eavesdrop* (E) on wireless communications. Thus, we get $\mathcal{S} = \{\mathcal{S}_\mathcal{N}^i, \mathcal{S}_\mathcal{A}^i\}_{i=1}^K$ with

Table 1. Normal form of game G at intersection i

$\mathcal{N}\backslash\mathcal{A}$	Eavesdrop (E)	Abstain (A)
Active mix zone (M)	$(\lambda_i m_i - c_p^i - c_q^i, \lambda_i(1 - m_i) - c_s)$	$(\lambda_i - c_p^i - c_q^i, 0)$
Passive mix zone (P)	$(-c_p^i, \lambda_i - c_s)$	$(\lambda_i - c_p^i, 0)$
Abstain (A)	$(0, \lambda_i - c_s)$	$(0, 0)$

$S_{\mathcal{N}}^i = \{M, P, A\}$ and $S_{\mathcal{A}}^i = \{E, A\}$. Finally, $\mathcal{U}$ is the payoffs' set, where utility u for each player is equal to benefit b minus cost c.

When a player *abstains*, it has neither benefits nor costs, its payoff being zero (Table 1). An eavesdropping station is worth c_s for the adversary, regardless of the intersection i. On the nodes' side, a passive mix zone (P) and an active mix zone (M) cost $c_p^i = \alpha\lambda_i$ and $c_m^i = c_p^i + c_q^i = (\alpha + \beta)\lambda_i$, respectively. Value c_p^i encompasses the cost of acquiring new pseudonyms, whereas c_q^i is the cost of remaining silent for a certain period. When the adversary plays E and the nodes play M, the benefit of nodes is proportional to the mixing effectiveness m_i and the traffic intensity at intersection i (i.e., $\lambda_i m_i$) whereas the attacker's benefit is proportional to $(1 - m_i)$ and the traffic intensity (i.e., $\lambda_i(1 - m_i)$). If the adversary plays A, m_i is equal to 1 because the nodes are not tracked. Thus, the nodes' benefit is λ_i and the adversary's benefit is zero. If the nodes play P or A while the adversary plays E, nodes lose all their privacy benefits and the attacker earns a maximal benefit (i.e., λ_i). Note finally that all players' costs (c_m^i and c_s) and benefits (λ_i and m_i) are normalized between zero and one.

In real life, nodes may not know the total amount of investment $\Gamma \cdot c_s$ (Γ being the number of eavesdropping stations that the attacker can afford) made by the adversary to eavesdrop on the communications, and thus its stations' number and position around the network. Nodes have *incomplete information* about the attacker's strategy and payoff. To solve this problem, Harsanyi [15] proposes to introduce a new player called *Nature* that turns an incomplete information game into an *imperfect information* game. To do so, Nature assigns a type θ to the adversary's power according to a *probability density function* $f(\theta)$ known to the nodes. We assume here that the adversary is aware of the nodes' costs c_p^i and c_q^i. We thus have an asymmetric information game, meaning that the information sets of the players differ in ways relevant to their behavior. The adversary has useful private information: an information partition that is different and not worse than that of the nodes [28]. Table 2 summarizes the notation used throughout the paper.

4   Game Results

In this section, we first analyze the complete information game, at one and then at K intersections ($\mathcal{C}_1$-game and $\mathcal{C}_K$-game). Then, we extend the analysis to the incomplete information $\mathcal{I}_1$-game and $\mathcal{I}_K$-game.

Table 2. List of symbols

Symbol	Definition
K	Number of intersections in the network
$\mathcal{N}$	Mobile nodes
$\mathcal{A}$	Adversary
λ_i	Normalized traffic intensity at intersection i
m_i	Mixing effectiveness of an active mix zone at intersection i
c_p^i	Nodes' cost of changing pseudonyms at intersection i
α	Cost of changing pseudonym per node
c_q^i	Nodes' cost of remaining silent at intersection i
β	Cost of being silent per node
c_m^i	Active mix zone cost: $c_p^i + c_q^i$
c_s	Adversary's cost of installing an eavesdropping station
θ	Nodes' belief in the type of the adversary
$f(\theta)$	Probability density function of the nodes' belief
$F(\theta)$	Cumulative distribution function of the nodes' belief
Γ	Total number of eavesdropping stations
$z_\mathcal{A}^i$	Nodes' belief in the presence of an eavesdropping station at intersection i
$u_\mathcal{N}^i$	Nodes' payoff function at intersection i
$u_\mathcal{A}^i$	Adversary's payoff function at intersection i
$s_{\mathcal{N},i}$	Nodes' strategy at intersection i
$s_{\mathcal{A},i}$	Adversary's strategy at intersection i
$s_{\mathcal{N},i}^*$	Nodes' best response at intersection i
$s_{\mathcal{A},i}^*$	Adversary's best response at intersection i
$u_{tot}^\mathcal{N}$	Nodes' global payoff function
$u_{tot}^\mathcal{A}$	Adversary's global payoff function

4.1 Complete Information Game

We begin the analysis with $\mathcal{C}_1$-game. The following theorem identifies all Nash equilibria (NE) of the game at one intersection with complete information.[1]

Theorem 1. *The $\mathcal{C}_1$-game has either a single pure-strategy Nash equilibrium:*

$$(s_{\mathcal{N},i}^*, s_{\mathcal{A},i}^*) = \begin{cases} (M, E) & \text{if} & (c_s < \lambda_i(1 - m_i)) \wedge (c_m^i < \lambda_i m_i) \\ (P, A) & \text{if} & (c_s > \lambda_i) \wedge (c_p^i < \lambda_i) \\ (A, E) & \text{if} & (c_s < \lambda_i) \wedge (c_m^i > \lambda_i m_i) \\ (A, A) & \text{if} & (c_s > \lambda_i) \wedge (c_p^i > \lambda_i) \end{cases}$$

or a single mixed-strategy NE:

$$(s_{\mathcal{N},i}^*, s_{\mathcal{A},i}^*) = (x_\mathcal{N}^i, x_\mathcal{A}^i) \quad \text{if} \quad (\lambda_i(1 - m_i) < c_s < \lambda_i) \wedge (c_m^i < \lambda_i m_i)$$

where $x_\mathcal{N}^i = \frac{\lambda_i - c_s}{\lambda_i m_i}$ is the probability of using an active mix zone at intersection i and $x_\mathcal{A}^i = \min(\frac{c_q^i}{\lambda_i m_i}, 1)$ is the probability of eavesdropping at intersection i. Moreover, $P(s_{\mathcal{N},i}^* = P) = 1 - x_\mathcal{N}^i$ and $P(s_{\mathcal{N},i}^* = A) = 0$.

---

[1] For convenience's sake, we focus in this paper on strict inequalities between benefits and costs.

Proof. We first distinguish five different cases that encompass all possible scenarios. For four of them, we get pure-strategy Nash equilibria, computed by finding both players' best responses in Table 1. In the last case, if

$$\begin{cases} \lambda_i(1 - m_i) < c_s < \lambda_i \\ c_m^i < \lambda_i m_i \end{cases}$$

there is no pure-strategy Nash equilibrium. However, we can derive a mixed-strategy Nash equilibrium. As nodes' strategy A is dominated by strategy M, it will never be used by the nodes. Then, we can find the mixed-strategy Nash equilibrium by simply finding the mixed-strategy Nash equilibrium of the 2-by-2 game shown in Table 3.

Table 3. Reduced $\mathcal{C}_1$-game for mixed-strategy Nash equilibrium

$\mathcal{N}\backslash\mathcal{A}$	Eavesdrop (E)	Abstain (A)
Active mix zone (M)	$(\lambda_i m_i - c_p^i - c_q^i, \lambda_i(1 - m_i) - c_s)$	$(\lambda_i - c_p^i - c_q^i, 0)$
Passive mix zone (P)	$(-c_p^i, \lambda_i - c_s)$	$(\lambda_i - c_p^i, 0)$

Assuming that

$$\begin{cases} Pr(s_{\mathcal{N},i} = M) = x_{\mathcal{N}}^i \\ Pr(s_{\mathcal{A},i} = E) = x_{\mathcal{A}}^i \end{cases},$$

we can solve

$$\begin{cases} x_{\mathcal{A}}^i(\lambda_i m_i - c_p^i - c_q^i) + (1 - x_{\mathcal{A}}^i)(\lambda_i - c_p^i - c_q^i) = -x_{\mathcal{A}}^i c_p^i + (1 - x_{\mathcal{A}}^i)(\lambda_i - c_p^i) \\ x_{\mathcal{N}}^i(\lambda_i(1 - m_i) - c_s) + (1 - x_{\mathcal{N}}^i)(\lambda_i - c_s) = 0 \end{cases},$$

and obtain the following mixed-strategy Nash equilibrium:

$$\begin{cases} P\{s_{\mathcal{N}}^i = M\} = \frac{\lambda_i - c_s}{\lambda_i m_i} \\ P\{s_{\mathcal{N}}^i = P\} = 1 - \frac{\lambda_i - c_s}{\lambda_i m_i} \\ P\{s_{\mathcal{A}}^i = E\} = \min(\frac{c_q^i}{\lambda_i m_i}, 1) \\ P\{s_{\mathcal{A}}^i = A\} = \max(1 - \frac{c_q^i}{\lambda_i m_i}, 0) \end{cases} \qquad \square$$

Theorem 1 shows that participants' strategies at NE are highly dependent on the traffic profiles at each specific intersection. The adversary plays E at NE either if the eavesdropping cost is low ($c_s < \lambda_i(1 - m_i)$), or if it is not too high ($c_s < \lambda_i$) and the nodes do not use an active mix zone at the same place. The nodes play M if c_m^i is small enough for given traffic intensity and mixing effectiveness ($c_m^i < \lambda_i m_i$). If the adversary abstains and the cost of changing pseudonym is not prohibitive ($c_p^i < \lambda_i$), they play P. Nodes abstain if the adversary is eavesdropping and c_m^i is not small enough to be beneficial for them. For a high pseudonym cost ($c_p^i > \lambda_i$), nodes abstain as well, regardless of the adversary's strategy. Finally, if c_s is neither too high nor too low and c_m^i small, players' best responses do not converge to a pure-strategy NE, leading to a mixed-strategy NE as defined in the theorem.

We will now extend the $\mathcal{C}_1$-game to the $\mathcal{C}_K$-game for K intersections. The $\mathcal{C}_K$-game can be viewed as a *supergame* with K simultaneous moves as defined in [24]. Because the strategy profiles are independent at different intersections and the set of strategies is not restricted by any constraints, both players can determine their best responses with $\mathcal{C}_1$-games at K intersections and aggregate them to get their $\mathcal{C}_K$-game best responses. This *supergame* NE can be defined by the union of the K NE of $\mathcal{C}_1$-games as follows:

$$(s_{\mathcal{N}}^*, s_{\mathcal{A}}^*) = \bigcup_{i=1}^{K}(s_{\mathcal{N},i}^*, s_{\mathcal{A},i}^*) \tag{3}$$

and the *supergame* payoff is the sum of payoffs provided by each $\mathcal{C}_1$-game:

$$\begin{cases} u_{tot}^{\mathcal{N}}(s_{\mathcal{N}}^*, s_{\mathcal{A}}^*) = \sum_{i=1}^{K} u_{\mathcal{N}}^i(s_{\mathcal{N},i}^*, s_{\mathcal{A},i}^*), \text{ for the nodes} \\ u_{tot}^{\mathcal{A}}(s_{\mathcal{N}}^*, s_{\mathcal{A}}^*) = \sum_{i=1}^{K} u_{\mathcal{A}}^i(s_{\mathcal{N},i}^*, s_{\mathcal{A},i}^*), \text{ for the adversary} \end{cases} \tag{4}$$

However, a local adversary cannot afford an unlimited number of eavesdropping stations. The total number of eavesdropping stations is thus assumed to be capped by an upper bound Γ. Consequently, the NE strategy profile $(s_{\mathcal{N}}^*, s_{\mathcal{A}}^*)$ of the $\mathcal{C}_K^{\Gamma}$-game can be defined as:

$$s_{\mathcal{N}}^* \in \arg\max_{s_{\mathcal{N}}} u_{tot}^{\mathcal{N}}(s_{\mathcal{N}}, s_{\mathcal{A}}^*) \tag{5}$$

$$\begin{cases} s_{\mathcal{A}}^* \in \arg\max_{s_{\mathcal{A}}} u_{tot}^{\mathcal{A}}(s_{\mathcal{N}}^*, s_{\mathcal{A}}) \\ \text{subject to } \sum_{i=1}^{K} \mathbf{1}_{s_{\mathcal{A},i}=E} \leq \Gamma \end{cases} \tag{6}$$

where the ith row of vectors $s_{\mathcal{N}}$ and $s_{\mathcal{A}}$ is $s_{\mathcal{N},i}$ and $s_{\mathcal{A},i}$, respectively.

Algorithm BoundedAdvCoverage copes with the new constraint on adversary's eavesdropping stations in the $\mathcal{C}_K^{\Gamma}$-game. This algorithm enables us to find the equilibrium of the game under the adversary's constraint for the whole network.

In BoundedAdvCoverage, we assume that $m_1 < m_2 < ... < m_K$, i.e. the first intersection has the lowest mixing effectiveness. Using Theorem 1, the algorithm first computes independently the Nash equilibria at each intersection (line 1). Then, if the total number of eavesdropping stations among the K intersections is larger than Γ, the adversary has to remove some of them.

First, the adversary changes strategy from $x_{\mathcal{A}}^i$ to A at the intersections where it has mixed strategies (lines 2 to 6). As the expected payoff of the adversary at mixed-strategy NE is equal to zero, it will not lose anything with this change. Note that the adversary starts with the intersection that has a mixed-strategy NE with smallest i (line 2), as this removes a mixed strategy with the highest probability of eavesdropping. If the first move is not sufficient, it considers the next intersection with a mixed-strategy NE. This continues until either the number of eavesdropping stations is smaller than Γ, or there are no more intersections with mixed-strategy NE. In the latter case, the adversary then moves to

Algorithm 1. BoundedAdvCoverage.

1: compute the Nash equilibria at each intersection $\Rightarrow (s_{\mathcal{N}}^*; s_{\mathcal{A}}^*)$
2: $i = 1$
3: **while** $(\sum_{i=1}^{K} \mathbf{1}_{s_{\mathcal{A},i}^*=E} > \Gamma) \wedge ((s_{\mathcal{N},i}^*; s_{\mathcal{A},i}^*) = (x_{\mathcal{N}}^i, x_{\mathcal{A}}^i))$ **do**
4:    $(s_{\mathcal{N},i}^*; s_{\mathcal{A},i}^*) = (P; A)$
5:    $i = i + 1$
6: **end while**
7: **while** $(\sum_{i=1}^{K} \mathbf{1}_{s_{A,i}^*=E} > \Gamma)$ **do**
8:    $j = \arg\min_{i,u_{\mathcal{A}}^i \neq 0} u_{\mathcal{A}}^i(s_{\mathcal{N},i}^*, s_{\mathcal{A},i}^*)$
9:    **if** $(c_p^j < \lambda_j)$ **then**
10:       $(s_{\mathcal{N},j}^*, s_{\mathcal{A},j}^*) = (P, A)$
11:    **else**
12:       $(s_{\mathcal{N},j}^*, s_{\mathcal{A},j}^*) = (A, A)$
13:    **end if**
14: **end while**

the second step of the algorithm (line 7) and removes its eavesdropping stations at intersections with pure-strategy NE, starting with the intersection where its payoff is the smallest (line 8). In this case, each time the adversary changes strategy, it reduces its number of eavesdropping stations by one. The adversary obviously stops this removal process when the constraint Γ is satisfied.

As nodes do not have any constraints on cost, they just concentrate on their best responses with respect to the new strategy of the adversary. The nodes' best response if the adversary does not have any eavesdropping station is to deploy a passive mix zone if and only if $c_p^i < \lambda_i$ (line 9). In this case, a new local equilibrium appears: $(s_{\mathcal{N},i}^*, s_{\mathcal{A},i}^*) = (P, A)$ (line 10). Whereas, if $c_p^i > \lambda_i$, the new NE is $(s_{\mathcal{N},i}^*, s_{\mathcal{A},i}^*) = (A, A)$ (line 12).

Theorem 2. *The $\mathcal{C}_K^\Gamma$-game has a single Nash equilibrium, provided by the K $\mathcal{C}_1$-games equilibria and the* BoundedAdvCoverage *algorithm.*

Proof. The BoundedAdvCoverage algorithm removes the eavesdropping stations in order to maximize the payoff of the adversary with the available eavesdropping stations, i.e. Γ. This algorithm also derives the nodes' best response with respect to the new adversary's strategy. Hence, the strategy profile $(s_{\mathcal{N}}^*, s_{\mathcal{A}}^*)$ is an equilibrium because no player is interested in unilaterally changing strategy. □

4.2 Incomplete Information Game

We extend the analysis to $\mathcal{I}$-games, where the mobile nodes have incomplete information about the adversary's payoff and strategy. Nodes must predict the attacker's best strategy based on the probability distribution $f(\theta)$ representing the nodes' belief in the adversary's type. For the purpose of analysis, we suppose that the nodes know Γ but do not know c_s that will be modeled by θ. Indeed, if c_s increases, the adversary will need more money if it wants to deploy the same number of eavesdropping stations. The power of the adversary is always relative to the cost of eavesdropping.[2]

Definition 1. *The strategy profile $(s^*_{\mathcal{N},i}, s^*_{\mathcal{A},i})$ is a pure-strategy Bayesian Nash equilibrium (BNE) of the $\mathcal{I}_1$-game at intersection i if and only if*

$$
\begin{cases}
s^*_{\mathcal{N},i} \in \underset{s_{\mathcal{N},i} \in \mathcal{S}^i_{\mathcal{N}}}{\arg\max}\; E_\theta[u^i_{\mathcal{N}}(s_{\mathcal{N},i}, s^*_{\mathcal{A},i}(\theta))] \\
s^*_{\mathcal{A},i} \in \underset{s_{\mathcal{A},i} \in \mathcal{S}^i_{\mathcal{A}}}{\arg\max}\; u^i_{\mathcal{A}}(s^*_{\mathcal{N},i}, s_{\mathcal{A},i})
\end{cases}
\tag{7}
$$

Let $z^i_{\mathcal{A}} = Pr\{s^*_{\mathcal{A},i} = E\}$ be the probability that the adversary installs an eavesdropping station at intersection i, in a given equilibrium. The following lemma provides the computation of $z^i_{\mathcal{A}}$ by the nodes, for any given distribution of adversary's type.

Lemma 1. *Supposing that $F(\theta)$ is the cumulative distribution function of the type of the eavesdropping station's cost, the nodes will assume that the adversary will play E at intersection i with probability*

$$
z^i_{\mathcal{A}} =
\begin{cases}
F(\lambda_i(1 - m_i)) + \min(\frac{c^i_q}{\lambda_i m_i}, 1)(F(\lambda_i) - F(\lambda_i(1 - m_i))) & \text{if } c^i_m < \lambda_i m_i \\
F(\lambda_i) & \text{if } c^i_m > \lambda_i m_i
\end{cases}
\tag{8}
$$

Proof. Nodes would like to express the probability that the adversary places an eavesdropping station based on the distribution probability $f(\theta)$ of the cost's type of such an eavesdropping station. First, let us define the cumulative distribution function of the cost's type:

$$
F(\theta) = P(\Theta < \theta) = \int_0^\theta f(u)du
$$

Moreover,

$$
P(a < \Theta < b) = \int_a^b f(u)du = F(b) - F(a)
$$

Assuming that nodes know the probability density function (and thus the cumulative distribution function), they can evaluate $z^i_{\mathcal{A}} = P(s^*_{\mathcal{A},i} = E)$ using the law of total probability:

$$
\begin{aligned}
P(s^*_{\mathcal{A},i} = E) = {} & P(s^*_{\mathcal{A},i} = E | \Theta < \lambda_i(1 - m_i))P(\Theta < \lambda_i(1 - m_i)) \\
& + P(s^*_{\mathcal{A},i} = E | \lambda_i(1 - m_i) < \Theta < \lambda_i)P(\lambda_i(1 - m_i) < \Theta < \lambda_i) \\
& + P(s^*_{\mathcal{A},i} = E | \Theta > \lambda_i)P(\Theta > \lambda_i)
\end{aligned}
$$

---

[2] It is similar to the purchasing power of consumers, which is relative to the level of goods/services' prices.

As $P(s^*_{\mathcal{A},i} = E|\Theta > \lambda_i) = 0$ and $P(s^*_{\mathcal{A},i} = E|\Theta < \lambda_i(1 - m_i)) = 1$, we get

$$
\begin{aligned}
P(s^*_{\mathcal{A},i} = E) &= P(\Theta < \lambda_i(1 - m_i)) \\
&\quad + P(s^*_{\mathcal{A},i} = E|\lambda_i(1 - m_i) < \Theta < \lambda_i)P(\lambda_i(1 - m_i) < \Theta < \lambda_i) \\
&= F(\lambda_i(1 - m_i)) \\
&\quad + P(s^*_{\mathcal{A},i} = E|\lambda_i(1 - m_i) < \Theta < \lambda_i)(F(\lambda_i) - F(\lambda_i(1 - m_i)))
\end{aligned}
$$

There remains to express $P(s^*_{\mathcal{A},i} = E|\lambda_i(1 - m_i) < \Theta < \lambda_i)$. Nodes can evaluate this probability using results of Theorem 1:

$$
P(s^*_{\mathcal{A},i} = E|\lambda_i(1 - m_i) < \Theta < \lambda_i) = \begin{cases} 1 & \text{if } c^i_m > \lambda_i m_i \\ \min(\frac{c^i_q}{\lambda_i m_i}, 1) & \text{if } c^i_m < \lambda_i m_i \end{cases} \qquad \square
$$

Using Lemma 1, the nodes can then find their best response that maximizes their payoff. This is shown with the following lemma.

Lemma 2. *The nodes' best response in $\mathcal{I}_1$-game is:*

$$
s^*_{\mathcal{N},i} = \begin{cases} M & \text{if} & (c^i_q < z^i_{\mathcal{A}}\lambda_i m_i) \wedge (c^i_m < \lambda_i(1 - z^i_{\mathcal{A}}(1 - m_i))) \\ P & \text{if} & (c^i_q > z^i_{\mathcal{A}}\lambda_i m_i) \wedge (c^i_p < (\lambda_i(1 - z^i_{\mathcal{A}}))) \\ A & \text{if} & (c^i_m > \lambda_i(1 - z^i_{\mathcal{A}}(1 - m_i))) \wedge (c^i_p > \lambda_i(1 - z^i_{\mathcal{A}})) \end{cases} \qquad (9)
$$

Proof. First, let us explicitly write the expected payoff:

$$
E_\theta[u^i_{\mathcal{N}}(s_{\mathcal{N},i}, s^*_{\mathcal{A},i}(\theta))] = z^i_{\mathcal{A}} u^i_{\mathcal{N}}(s_{\mathcal{N},i}, s^*_{\mathcal{A},i} = E) + (1 - z^i_{\mathcal{A}})u^i_{\mathcal{N}}(s_{\mathcal{N},i}, s^*_{\mathcal{A},i} = A)
$$

In order to get $s^*_{\mathcal{N},i} = M$, we must verify both conditions below:

$$
\begin{cases} E_\theta[u^i_{\mathcal{N}}(s_{\mathcal{N},i} = M, s^*_{\mathcal{A},i}(\theta))] > E_\theta[u^i_{\mathcal{N}}(s_{\mathcal{N},i} = P, s^*_{\mathcal{A},i}(\theta))] \\ E_\theta[u^i_{\mathcal{N}}(s_{\mathcal{N},i} = M, s^*_{\mathcal{A},i}(\theta))] > E_\theta[u^i_{\mathcal{N}}(s_{\mathcal{N},i} = A, s^*_{\mathcal{A},i}(\theta))] \end{cases}
$$

or, explicitly:

$$
\begin{cases} z^i_{\mathcal{A}}(\lambda_i m_i - c^i_p - c^i_q) + (1 - z^i_{\mathcal{A}})(\lambda_i - c^i_p - c^i_q) > -z^i_{\mathcal{A}}c^i_p + (1 - z^i_{\mathcal{A}})(\lambda_i - c^i_p) \\ z^i_{\mathcal{A}}(\lambda_i m_i - c^i_p - c^i_q) + (1 - z^i_{\mathcal{A}})(\lambda_i - c^i_p - c^i_q) > 0 \end{cases}
$$

or, by simplifying:

$$
\begin{cases} c^i_q < z^i_{\mathcal{A}}\lambda_i m_i \\ c^i_p + c^i_q = c^i_m < \lambda_i(1 - z^i_{\mathcal{A}}(1 - m_i)) \end{cases}
$$

We can prove in the same way both other best responses. For $s^*_{\mathcal{N},i} = P$,

$$
\begin{cases} E_\theta[u^i_{\mathcal{N}}(s_{\mathcal{N},i} = P, s^*_{\mathcal{A},i}(\theta))] > E_\theta[u^i_{\mathcal{N}}(s_{\mathcal{N},i} = M, s^*_{\mathcal{A},i}(\theta))] \\ E_\theta[u^i_{\mathcal{N}}(s_{\mathcal{N},i} = P, s^*_{\mathcal{A},i}(\theta))] > E_\theta[u^i_{\mathcal{N}}(s_{\mathcal{N},i} = A, s^*_{\mathcal{A},i}(\theta))] \end{cases}
$$

must be verified, and

$$\begin{cases} E_\theta[u^i_\mathcal{N}(s_{\mathcal{N},i} = A, s^*_{\mathcal{A},i}(\theta))] > E_\theta[u^i_\mathcal{N}(s_{\mathcal{N},i} = M, s^*_{\mathcal{A},i}(\theta))] \\ E_\theta[u^i_\mathcal{N}(s_{\mathcal{N},i} = A, s^*_{\mathcal{A},i}(\theta))] > E_\theta[u^i_\mathcal{N}(s_{\mathcal{N},i} = P, s^*_{\mathcal{A},i}(\theta))] \end{cases}$$

for $s^*_{\mathcal{N},i} = A$. □

Note that the adversary has complete information, and consequently can obtain its best response using the calculated payoffs in Table 1. This is shown by the following lemma.

Lemma 3. *The adversary's best response of the $\mathcal{I}_1$-game is*

$$s^*_{\mathcal{A},i} = \begin{cases} E & if & (c_s < \lambda_i(1 - m_i)) \vee ((\lambda_i(1 - m_i) < c_s < \lambda_i) \wedge (s^*_{\mathcal{N},i} \neq M)) \\ A & if & (c_s > \lambda_i) \vee ((\lambda_i(1 - m_i) < c_s < \lambda_i) \wedge (s^*_{\mathcal{N},i} = M)) \end{cases}$$

(10)

Considering Lemmas 1, 2 and 3, we immediately have the following theorem.

Theorem 3. *The $\mathcal{I}_1$-game has at least one pure-strategy Bayesian Nash equilibrium.*

Proof. As the Bayesian NE is defined by the players' mutual best responses (Definition 1), the result follows from Lemmas 1, 2, and 3. □

Note that, comparing to the $\mathcal{C}_1$-game, (M, A) and (P, E) can also be pure-strategy BNE for the $\mathcal{I}_1$-game. For example, (M, A) is a BNE if the nodes believe that the cost of an eavesdropping station is small, whereas in reality the actual cost of an eavesdropping station is high (typically greater than λ_i). If the mobile nodes had perfect knowledge about the adversary's payoff, they would have deployed a passive mix zone instead of an active mix zone. Similarly, the nodes deploy passive mix zone at (P, E) BNE due to incomplete information about the adversary, which degrades their location privacy.

We now generalize our $\mathcal{I}_1$-game to the $\mathcal{I}_K^\Gamma$-game by aggregating all the equilibria at each intersection and sum the payoffs of all intersections to obtain the *supergame* payoffs for both participants. Similarly, the BNE strategy profile $(s^*_\mathcal{N}, s^*_\mathcal{A})$ can be expressed as:

$$s^*_\mathcal{N} \in \arg\max_{s_\mathcal{N}} \sum_{i=1}^K E_\theta[u^i_\mathcal{N}(s_{\mathcal{N},i}, s^*_{\mathcal{A},i}(\theta))]$$

(11)

$$\begin{cases} s^*_\mathcal{A} \in \arg\max_{s_\mathcal{A}} \sum_{i=1}^K u^i_\mathcal{A}(s^*_{\mathcal{N},i}, s_{\mathcal{A},i}) \\ \text{subject to } \sum_{i=1}^K \mathbf{1}_{s_{\mathcal{A},i}=E} \leq \Gamma \end{cases}$$

(12)

BayesianBoundedAdvCoverage algorithm enables the players to find the BNE of the $\mathcal{I}_K^\Gamma$-game.

The algorithm first computes the BNE at each intersection independently, using Theorem 3. Then, the adversary removes eavesdropping stations at intersections where they provide the smallest payoffs (lines 3 and 4), until its total

---

Algorithm 2. BayesianBoundedAdvCoverage.

1: compute the Bayesian Nash equilibria at each intersection $\Rightarrow (s_\mathcal{N}^*; s_\mathcal{A}^*)$
2: **while** $(\sum_{i=1}^{K} \mathbf{1}_{s_{\mathcal{A},i}=E}^* > \Gamma)$ **do**
3:     $j = \underset{i, u_\mathcal{A}^i \neq 0}{\arg\min} \; u_\mathcal{A}^i(s_{\mathcal{N},i}^*, s_{\mathcal{A},i}^*)$
4:     $s_{\mathcal{A},j}^* = A$
5: **end while**
6: **for** $i = 1 : K - \Gamma$ **do**
7:     $j = \underset{i \neq k, \forall k < i}{\arg\min} \; E[u_\mathcal{A}^i(s_{\mathcal{N},i}^*, s_{\mathcal{A},i}^* = E)]$
8:     **if** $(c_p^j < \lambda_j)$ **then**
9:       $s_{\mathcal{N},j}^* = P$
10:     **else**
11:       $s_{\mathcal{N},j}^* = A$
12:     **end if**
13: **end for**

Table 4. Number of NE (among all intersections)

scenario\NE	$(\mathbf{M}, \mathbf{E})$	$(\mathbf{A}, \mathbf{E})$	$(\mathbf{P}, \mathbf{A})$	mixed
$c_s = 0.1, \Gamma = 23$	17	6	0	0
$c_s = 0.1, \Gamma = 5$	2	3	18	0
$c_s = 0.5, \Gamma = 23$	2	3	5	13
$c_s = 0.5, \Gamma = 5$	2	3	18	0

number of eavesdropping stations satisfies the upper bound Γ. The mobile nodes find the $K - \Gamma$ intersections where the expected payoff of the adversary playing E is the smallest (line 7). Indeed, these intersections are those where there is the highest probability that the adversary removes its eavesdropping stations. The nodes then play P at these intersections if $c_p^j < \lambda_j$ or play A if c_p^j is prohibitive.

5   Numerical Results

In this section, we evaluate our game-theoretic model by means of numerical results based on traffic data[3] from Lausanne [22]. For convenience, we concentrate on the $K = 23$ main intersections of Lausanne and use Matlab to numerically evaluate the results. We test both the $\mathcal{C}_K^\Gamma$-game and the $\mathcal{I}_K^\Gamma$-game, with different costs. Benefits depend on the traffic parameters λ_i and m_i.

5.1   Complete Information Game

Table 4 summarizes the results with different players' costs in the complete information scenario. In all of the four cases, nodes' costs are fixed: $c_p^i = \alpha\lambda_i = 0.1\lambda_i$ and $c_q^i = \beta\lambda_i = 0.1\lambda_i$. We sum the different NE at each intersection and

---

[3] The data are publicly available on http://icapeople.epfl.ch/mhumbert/tracking.

Table 5. Number of Bayesian Nash equilibria (BNE) among all intersections

scenario\BNE	(M, E)	(P, E)	(A, E)	(M, A)	(P, A)	(A, A)
$\theta \sim U(0,1), c_s = 0.2, \Gamma = 23$	10	13	0	0	0	0
$\theta \sim U(0,1), c_s = 0.2, \Gamma = 5$	1	4	0	0	18	0
$\theta \sim \beta(2,5), c_s = 0.2, \Gamma = 23$	16	3	4	0	0	0
$\theta \sim \beta(2,5), c_s = 0.2, \Gamma = 5$	1	0	4	0	18	0
$\theta \sim \beta(2,5), c_s = 0.5, \Gamma = 23$	2	0	2	14	3	2
$\theta \sim \beta(2,5), c_s = 0.5, \Gamma = 5$	1	1	2	0	17	2

provide the results for two values of c_s (0.1 and 0.5). For each case, we solve the game with an unlimited and a limited number of stations ($\Gamma = 23$ and $\Gamma = 5$).

In the first scenario, as c_s is very low, the adversary plays E at each intersection. On the contrary, the nodes decide to abstain at six intersections, where m_i is too low to get a significant benefit, despite the relatively low price of an active mix zone (Figure 2(a)). In the second scenario, the adversary keeps eavesdropping stations at two intersections where there are active mix zones, instead of placing them at intersections free of mix zones (Figure 2(b)). This is due to the fact that, at those two intersections, the number of vehicles per hour is quite high, with a mixing effectiveness that does not confuse the adversary too much ($m_i < 0.5$). Finally, we notice that the nodes take advantage of their complete knowledge of the adversary's payoff to use passive mix zones wherever the attacker ceases eavesdropping.

If c_s increases to 0.5 (third and fourth scenarios), the adversary deploys fewer eavesdropping stations, five in total without any limit on the stations' number (Figure 3). The eavesdropping stations tend to be placed at intersections with the lowest mixing effectiveness. Most surprising here is that the nodes' best responses change as well, showing that they are not independent of the adversary's strategies. Except for two intersections, the nodes and the adversary adopt complementary strategies. If the adversary places an eavesdropping station, the nodes abstain, whereas, if the adversary abstains, the nodes place a (passive) mix zone. If we limit the number of stations to five, we get the same resulting equilibrium as in Figure 2(a) and reach the same conclusions.

5.2   Incomplete Information Game

We model the imperfect nodes' knowledge of c_s by using two different probability distributions. First, the uniform distribution $U(0,1)$ represents the case when mobile nodes have no idea about c_s. Second, the beta distribution $\beta(2,5)$[4] models the case when the nodes' belief in c_s is more accurate. Table 5 summarizes the results of the $\mathcal{I}_K^\Gamma$-game.

In the first scenario, we notice that there are 13 intersections where the nodes deploy passive mix zones while the adversary is eavesdropping at the same places.

---

[4] The beta distribution is a family of continuous probability distributions defined on the interval $[0,1]$. $\beta(2,5)$ is maximal in 0.2 and its mean is equal to $2/7$.

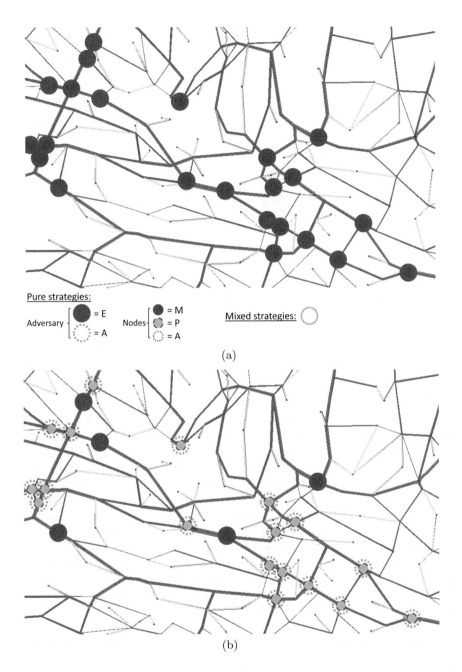

Fig. 2. Maps of Lausanne downtown and strategies chosen (at the main 23 intersections) with $\alpha = 0.1$, $\beta = 0.1$ and $c_s = 0.1$. (a) Equilibrium with an unlimited number of eavesdropping stations, (b) Equilibrium with a limited number of eavesdropping stations (equal to five).

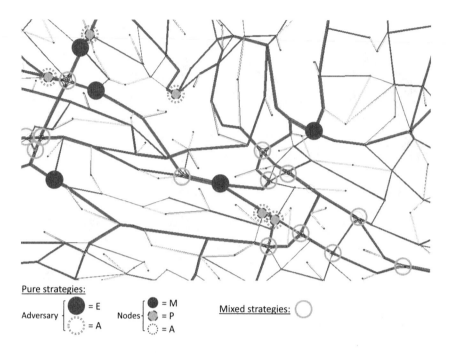

Fig. 3. Map of downtown Lausanne with $\alpha = 0.1$, $\beta = 0.1$ and $c_s = 0.5$. Equilibrium with an unlimited number of eavesdropping stations.

Nodes lose all their location privacy and pay the cost of changing pseudonyms, which leads to a negative payoff. The nodes have no clue about c_s, and thus must lay a bet on the adversary's payoff. Nodes believe that c_s is close to $E[\Theta] = 0.5$, whereas $c_s = 0.2$. Thus, the nodes think that the adversary will not play E everywhere, whereas it will, because of low actual c_s. We also notice that the nodes privilege passive mix zones at intersections with low m_i and active mix zones where m_i is higher. This is surprising because nodes should expect that the adversary places eavesdropping stations where the m_i is low, and thus deploy active mix zones at these intersections, instead of passive ones. In the second scenario, the nodes take advantage of the limited number of eavesdropping stations to deploy more passive mix zones. However, the nodes still have passive mix zones at four intersections out of five where the adversary keeps eavesdropping. Hence, in this case either, the BNE is not optimal for the nodes.

In the third and fourth scenarios, we observe that if nodes' knowledge about c_s becomes more accurate, the nodes' strategy at equilibrium leads to a higher payoff. There are three and no (P, E) in the third and fourth cases, respectively. The nodes know that $E[\Theta] = 2/7 \approx 0.29$, which is quite close to $c_s = 0.2$, leading to a much better strategy than with the uniform distribution.

The last two cases depict a nodes' wrong belief in c_s. Their belief is the same as in cases 3 and 4, but the real c_s is higher. This inaccuracy leads to a decrease on the nodes' payoff at BNE but not as significant as with a uniform distribution.

We can observe this especially in the fifth scenario. In this case, there are 14 (M, A) at BNE, whereas with a good knowledge on c_s, the nodes would have played P instead of M. Thus, nodes adopt non-optimal strategies, leading to a decrease in payoff equal to c_q^i (for intersection i). We also notice in the last case a single (P, E) and a single (M, E). The difference between these two intersections is in the value of m_i (both values of λ_i are high). In the former intersection, $m_i = 0.42$, whereas in the latter $m_i = 0.35$. Thus, nodes probably believe that the adversary ceases eavesdropping at the intersection with highest m_i, whereas it does not.

6   Conclusion

We have considered the problem of deploying mix zones in the presence of a passive adversary equipped with a limited number of eavesdropping stations. We have proposed a game-theoretic model to evaluate the strategic behaviors of both players in such *tracking games*. First, we analyze the complete information game and derive an algorithm to obtain NE strategy profiles for a large network. Second, we evaluate the incomplete information game where mobile nodes are uncertain about the placement of eavesdropping stations. We obtain a single pure-strategy Bayesian NE at one intersection. We also describe an algorithm to obtain the equilibrium in a large network. Finally, we evaluate using real road traffic statistics both the complete information and incomplete information games. Among other results, the numerical evaluations show that the adversary and mobile nodes often adopt complementary strategies when they have complete information: nodes place (passive) mix zones at locations where there are no eavesdropping stations, whereas the adversary deploys eavesdropping stations at places where there are no (active) mix zones. In the incomplete information case, we notice that mobile nodes' strategy (and thus payoff) highly depends on their belief about the type of adversary. Our results quantify how the lack of information by mobile nodes about the attacker's strategy leads to a significant decrease in the achievable location privacy level at BNE. In summary, our results enable system designers to predict the strategy of a local adversary and mobile nodes with limited capabilities in tracking games.

For future work, we intend to test our results by using traffic data from other cities and more precisely measure the mixing effectiveness using the sojourn times and the evolution of traffic over time. Moreover, we would like to extend our results to other kinds of mobile networks, such as pedestrian ones. We would also like to enrich our analysis by developing a scenario where the attacker leverages on the geographical positions and the interdependence of the intersections to improve his tracking power. This approach would require more complex strategies and utility functions, and the games at different intersections would no longer be independent [19]. Another extension of this work is the evaluation of the interactions between the attacker and the defenders by using repeated games.

Acknowledgements

We would like to thank Tansu Alpcan, Igor Bilogrevic, Joseph Y. Halpern and Georgios Theodorakopoulos for their insights and discussions about the game-theoretic analysis. We also thank Nevena Vratonjic and the anonymous reviewers for their helpful feedback. We are very grateful to Jean-Pierre Leyvraz for providing us with the traffic data of Lausanne.

References

1. Acquisti, A., Dingledine, R., Syverson, P.: On the economics of anonymity. In: Wright, R.N. (ed.) FC 2003. LNCS, vol. 2742, pp. 84–102. Springer, Heidelberg (2003)
2. Aka-aki, http://www.aka-aki.com
3. Beresford, A.R., Stajano, F.: Location privacy in pervasive computing. IEEE Pervasive Computing 2, 46–55 (2003)
4. Beresford, A.R., Stajano, F.: Mix zones: User privacy in location-aware services. In: Proceedings of the Second IEEE Annual Conference on Pervasive Computing and Communications Workshops (2004)
5. Buchegger, S., Alpcan, T.: Security games for vehicular networks. In: 46th Annual Allerton Conference on Communication, Control, and Computing (2008)
6. Buttyán, L., Holczer, T., Vajda, I.: On the effectiveness of changing pseudonyms to provide location privacy in VANETs. In: Stajano, F., Meadows, C., Capkun, S., Moore, T. (eds.) ESAS 2007. LNCS, vol. 4572, pp. 129–141. Springer, Heidelberg (2007)
7. Buttyán, L., Hubaux, J.P.: Security and Cooperation in Wireless Networks. Cambridge University Press, Cambridge (2008)
8. Freudiger, J., Manshaei, M.H., Hubaux, J.P., Parkes, D.C.: On non-cooperative location privacy: a game-theoretic analysis. In: Proceedings of the 16th ACM Conference on Computer and Communications Security (2009)
9. Freudiger, J., Raya, M., Felegyhazi, M., Papadimitratos, P., Hubaux, J.P.: Mix zones for location privacy in vehicular networks. In: Proc. 1st Intl. Wksp. Wireless Networking for Intelligent Transportation Systems (Win-ITS) (2007)
10. Freudiger, J., Shokri, R., Hubaux, J.P.: On the optimal placement of mix zones. In: Goldberg, I., Atallah, M.J. (eds.) Privacy Enhancing Technologies. LNCS, vol. 5672, pp. 216–234. Springer, Heidelberg (2009)
11. Friedman, J.W.: A non-cooperative equilibrium for supergames. The Review of Economic Studies 38, 1–12 (1971)
12. Grossklags, J., Johnson, B., Christin, N.: The price of uncertainty in security games. In: Proceedings (online) of the Eighth Workshop on the Economics of Information Security (WEIS), London, UK (2009)
13. Gruteser, M., Grunwald, D.: Anonymous usage of location-based services through spatial and temporal cloaking. In: Proceedings of the 1st International Conference on Mobile Systems, Applications and Services (2003)
14. Halpern, J.Y., Teague, V.: Rational secret sharing and multiparty computation: extended abstract. In: Proceedings of the Thirty-Sixth Annual ACM Symposium on Theory of Computing, pp. 623–632 (2004)
15. Harsanyi, J.: Games with incomplete information played by Bayesian players. Management Science 14, 159–182 (1967)

16. Hartenstein, H., Laberteaux, K.: A tutorial survey on vehicular ad hoc networks. IEEE Communications Magazine 46(6), 164–171 (2008)
17. Huang, L., Matsuura, K., Yamane, H., Sezaki, K.: Enhancing wireless location privacy using silent period. In: IEEE Wireless Communications and Networking Conference (2005)
18. Huang, L., Matsuura, K., Yamane, H., Sezaki, K.: Towards modeling wireless location privacy. In: Danezis, G., Martin, D. (eds.) PET 2005. LNCS, vol. 3856, pp. 59–77. Springer, Heidelberg (2006)
19. Humbert, M.: Location privacy amidst local eavesdroppers. Master's thesis, EPFL (2009)
20. Isaacs, R.: Differential games: a mathematical theory with applications to warfare and pursuit, control and optimization. Dover Publications, New York (1999)
21. Katz, J.: Bridging game theory and cryptography: recent results and future directions. In: Canetti, R. (ed.) TCC 2008. LNCS, vol. 4948, pp. 251–272. Springer, Heidelberg (2008)
22. Leyvraz, J.P., Mattenberger, P., Robert-Grandpierre, A.: Mise à jour majeure de la modélisation EMME2 de l'agglomération Lausanne-Morges. Technical Report TRANSP-OR 061208, EPFL (2006)
23. Li, M., Sampigethay, K., Huang, L., Poovendra, R.: Swing & swap: User centric approaches towards maximizing location privacy. In: Proceedings of the 5th ACM Workshop on Privacy in Electronic Society (2006)
24. Luce, R.D., Raiffa, H.: Games and Decisions. Wiley, Chichester (1957)
25. MIT Media Lab: Reality Mining,
    http://reality.media.mit.edu/serendipity.php
26. Nash, J.: Non-cooperative games. Annals of Mathematics 54 (1951)
27. Pfitzmann, A., Köhntopp, M.: Anonymity, unobservability, and pseudonymity - a proposal for terminology. In: Federrath, H. (ed.) Designing Privacy Enhancing Technologies. LNCS, vol. 2009, p. 1. Springer, Heidelberg (2001)
28. Rasmusen, E.: Games and information. Blackwell, Malden (1989)
29. Raya, M., Manshaei, M.H., Felegyhazi, M., Hubaux, J.P.: Revocation games in ephemeral networks. In: Proceedings of the 16th ACM Conference on Computer and Communications Security (2008)
30. Sampigethaya, K., Huang, L., Li, M., Poovendran, R., Matsuura, K., Sezaki, K.: CARAVAN: providing location privacy for VANET. In: Proceedings of Embedded Security in Cars (ESCAR) (2005)
31. Schoch, E., Kargl, F., Leinmüller, T., Schlott, S., Papadimitratos, P.: Impact of pseudonym changes on geographic routing in VANETs. In: Buttyán, L., Gligor, V.D., Westhoff, D. (eds.) ESAS 2006. LNCS, vol. 4357, pp. 1–2. Springer, Heidelberg (2006)
32. Serjantov, A., Danezis, G.: Towards an information theoretic metric for anonymity. In: Dingledine, R., Syverson, P.F. (eds.) PET 2002. LNCS, vol. 2482, pp. 41–53. Springer, Heidelberg (2003)
33. Varian, H.R.: Economic aspects of personal privacy. In: Internet Policy and Economics (2009)

gPath: A Game-Theoretic Path Selection Algorithm to Protect Tor's Anonymity

Nan Zhang[1], Wei Yu[2], Xinwen Fu[3], and Sajal K. Das[4]

[1] George Washington University
nzhang10@gwu.edu
[2] Towson University
wyu@towson.edu
[3] University of Massachusetts, Lowell
xinwenfu@cs.uml.edu
[4] The University of Texas at Arlington
das@cse.uta.edu

Abstract. In this paper, we address the problem of defending against entry-exit linking attacks in Tor, a popular anonymous communication system. We formalize the problem as a repeated non-cooperative game between the defender and the adversary (i.e., controller of the compromised Tor nodes to carry out entry-exit linking attacks). Given the current path selection algorithm of Tor, we derive an optimal attack strategy for the adversary according to its utility function, followed by an optimal defensive strategy against this attack. We then repeat such interactions for three additional times, leading to three design principles, namely *stratified path selection*, *bandwidth order selection*, and *adaptive exit selection*. We further develop *gPath*, a path selection algorithm that integrates all three principles to significantly reduce the success probability of linking attacks. Using a combination of theoretical analysis and experimental studies on real-world Tor data, we demonstrate the superiority of our algorithm over the existing ones.

1 Introduction

Anonymous Communication Systems: In this paper, we address the problem of defending against entry-exit linking attacks in Tor, a popular anonymous communication system [1, 2]. Tor is an overlay system over the existing Internet. It consists of thousands of computers distributed around the world, each of which is called a *Tor node* and donates its bandwidth to relay traffic in the Tor network. Along with the Tor nodes, the Tor network maintains a *Tor directory server* which holds descriptive information about all Tor nodes.

A *Tor client* can be a computer or mobile device which makes use of Tor - it does not have to be a Tor node. To achieve anonymous communication with a destination host, a Tor client in general chooses three Tor nodes (namely entry, middle, and exit nodes) based on information stored in the directory server (we shall describe the details of node selection later in the paper), and then constructs a path (called *Tor circuit*) with these three nodes to relay packets to/from the destination. To construct the path, the Tor client needs to negotiate session keys with all three Tor nodes on the path. This is because Tor uses layered encryption - i.e., at each hop, a Tor node removes a layer of encryption to

T. Alpcan, L. Buttyan, and J. Baras (Eds.): GameSec 2010, LNCS 6442, pp. 58–71, 2010.

learn where to forward the packet. The destination, after receiving the packet, sees it as coming from the exit Tor node on the path. Note that each Tor node on the path also has the knowledge of the preceding node. Thus, a return packet (from the destination) can take an exact reverse of the path and arrive at the Tor client. One can see that the destination is oblivious to the fact that it is actually communicating with the Tor client, hence achieving anonymous communication.

Defense Against Entry-Exit Linking Attacks: While the volunteer-based nature of Tor reduces its chance of being controlled/manipulated by a single organization or government, recent studies discovered powerful techniques for a small number of malicious Tor nodes to collude with each other and identify the real source and destination of a packet. There are various types of such linking attacks. For example, one type of attacks manipulates the packet inter-arrival time with a secret pattern at the entry node, and then observes whether such pattern emerges at the exit node [3, 4]. Another type intentionally replays a packet at the entry node [5]. Since Tor uses counter mode AES (AES-CTR) for encryption, the duplicate packet will incur a decryption error the exit node, causing a pair of colluding entry-exit nodes to identify the real source and destination of a packet. A common property of these attacks is that they require the colluding nodes to be selected as both the entry and the exit nodes of a Tor path.

There are two possible strategies to defend against such entry-exit linking attacks. One is to revise the design of Tor communication protocol specifically counteract each entry-exit linking attack (e.g., use padding to counteract timing-based attacks [6, 7]). The other is to revise the path selection algorithm to minimize the probability for an adversary to control both entry and exit of a Tor path. Comparing these two strategies, the first one may completely eliminate a linking attack while the second one naturally cannot - there is always a small albeit nonzero probability for the entry and exit nodes to collude with each other. Nonetheless, the second strategy produces a more generic solution that may significantly reduce the success probability of entry-exit linking attacks. We focus on the second strategy in this paper.

Various path selection rules have already been adopted by Tor to reduce the chance of selecting two colluding nodes as entry and exit. For example, to construct a path, Tor picks at most one node from a /16 subnetwork (i.e., with a common 16-bit routing prefix). While this rule may limit the success of a resource-limited adversary, an adversary with sufficient resources may soon learn to bypass it by spreading the malicious nodes across different /16 networks. Similar to the arm races between the adversary and the defender in many other domains (e.g., email spam detection [8], intrusion detection [9, 10]), once the path selection algorithm is updated to detect certain types of colluding nodes, adversaries start using new ones to evade such detections.

To the best of our knowledge, no systematic approach has been developed to formally analyze the arm race between the design of the path selection algorithm and the adversarial strategies for deploying collusive Tor nodes. Indeed, it is unclear how to counteract entry-exit linking attacks with the presence of a smart adversary that strategically deploys a set of malicious Tor nodes into the system.

Outline of Technical Results: In this paper, we present the first game-theoretic study on the design of path selection algorithm against entry-exit linking attacks. Our goal is to provide a set of principles for the design of path selection algorithms.

To this end, we first observe that a path selection algorithm has to make a proper tradeoff between the *privacy* and *utility* of the constructed path. The key privacy-preserving measure we consider in the paper is *the success rate of an entry-exit linking attack*. The key utility factor we consider is the *bandwidth* of the constructed Tor path. To achieve a higher utility, the path selection algorithm should use high-bandwidth nodes as much as possible for path construction. However, an adversary may adapt to this strategy and inject high-bandwidth nodes to facilitate linking attacks, leading to a dilemma on the design of path selection algorithms. On the other hand, the adversary is also bandwidth-sensitive in that the bandwidth it can assign to malicious nodes is not unlimited. Thus, the adversary must determine an optimal distribution of bandwidth to the colluding nodes under its control.

We formalize the problem as a repeated non-cooperative game between a bandwidth-sensitive path selection algorithm and a bandwidth-sensitive adversary which controls a number of colluding nodes. Given the current path selection algorithm of Tor, we derive the optimal strategy for the adversary. Then, we produce the optimal path selection algorithm against this adversarial strategy. We repeat such interactions for three times, leading to three design principles that we then prove to significantly reduce the success probability of linking attacks without substantially affecting the bandwidth of the constructed Tor paths. The three design principles are listed as follows:

- The path selection algorithm should *draw the entry and exit nodes from separate pools*. That is, an exit-eligible node should not be considered for the entry.
- While choosing the entry and middle nodes, the path selection algorithm should ensure that *the middle node has equal or higher bandwidth than the entry node*.
- The path selection algorithm should *draw the entry and middle nodes before choosing the exit node based on the bandwidths of the entry and middle nodes*. In particular, let entry and middle nodes' bandwidth be b_i and b_j, respectively. To draw the exit node, all nodes with bandwidth not exceeding $\min(b_i, b_j)$ should be chosen with equal probability[1].

Next, we develop gPath, a path selection algorithm that integrates all three principles to effectively reduce the success probability of linking attacks. We derive theoretical bounds on the performance of gPath and validate the analytical results with experiments on real-world data of Tor nodes. Both theoretical and experimental results demonstrate the effectiveness of our algorithm. In particular, while our algorithm incurs virtually no loss on the utility (i.e., bandwidth) of constructed Tor paths, it is capable of reducing the success probability of entry-exit linking attacks by orders of magnitude.

The rest of the paper is organized as follows. We introduce the system framework in Section 2, followed by a game-theoretic model in Section 3. Then, we present an analysis of the repeated interactions in Section 4. Also in this section, we derive the three design principles and describe gPath, our path selection algorithm. We present the experimental evaluation in Section 5, followed by a review of the related work in Section 6 and the final remarks in Section 7.

---

[1] Since we focus on the bandwidth factor in the paper, here we assume that the entry, middle, and exit nodes are from different subnetworks.

2   System Model

2.1   Model of Tor

Consider a system consisting of n Tor nodes $\Omega = \{v_1, \ldots, v_n\}$. Each node v_i specifies two parameters: the donated bandwidth b_i and whether it is willing to serve as an exit node, denoted by c_i. Here $c_i = 1$ if v_i is willing to serve, and 0 otherwise. Without loss of generality, normalize (and discretize, if necessary) the bandwidth to an integer $b_i \in [1, 100]$. Let $\Omega_1 \subseteq \Omega$ be the subset of nodes that are willing to serve as exit nodes, i.e., $\Omega_1 = \{v_i | c_i = 1\}$.

Each path $P : \langle v_i, v_j, v_k \rangle$ constructed by Tor consists of the following three nodes: the entry v_i, the middle v_j, and the exit v_k. There must be $v_k \in \Omega_1$. For our purpose, it suffices to consider the bandwidth of a path to follow a simple model of $b_P = \min(b_i, b_j, b_k)$.

2.2   Adversary Model

We consider an adversary which launches entry-exit linking attacks to compromise the communication anonymity of Tor users. In the following, we first describe the model of such an attack, and then define the adversarial strategy to launch it.

Let Ω_A be the set of nodes in Ω which the adversary has compromised. As we described in Section 1, the entry-exit linking attack succeeds for a constructed path $P : \langle v_i, v_j, v_k \rangle$ if and only if $v_i \in \Omega_A$ and $v_k \in \Omega_A$.

To launch such an entry-exit linking attack, an adversary has three possible strategies: (i) add nodes controlled by the adversary into Tor network, (ii) compromise existing Tor nodes, or (iii) a combination of these two strategies. Barring the difference between the cost associated with adding a new Tor node and compromising an existing one, these three strategies are essentially equivalent for the purpose of this paper because of two reasons: First, in practice, the number of Tor nodes that can be controlled by an adversary is significantly smaller than the total number of Tor nodes. Thus, adding or compromising the same number of nodes will lead to almost equal percentage of compromised Tor nodes in the system. Second, once compromised, an existing Tor node can change its bandwidth and exit-eligibility, i.e., b_i and c_i. Thus, the space of parameter settings available to the adversary remains the same for all strategies.

Therefore, without loss of generality, we shall consider an universal setting where the adversary is in control of a subset of nodes Ω. The adversarial strategy is to set b_i and c_i for each compromised node such that the entry-exit linking attack has the maximum success probability $\Pr\{\{v_i, v_k\} \subseteq P\}$. We shall introduce the precise semantics of the adversarial strategy in the game-theoretic setting presented in the next subsections.

2.3   Performance of Path Selection

The performance of path selection should be measured in terms of two metrics: (1) *Privacy* (resilience against linking attacks): The construction of $P : \langle v_i, v_j, v_k \rangle$ should minimize the success probability of linking attack, i.e., $\Pr\{\{v_i, v_k\} \subseteq \Omega_A\}$. (2) *Utility* (efficiency of communication): The construction of $P : \langle v_i, v_j, v_k \rangle$ should maximize the path bandwidth $b_P = \min(b_i, b_j, b_k)$.

3   Proposed Game-Theoretic Framework

We consider a non-cooperative game between the defender (i.e., the designer of the path selection algorithm) and the adversary (i.e., controller of the compromised Tor nodes to carry out entry-exit linking attacks). In this section, we define the strategies and the utility functions for the adversary and the defender, respectively. After that, we discuss the objective of our game-theoretic analysis.

3.1   Adversary and Defender's Strategies

Adversary's Strategy: The adversary is controlling a subset of nodes in $\{v_1, \ldots, v_n\}$. Suppose that each adversary can only control (up to) n_A nodes where $1 \ll n_A \ll n$. Let the adversary-controlled subset be denoted as Ω_A where $|\Omega_A| = n_A$. The strategy of the adversary can be formalized as choosing b_i and c_i for all $v_i \in \Omega_A$.

Defender's Strategy: Since the adversarial nodes may be distributed not only over different subnetworks but also geographically over different locations, we consider the worst-case scenario where the defender does not have the ability to distinguish between honest and malicious Tor nodes. As such, the defensive strategy is essentially the path selection algorithm, i.e., how to construct a path with three parties: the entry N_1, the middle node N_2, and the exit node N_3. Such a construction algorithm can be modeled as selecting the probability for any three given nodes to be selected in oder - i.e., selecting a function $f(b_i, c_i, b_j, c_j, b_k) \in [0, 1]$ for three nodes v_i, v_j, v_k which is the probability for the three nodes to be chosen as N_1, N_2, N_3, respectively, when $c_k = 1$.

3.2   Adversary and Defender's Utility Functions

Adversary's Objective: The adversary only has one objective: to compromise the anonymity of as many paths as possible. Recall that the adversary breaks the anonymity of a path if and only if it compromises both the entry and exit nodes. Thus, the adversary's objective can be formalized as maximizing its utility function

$$u_A(f, \Omega_A) = \sum_{v_i, v_k \in \Omega_A, v_j \in \Omega \text{ and } c_k=1} f(b_i, c_i, b_j, c_j, b_k), \tag{1}$$

i.e., the probability for the adversary to compromise a path generated by f.

Defender's Objective: The objective of the defender is two-fold: One is to minimize $u_A(f, \Omega_A)$ to protect the anonymity of communication. The other is to maximize the bandwidth of constructed path. In particular, consider

$$\beta(f) = \sum_{v_i, v_j, v_k \in \Omega, c_k=1} (f(b_i, c_i, b_j, c_j, b_k) \cdot \min(b_i, b_j, b_k)), \tag{2}$$

which is the expected bandwidth of a path constructed by the defender's strategy f. The defender's objective can be formalized as maximizing its utility function

$$u_D(f, \Omega_A) = r_D \cdot \beta(f) - (1 - r_D) \cdot u_A(f, \Omega_A), \tag{3}$$

where $r_D \in [0, 1]$ is the *preference parameter* of the defender while capturing the defender's preference between protecting anonymity and maximizing bandwidth.

3.3 Objective of Game-Theoretic Analysis

Note that both the attacking and defensive strategies form finite sets. Thus, Nash equilibrium (i.e., a state in which no player can benefit by unilaterally changing its strategy) always exist for the game. Nonetheless, it may not be practical to expect that the defender and adversary can actually reach this state due mainly to the possible irrationality of players and the intractability of computing the Nash equilibrium. Thus, in this paper we do *not* focus on the derivation of the Nash equilibrium. Instead, we consider the problem from a practical standpoint and analyze the strategies of the defender and the adversary in an iterative fashion. In particular, we consider a repeated game where each player makes a move sequentially, starting from the current path selection algorithm used by Tor. In the next section, we shall analyze the first four rounds of this repeated game, and derive three design principles that will not only hold for these iterations, but also in general can be applied to the design of any Tor path selection algorithm. Note that while the analysis of future rounds (i.e., fifth and later) of interactions is possible, we did not find additional design principles from such analysis which can be intuitively explained and applied to the design of path selection algorithms in practice.

4 Interactions

In this section, we consider four rounds of interactions between the defender and the adversary. Each round consists of two steps: (i) the defender makes a move on its path selection algorithm by devising an optimal countermeasure to the adversarial strategy in the previous round (in the first round, the defender will devise the current Tor algorithm), and (ii) the adversary will respond by changing its strategy to the optimal one against the defender's new algorithm. The defensive algorithm used in the first step is a simplified version of the current algorithm used in Tor [1, 2]. For each subsequent round, we shall first describe the main idea of our defensive strategy, and then derive the optimal attacking strategy against it. Finally, we draw observations from the defensive strategy design in each step to form the three design principles that we propose.

4.1 Round 1: Basic Design

Defender (Tor)'s Current Strategy: We start with a simplified version of the path selection strategy currently used in Tor. With this strategy, the exit node is first selected from all nodes with $c_i = 1$. The selection is preferential according to a node's (claimed) bandwidth. In particular, suppose that the probability of selecting a node v_i is proportional to $w(b_i) \in [0, 1]$. According to the current strategy adopted by Tor, $w(b_i)$ satisfies the following two properties: (1) When $b_i < b_L$ where b_L is a pre-determined threshold, $w(b_i)$ monotonically increases with an increasing b_i, because a larger bandwidth is preferred by the path selection algorithm. (2) When $b_i \geq b_L$, $w(b_i) = w(b_L)$.

The introduction of the threshold b_L is to prevent a Tor node from claiming an arbitrarily large bandwidth to maximize its selection probability. After the exit node is selected, the entry and the middle nodes are selected, in turn, from all remaining nodes with probability proportional to the weight function $w(\cdot)$. Besides the natural rule that the entry, middle, and exit nodes must be unique, the selections of these three nodes are independent of each other.

Adversary's Strategy: We now analyze the optimal attacking strategy against the current Tor path selection algorithm. Note that the attack success probability is monotonic with b_i and b_k. Thus, the optimal strategy for the adversary is to make the bandwidths of compromised nodes as large as b_L. In addition, since an exit-eligible node is considered in the selection of all three nodes, the adversary should make all compromised nodes exit-eligible. In other words, $c_i = 1$ for all $v_i \in \Omega_A$. In this case,

$$u_A = \frac{n_A \cdot w(b_L)}{w_{1R} + n_A \cdot w(b_L)} \cdot \frac{(n_A - 1) \cdot w(b_L)}{w_R + (n_A - 1) \cdot w(b_L)}, \tag{4}$$

where $w_R = \sum_{i \in \Omega \setminus \Omega_A} w(b_i)$ and $w_{1R} = \sum_{i \in \Omega \setminus \Omega_A, c_i=1} w(b_i)$ are the total weights for all bona fide nodes not controlled by the adversary and eligible to serve as the exit node, respectively. Note that the value of u_A - i.e., the probability for the adversary to successfully launch the entry-exit linking attack - can be high even with a small value of n_A. For example, in our experiments with the current Tor node bandwidth distribution, the probability for an adversary to compromise both the entry and the exit nodes of a path is greater than 5% when only 1% of all Tor nodes (i.e., 15 nodes) are compromised.

4.2   Round 2: Stratified Path Selection

Defender's Strategy: Given the optimal adversarial strategy in Round 1, let us now consider the defender's next move to best block the attack. We consider a simple way to defend against this adversarial strategy - to select the entry node from only the nodes that *refuse* to serve as the exit - i.e., we ensure $f(v_i, v_j, v_k) = 0$ if $v_i \in \Omega_1$. Note that this defensive strategy chooses v_i and v_k from mutually exclusive subsets of Ω. Thus, we call it *stratified path selection*.

   If the adversary still uses the first round's optimal strategy, its entry-exit linking attack will never succeed because no compromised node can serve as the entry node. Thus, stratified path selection will force the adversary to change its strategy, as shown in our following discussion of the adversary's optimal response in Round 2.

   The expected bandwidth of a constructed path is barely affected by stratified path selection because the average bandwidth of current Tor nodes with $c_i = 0$ is actually higher than that of nodes with $c_i = 1$. For example, with the Tor node bandwidth distribution we recorded in January 10, 2010, nodes with $c_i = 0$ and 1 have average bandwidth of 341.87 and 257.27 KB/s, respectively.

Adversary's Strategy: What stratified path selection does *not* change is the optimal value of the bandwidth to assign to the compromised nodes. The adversary can still increase the success probability of its attack by choosing a bandwidth as large as b_L for all compromised nodes. The defender's adoption of stratified path selection only affects the adversary's assignments of c_i. In particular, we derive the following theorem.

Theorem 1. *The optimal strategy of the adversary is to assign bandwidth b_L to all n_A compromised nodes, and to assign $c_i = 0$ to $p_c \cdot n_A$ of them, where*

$$p_c = \frac{w_{1R} \cdot d_0 - \sqrt{w_{1R}^2 \cdot d_0^2 - d_0 \cdot w_{1R} n_A w(b_L)(d_1 - d_0)}}{d_1 - d_0}, \tag{5}$$

where $d_0 = w_{0R}+n_A \cdot w(b_L)$, $d_1 = w_{1R}+n_A \cdot w(b_L)$, and $w_{0R} = \sum_{i \in \Omega \setminus \Omega_A, c_i=0} w(b_i)$ (i.e., the total weight for all bona fide exit-ineligible nodes).

All proofs in the paper are not included due to the space limitation.

As an example of the optimal strategy derived in Theorem 1, when $w_{1R} = \hat{w}_{0R}$, the optimal strategy for the adversary is to specify $p_c = 0.5$. As we shall show in the experimental evaluation section, given the current distribution of Tor node bandwidth, the derived optimal strategy yields a success probability of 1% for the attack when the adversary is capable of compromising 1% of all Tor nodes. One can see that this is significantly lower than that after Round 1 (i.e., $\approx 5\%$).

Observation from Round 2: A key observation from Round 2 is that the anonymous routing system should draw the entry and exit nodes from separate pools, such that the adversary cannot "double-dip" a compromised node by making it exit-eligible. While the separation is done based on c_i in the above discussions, it may also be conducted based on other criteria. For example, when the majority of nodes are exit-eligible (i.e., with $c_i = 1$), the system may designate a subset of them as the pool for exit node selection, and use the remaining ones along with the exit-ineligible nodes to choose the entry node. We will not elaborate on this option because it does not represent the current distribution of Tor nodes.

4.3   Round 3: Bandwidth Order Selection

Defender's Strategy: Round 2 focuses on the adversary's selection of c_i, and devises a defensive strategy that forces the adversary to separate the compromised nodes into two groups, with $c_i = 0$ and 1, respectively. The adversary, in response, computes an optimal allocation of compromised nodes between the two groups. In Round 3, we focus on the adversary's bandwidth selection strategy and aim at forcing it to reduce b_i and therefore settle with a smaller success probability of the entry-exit linking attacks.

The key idea for the defensive strategy is to add one step *after* the node selections: if the entry node has higher bandwidth than the middle node, then swap the entry and middle nodes to ensure that the entry has a lower bandwidth. We call this method *bandwidth order selection*. The premise of bandwidth order selection is that the adversary is forced to reduce the bandwidth of compromised nodes with $c_i = 0$, because otherwise it can compromise the entry node only if all three chosen nodes have $b_i \geq b_L$.

Adversary's Strategy: Given the defender's updated strategy, what does not change for the adversary is to assign bandwidth of at least b_L to all exit-eligible nodes (with $c_i = 1$). On the other hand, the adversary has to reduce b_i for an exit-ineligible node in order to increase its probability of being chosen as the entry node. In particular, we derive the following theorem:

Theorem 2. *With stratified path selection and bandwidth order selection, the optimal strategy for the adversary is to assign $b = b_{entry}$, where*

$$b_{entry} = \frac{w_{0R} \cdot 100}{n_A \cdot (1 - p_c) \cdot (1 - 2w_{0R}) \cdot 100 + 2w_{0R}}. \tag{6}$$

Observe from the experimental evaluation section that, given the current Tor node bandwidth distribution, an adversary with 1% compromised node now only has a success probability of less than 0.7% for launching entry-exit linking attacks.

Observation from Round 3: In the first two rounds, the selection of the entry, middle, and exit nodes are independent. A key observation from Round 3 is that the defender should draw the entry and middle nodes in a correlated fashion - e.g., to ensure that the middle node has higher bandwidth than the entry node. By doing so, the path bandwidth is not influenced, while the attack success probability is significantly reduced.

4.4 Round 4: Adaptive Exit Selection

Defender's Strategy: Round 4 focuses on answering a simple question for the defender: Since Round 2 dictates the selection of entry and exit nodes to be from mutually exclusive subsets, should the defender select the entry or the exit first?

Given the optimal strategy of the adversary in Round 3, the key idea for the defender's response in Round 4 can be stated as follows: First, the defender selects the entry and middle nodes v_i and v_j. Then, while selecting the exit node v_k, the defender no longer needs to distinguish between nodes with bandwidth $b \geq \min(b_i, b_j)$, because if one of these nodes is chosen as the exit, the bottleneck on bandwidth will be on the entry or middle node. In particular, while choosing the exit node, we set $w(b) = w(\min(b_i, b_j))$ if and only if $b \geq \min(b_i, b_j)$. To maintain the same probability of $b_k \geq \min(b_i, b_j)$, we also adjust the weight for $b < \min(b_i, b_j)$ accordingly:

$$w_{\text{new}}(b) = w(b) \cdot \frac{\sum_{h=1}^{n} w(b_i) - \sum_{h:b_h \geq \min(b_i,b_j)} w(\min(b_i, b_j))}{\sum_{h:b_h < \min(b_i,b_j)} w(b_h)}. \tag{7}$$

We call this strategy *adaptive exit selection*. It has no impact on the path bandwidth. Meanwhile, if the adversary still uses Round 3 strategy, the attack success probability will decrease because a compromised node becomes less likely to be chosen as the exit.

Adversary's Strategy: The optimal adversarial strategy remains the same as Round 3.

Observation from Round 4: In the first three rounds, the order of selection for the nodes is not a factor. A key observation from Round 4 is that the defender should draw the entry and middle nodes first. Then, while drawing the exit node, it should weigh equally all nodes with bandwidth no less than $\min(v_i, v_j)$. By doing so, the path bandwidth is not affected, while the attack success probability can be significantly reduced.

4.5 gPath: Our Proposed Algorithm

1: Randomly choose two nodes v_i, v_j from Ω such that the probability for a node $v : \langle b, c \rangle$ to be picked up is proportional to $w(b)$ if $c = 0$, and 0 if $c = 1$.
2: Choose the node with smaller bandwidth as the entry node, the other one as the middle node. Break tie arbitrarily.
3: Randomly choose node v_k from Ω such that the probability for a node $v : \langle b, c \rangle$ to be picked up is proportional to $\min(w(b), w(\min(b_i, b_j)))$ if $c = 1$, and 0 if $c = 0$.

5   Experimental Evaluation

Experimental Setup: We used the configuration of Tor nodes according to a snapshot captured on Jan 10, 2010. There are 1541 Tor nodes with an average bandwidth of 308.82KB. Among them, 602 nodes are exit-eligible with an average bandwidth of 257.27KB. The average bandwidth of an exit-ineligible node is 341.87KB. The lowest and highest bandwidth are 0 and 9995KB, respectively. We set the number of nodes compromised by the adversary to be 1% of total nodes, i.e., 15 nodes.

In this section, we focus on presenting experimental results on the *privacy* measure, i.e., how gPath reduces the probability for an entry-exit attack to succeed. The utility measure, i.e., the bandwidth of a constructed Tor path, is barely affected in the experiments, as discussed in the theoretical analysis in Section 4. Due to the space limitation, we do not include experimental results on the utility measure in the paper.

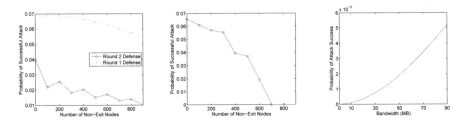

Fig. 1. Effect of Stratified Path Selection   **Fig. 2.** Effect After Bandwidth Adjustment   **Fig. 3.** Effect of Bandwidth Order Selection

Round 2 (Stratified Path Selection): Round 2 defense chooses the entry/middle nodes and the exit node from disjoint sets. Figure 1 shows the success probability of the *optimal* adversarial strategy given the stratified path selection defense. We conducted experiments with a fixed number (602) of exit-eligible nodes while varying the number of exit-ineligible nodes from 1 to 939, the actual number of exit-ineligible nodes in our snapshot of Tor, in order to demonstrate the change of attack performance given different exit-eligible/ineligible ratios. Figure 1 demonstrates that, compared with the current path selection algorithm, the stratified path selection algorithm significantly reduces the attack success probability from more than 5% to less than 1%. Also, the attack success probability decreases when more (honest) exit-ineligible nodes exit in the system.

Recall from Section 4 that if the number of exit-ineligible nodes is too small, then we have to place exit-eligible nodes into the pool for entry and the middle node in order to maximize the path bandwidth. We tested the performance of stratified path selection while maintaining equal total bandwidth for the exit and entry/middle pools. Figure 2 depicts the results. One can see that the probability of a successful attack remains substantially lower than that according to the defensive strategy in Round 1.

Round 3 (Bandwidth Order Selection): Compared with Round 2 defense, the defensive strategy in Round 3 further requires the middle node to have the highest bandwidth. We tested the performance of Round 3 defense while varying the attacking strategy with bandwidth from 0MB to 90MB. Figure 3 depicts the results which show that while a larger bandwidth still makes an attack more likely to succeed (as in Rounds 1 and 2), the probability of a successful attack is an order of magnitude lower than that in Round 2. For example, even the highest bandwidth only yields a success probability of 0.6% in Round 3, as compared with more than 6% in Round 2.

We also tested the performance of Round 3 defense given the optimal attacking strategy while varying the number of non-exit nodes from 1 to 939. The results are depicted in Figure 4. Observe that, for any number of non-exit nodes, the attack success probability is consistently an order of magnitude lower than that in Round 2.

Round 4 (Adaptive Exit Selection): Finally, Round 4 defense further requires the entry and middle nodes to be chosen before the exit node, so that any candidate for the exit node can be treated evenly as long as its bandwidth is greater than the smaller of the entry and middle nodes. Figure 5 depicts the success probability for the optimal adversarial strategy when the number of non-exit nodes varies from 1 to 939. From this figure we conclude that the probability is further reduced by an order of magnitude as compared with Round 3 (e.g., from 0.6% to less than 0.04%).

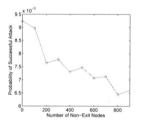   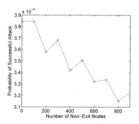

Fig. 4. Effect of Bandwidth Order Selection: Optimal Attacking Strategy

Fig. 5. Effect of Adaptive Exit Selection

6   Related Work

Chaum [11] pioneered the idea of anonymous communication. A survey of various anonymous communication systems can be found in [1, 12]. There exists a lot of research on compromising anonymous communication systems based on the network traffic analysis, most of which falls into the category of entry-exit linking attacks studied in this paper. Technically, these entry-exit linking attacks can largely be categorized into two groups: passive traffic analysis and active watermarking.

With passive traffic analysis, the attacker eavesdrops traffic passively and aims to identify the similarity between the entry's inbound traffic and the exit's outbound traffic [13, 14]. For example, a cross-correlation based similarity measure was studied in [14], whereas similarity measures based on packet size and timing were studied in [15,16,17].

With active watermarking, the attacker designs a secret signal (i.e., watermark) and embeds it into the inbound traffic of the entry node, aiming to identify the secret pattern from the observed outbound traffic of the exit node [4, 18]. Also in this category, Murdoch *et al.* [19] investigated a timing-based attacks on Tor by using a number of compromised Tor node. Overlier *et al.* [20] studied a scheme using one compromised node to identify the "hidden server" anonymized by Tor. The attacks studied in [3, 5] exploit the design of cell packet delivery in Tor to carry out linking attacks.

The application of game theory has also been extensively studied in many aspects of network security [21, 22, 23, 24, 25, 26, 27]. The interactions between defender and adversary, in particular, have been studied with various game-theoretic models such as static [28], stochastic [10, 29], and repeated games [30, 31]. Based on such models, most existing work focused on the modeling of adversarial intent, objectives, and strategies, as well as the corresponding defensive countermeasures. More closely related to our work is [32] which analyzed the initial rounds of interactions between email spammers and spam email detectors to produce a spam-email classification algorithm. Compared with the above existing work, our work is the first to draw design principles of an anonymous communication system from a novel game-theoretic analysis.

7 Final Remarks

In this paper, we have presented the first game-theoretic study on the design of path selection algorithm against entry-exiting linking attack in anonymous communicating systems such as Tor. Based on the model of a non-cooperative game between the defender (i.e., the designer of the path selection algorithm) and the adversary (i.e., controller of the compromised Tor nodes to carry out entry-exit linking attacks), we analyzed three rounds of the repeated arm race between the two players, and derived three design principles. These principles are *stratified path selection, bandwidth order selection*, and *adaptive exit selection*, which can effectively reduce the success probability of the attack without substantially affecting the bandwidth of the constructed Tor paths. Using a combination of theoretical analysis and experimental studies, we demonstrate the superiority of our developed algorithm over the existing ones.

References

1. Dingledine, R., Mathewson, N., Syverson, P.: Tor: The second-generation onion router. In: Proceedings of the 13th USENIX Security Symposium (2004)
2. Dingledine, R., Mathewson, N., Syverson, P.: Tor: anonymity online (2008), http://tor.eff.org/index.html.en
3. Ling, Z., Luo, J., Yu, W., Fu, X., Xuan, D., Jia, W.: A new cell counter based attack against tor. In: ACM Conference on Computer and Communications Security (CCS) (2009)
4. Wang, X., Chen, S., Jajodia, S.: Tracking anonymous peer-to-peer VoIP calls on the internet. In: ACM Conference on Computer Communications Security (CCS) (2005)
5. Pries, R., Yu, W., Fu, X., Zhao, W.: A new replay attack against anonymous communication networks. In: IEEE International Conference on Communications (ICC) (2008)
6. Johnson, A., Feigenbaum, J., Syverson, P.: Preventing active timing attacks in low-latency anonymous communication. In: Privacy Enhancing Technologies Symposium (PETS) (2010)
7. Diaz, C., Murdoch, S.J., Troncoso, C.: Impact of network topology on anonymity and overhead in low-latency anonymity networks. In: Privacy Enhancing Technologies Symposium (PETS) (2010)

8. Xie, M., Yin, H., Wang, H.: An effective defense against email spam laundering. In: ACM Conference on Computer Communications Security (CCS) (2006)
9. Debar, H., Dacier, M., Wespi, A.: Towards a taxonomy of intrusion detection systems. Computer Networks 31(8), 805–822 (1999)
10. Alpcan, T., Basar, T.: An intrusion detection game with limited observations. In: Proceeding of the 12th Intl. Symposium on Dynamic Games and Applications (2006)
11. Chaum, D.: Untraceable electronic mail, return addresses, and digital pseudonyms. Communications of the ACM 4(2) (February 1981)
12. Danezis, G., Dingledine, R., Mathewson, N.: Mixminion: design of a type iii anonymous remailer protocol. In: IEEE Symposium on Security and Privacy (S&P) (2003)
13. Zhu, Y., Fu, X., Graham, B., Bettati, R., Zhao, W.: On flow correlation attacks and countermeasures in mix networks. In: Martin, D., Serjantov, A. (eds.) PET 2004. LNCS, vol. 3424, pp. 207–225. Springer, Heidelberg (2005)
14. Levine, B.N., Reiter, M.K., Wang, C., Wright, M.: Timing attacks in low-latency mix-based systems. In: Juels, A. (ed.) FC 2004. LNCS, vol. 3110, pp. 251–265. Springer, Heidelberg (2004)
15. Song, D.X., Wagner, D., Tian, X.: Timing analysis of keystrokes and timing attacks on ssh. In: Proceedings of 10th USENIX Security Symposium (August 2001)
16. Sun, Q.X., Simon, D.R., Wang, Y., Russell, W., Padmanabhan, V.N., Qiu, L.L.: Statistical identification of encrypted web browsing traffic. In: Proceedings of IEEE Symposium on Security and Privacy (S&P) (May 2002)
17. Liberatore, M., Levine, B.N.: Inferring the source of encrypted http connections. In: ACM conference on Computer and Communication Security (CCS) (2006)
18. Yu, W., Fu, X., Graham, S., Xuan, D., Zhao, W.: DSSS-based flow marking technique for invisible traceback. In: IEEE Symposium on Security and Privacy (S&P) (2007)
19. Murdoch, S.J., Danezis, G.: Low-cost traffic analysis of tor. In: Proceedings of the IEEE Security and Privacy Symposium (S&P) (May 2006)
20. Overlier, L., Syverson, P.: Locating hidden servers. In: Proceedings of the IEEE Security and Privacy Symposium (S&P) (May 2006)
21. Xu, J., Lee, W.: Sustaining availability of web services under distributed denial of service attacks. IEEE Transactions on Computer 52(4), 195–208 (2003)
22. Gordon, L.A., Loeb, M.P.: Using information security as a response to competitor analysis systems. Communications of the ACM 44(9), 70–75 (2001)
23. Liu, P., Zang, W., Yu, M.: Incentive-based modeling and inference of attacker intent, objectives, and strategies. ACM Transactions on Information System and Security 8(1), 78–118 (2005)
24. Buttyan, L., Hubaux, J.P.: Security and Cooperation in Wireless Networks. Cambridge University Press, Cambridge (2007)
25. Yu, W., Liu, K.J.R.: Game theoretic analysis of cooperation stimulation and security in autonomous mobile ad hoc networks. IEEE Transactions on Mobile Computing 6(5), 459–473 (2007)
26. Alpcan, T., Basar, T.: A game theoretic analysis of intrusion detection in access control systems. In: Proceedings of the 43th IEEE Conference on Decision and Control (2004)
27. Liu, Y., Comaniciu, C., Man, H.: A bayesian game approach for intrusion detection in wireless ad hoc networks. In: Proceedings of the 2006 Workshop on Game Theory for Communications and Networks (2006)
28. Zhang, N., Zhao, W.: Distributed privacy preserving information sharing. In: Proceedings of the 31th International Conference on Very Large Data Bases (VLDB) (2005)

29. Lye, K., Wing, J.M.: Game strategies in network security. In: Proceedings of the 15th IEEE Computer Security Foundations Workshop (CSFW) (2002)
30. Buike, D.: Towards a game theory model of information warfare. Airforce Institute of Technology, Master Thesis (1999)
31. Zhang, N., Yu, W., Fu, X., Das, S.K.: Maintaining defender's reputation in anomaly detection against insider attacks. IEEE Transactions on Systems, Man, and Cybernetics, Part B: Cybernetics 40 (2010)
32. Dalvi, N., Domingos, P., Mausam, S.S., Verma, D.: Adversarial classification. In: SIGKDD (2004)

When Do Firms Invest in Privacy-Preserving Technologies?

Murat Kantarcioglu[1], Alain Bensoussan[2], and SingRu(Celine) Hoe[1]

[1] University of Texas at Dallas, USA
[2] University of Texas at Dallas, USA and The Hong Kong Polytechnic University, HK
muratk@utdallas.edu, alain.bensoussan@utdallas.edu, hoceline02@yahoo.com

Abstract. Privacy is a central concern in the information age. In some circumstances, customers' decisions whether to use firms' services rely on the extent of privacy that firms are able to provide, for example, the use of certain banking services, health care information technology [7]...etc. Firms thus face crucial assessment of investment on privacy-preserving technologies. Two important factors affect firms' valuation: (1) a customer's valuation of his private information and (2) a customer's profitability to the firm. The former determines the potential customer base that a firm can exploit given certain privacy protection, and the latter establishes profits that a firm can make. Both factors have some random components which can be best described by their descriptive probability distributions. We view firms' evaluation processes as a variant of Stackelberg type leader-follower game under complete information with customers taking the role of the follower. Firms integrate customers' optimal decisions into their valuation. Rational utility maximizing customers optimally decide whether to use firms' services by linking to their own decision threshold. The threshold is their own fair valuation of privacy connected to their private information. This fair privacy valuation is determined by a standardized premium over a fixed privacy rank related to values of private information common to the general population. This assertion is motivated by a recent research study [2]. We explore how the two underlying distributions and their dependence structures impact firms' investment valuation. Copulas, useful tools to study the relationship between random variables, are used to allow great flexibility on constructing bivariate distribution functions from arbitrarily univariate marginals with various dependence structures. We find that dependence structures and underlying univariate distributions have significant impacts on valuation. This suggests that, for appropriate investment decision making, firms shall be cautious on estimating underlying univariate distributions and their dependence structures. If distribution validation is not empirically possible, firms shall proceed with distributions and dependence structures which are practically justifiable for their market segments/industries. Our results identify several cases where the government intervention may be required to have firms invest in privacy-preserving technologies.

Keywords: Copulas, Privacy-Preserving Technology, Stackelberg Game.

T. Alpcan, L. Buttyan, and J. Baras (Eds.): GameSec 2010, LNCS 6442, pp. 72–86, 2010.
© Springer-Verlag Berlin Heidelberg 2010

1   Introduction

Consumers place a positive value on their private information and in many cases act upon their values of privacy to decide whether to use a firm's services. [7] shows that a major barrier to consumer acceptance of a health information technology, in particular personal health records, is that private information may not be adequately protected. On the other hand, firms make investment decisions about privacy-preserving technologies, which require the technology adoption to yield them enough compensation for making investment. We are interested in assessing privacy-preserving technologies to formulate useful guidelines for firms' investment decisions and to make inferences, if any, about the necessity of government intervention to encourage firms' investment in privacy-preserving technologies. This job distinguishes our work from other privacy related works.

In general, most privacy related works focus on issues related to clarifying the privacy trade-offs that individuals will make to gain access to specific services or quantifying individuals' privacy values. [2] seeks to quantify how general privacy attitudes impact the price participants set for revealing private information. Through experimental auctions, they show that a trait's desirability in relation to the group plays a key role in the amount people demand to publicize private information. [4] suggests that the information-seeking organization has to offer financial incentives and convenience (i.e., privacy mitigation strategies) in exchange for individuals to relinquish personal information. They apply the expectancy theory of motivation, and find that benefits – monetary reward and future convenience – significantly affect individuals' preferences over Web sites with different privacy policies. Using the technique of conjoint analysis on survey data collected from participating institutions, they are also able to quantify the value of Web site privacy protection. Observing that research has uncovered a dichotomy between stated attitudes and actual behavior of individuals facing decisions, privacy and personal information security, [1] provides an analysis of the dichotomy, outlining an experimental design to test their hypotheses about the observed inconsistency. More recently, multidisciplinary fields of human-computer interaction (HCI) have emerged with a raft of work on privacy in computing. [3] updates the development of HCI from psychology aspects, giving explicit attention to the emergence of computer-supported cooperative work and pointing out that having both "useful and usable" computing systems are of paramount importance. In accordance with the "useful and usable" criterion postulated by [3], [6] proposes a privacy expectations and security assurance offer system satisfying this criterion. Under the proposed system, the on-line organization offers consumers choices of privacy preferences and security levels with fee schedules. Consumers will get compensated if their designated privacy is violated. This proposed system thus bears the benefits of enhancing consumer privacy choices, creating a market for privacy preferences, and providing direct incentives for privacy offering organizations to care about the security of personal information.

Two unique factors affect firms' investment valuation in our model: (1) a customer's valuation of his private information and (2) a customer's profitability to a firm. The former determines the potential customer base that firms can exploit given certain privacy protection and the latter establishes profits that firms can make from customers using firms' services. For example, a customer who wants to use banking services may consider his total wealth as private information. At the same time, it may be the case that the wealthier the customer, the more profitable it is to the bank. Clearly, different customers could demand different levels of privacy protection. For example, a very wealthy individual would care more about keeping his investments private compared to an average customer. A similar phenomenon can also be observed in medical research domain. A person who is HIV positive could be more valuable to medical researchers working on a vaccine for HIV. On the other hand, such HIV positive patients may be more sensitive about their privacy. In this paper, our goal is to analyze how the relationship between these two factors affects firms' decisions to invest in privacy-preserving technologies.

In our valuation process, we view firms' evaluation processes as a variant of a Stackelberg leader-follower type game under complete information with costumers taking the role of the follower. We assume homogenous customers. Each costumer chooses whether to use a firm's services by solving his own utility maximization problem. In our model, customers' utilities link directly to the degree of privacy which they believe that the firm can provide. Rational utility maximizing customers will use a firm's services if and only if the firm's privacy-preserving technology can provide a level of privacy protection no less than their thresholds. The threshold is their own "fair" rank/level of privacy connected to their valuation of private information. When individuals are more sensitive about their private information, they will assign higher values for their private information. This fair privacy rank is determined by a standardized premium over a fixed rank of privacy related to values of private information common to the general population. In the context of this paper, a privacy-preserving technology could be anything ranging from specialized tools to protect individual privacy to the simple promise of not sharing customer data with outside companies. Clearly, each of these cases can have different costs to the firm. For example, a firm may loose some potential income if it is not selling its customers' information.

The specification of fair privacy rank is motivated by the study of [2] which shows that, with more than 95 percent statistical confidence, a linear relationship exists between an individual's belief about a trait and the private value he places on it. For example, overweight people tend to valuate their private weight information with higher prices. By using standard tools for analyzing Stackelberg games, we find the equilibrium behavior using backward induction. Customers (the follower) decide whether to use firms' services by maximizing their utility functions given firms' privacy-preserving technologies adopted. Firms (the leader) then integrate customers' decisions into their valuation and will adopt the technology if and only if technology adoption yields them nonnegative profits. We simplify the game setting without considering competition among firms, which could lead to potentially dynamic movements among customers. In addition, we do not consider the existence of certain privacy regulations. Instead, we focus on analyzing the impacts of two underlying univariate distributions and their dependence structures. We leave such extensions to future work.

To allow sufficient flexibility in formulating bivariate distribution functions, we employ copula functions. Copula functions describe the interrelation of several random variables separated from the marginals to arrive at the joint distribution function. There are several advantages to use copulas, including: (1) We may know a great deal about (or it's easier to estimate) marginal distributions of two underlying individual variables, but we may know little about (or it's difficult to estimate) their joint behaviors. Copulas allow us to piece together joint distributions when only the marginals are known with certainty (or can be best guessed). (2) Using a copula as a basis for constructing bivariate joint distribution functions is flexible because no restrictions are placed on the marginal distributions, i.e., marginals can come from different families. (3) The separation of dependence structure and marginals permits great flexibility in deriving a richer class of joint distribution functions, in which the level of dependence among variables can go beyond standard ones.

Thanks to copulas, we are able to explore values of privacy-preserving technologies with a richer class of joint distribution. In majority cases, the mean of distribution of customers' profitability to firms is positively related to firms' revenues. On the other hand, the impact of its volatility is inconclusive. We caution that this does not mean "volatility" doesn't really matter if for instance the confidence interval of expected profits is discussed. Furthermore, the distribution of customers' valuation of private information affects firms' investment decision making in accordance with two underlying marginals. In some cases, the investment evaluation is irrelevant to the mean and the standard deviation of customers' valuation of private information. In these cases, it suggests that the customer base is implicitly given by the adjusted excess level of privacy protection from technology adoption (i.e., $\frac{\alpha_p - b}{a}$ to be formally defined in Sect.2.2). This in turn implicitly suggests that, given these situations, firms would likely invest in privacy-preserving technologies requiring significant costs if government regulation provides additional motivation. Finally, we find that the impact of dependence structure on firms' investment decision making hinges on underlying univariate marginals. The Pearson correlation, ρ, measures the dependence structure of two underlying marginals under the Gaussian copula. Depending on the two univariate underlying marginals, ρ may be independent of, negatively related to, or positively related to the possibility of firms' privacy-preserving technology adoption. In some cases, the necessity of government is positively related to the correlation of the two underlying marginals. Our results draw attention to the importance of thorough estimation when obtaining the underlying marginals and dependence structures for optimal decision making.

2   The Model

2.1   Background Information

We consider a firm facing a privacy-preserving technology investment problem. The adoption of this privacy-preserving technology, P, will pose a fixed investment cost K.[1] Alternatively, we can view that the technology adoption would cause a firm an opportunity cost K due to preventing the firm from using certain private information (eg. facebook's new privacy controls). In return for the cost K, the firm will make profits from customers who use firm's services given that the privacy-preserving technology can provide them with the required level of privacy protection. For a firm's potential customer group, I, each individual is assigned a customer profile characterized by his valuation of private information and his profitability to the firm. For example, for customer $i \in I$, his profile revealed to the firm is (x_i, y_i), where x_i represents customer i's valuation of private information and y_i represents customer i's profitability to the firm. We assume complete information, that is, x_i and y_i are publicly available information[2]. The random variables associated with customers' valuation of private information, X, and customers' profitability to a firm, Y, can be best described by their descriptive probability distributions, which can be characterized by corresponding cumulative distribution functions, $F_X(x)$ and $F_Y(y)$ respectively. We proceed with the firm's valuation as a Stackelberg leader-follower type game with customers as the follower. In our simplified game setting, a customer's strategy is defined to be a set of $D = \{0, 1\}$ with 0 representing not to use firm's services and 1 otherwise. Following standard procedures, we solve the problem backwards. Given the firm's adoption of the technology, P, customers (the follower) choose their strategies, which are to use or not to use the firm's services, by solving their utility maximization problem. In our specification, customers' utility maximization links directly to their valuation of private information and the level of privacy protection that the firm's privacy-preserving technology can provide. The firm (the leader) then integrates customers' decisions (i.e., the best response function in the terminology of game theory) into their valuation. Given the customers' decision function, the firm's expected revenues from technology adoption is the expected values it can receive from customers under the joint distribution function $F_{X,Y}(x, y)$. Clearly, the market mechanism makes it the

---

[1] It does not necessarily mean that this analysis work is only appropriate for valuating a single privacy-preserving technology adoption. Rather, we can consider the privacy-preserving technology P as any possible combination of available technologies resulting in different services and costs. That is, P can be considered as an element of the power set $\mathcal{P}(S)$, $S = \{s_1, s_2, ..., s_n\}$, $n \geq 1$ where s_i, $i = 1, 2...n$ represents different technologies. Apparently, there would be at most 2^n possible combinations as well as associated costs. When $P = \emptyset$, it indicates that no privacy-preserving technology is evaluated; hence $K = 0$ in this scenario. When $n = 1$, it resorts to the valuation of single technology adoption. When $n > 1$, we can valuate all potential combinations and rank them in order to make the best investment decision.

[2] In practice, we can identify these information through marketing research for example.

case that the firm will make the investment if and only if the investment yields non-negative profits.

2.2 Customer's (Follower's) Decision Function

We solve the problem backwards. Given the firm's adoption of the technology, P, an individual customer chooses his strategy, D, which maximizes his utility. We assume homogenous individuals. For each individual customer $i \in I$, we define his utility function as:

$$U(x_i, D) = (2D - 1)\left(\alpha_P - \left(a \times \frac{|x_i - \mu_X|}{\sigma_X} + b\right)\right)$$

where $D = \{0, 1\}$, μ_X and σ_X represent the mean and the standard deviation of X respectively, both $a > 0$ and $b > 0$ are constants, and $\alpha_P > b$ is a constant representing the level of privacy protection that the privacy-preserving technology P can provide[3]. Obviously, rational customers would choose $D = 1$ if $\alpha_P \geq \left(a \times \frac{|x_i - \mu_X|}{\sigma_X} + b\right)$ and $D = 0$ otherwise. That is, a customer's optimal strategy in response to his utility maximization solution relies solely on his value of private information and privacy protection that the firm's privacy-preservation technology can provide. That is, a customer's best response exclusively depends on the threshold, which is his own fair valuation of privacy connected to his private information. For facilitating our further exposition, we define such a rule as a customer's decision function given:

$$D(x_i) = \mathbb{1}_{a \times \frac{|x_i - \mu_X|}{\sigma_X} + b \leq \alpha_P},$$

The basic intuition behind this decision function is that a customer would use a firm's services if the level of privacy protection provided by the technology is no less than his "fair" level of privacy connected to his valuation of private information. The fair level of privacy consists of two components: (1) a basic level of privacy related to values of private information common to the population, b (exogenously determined in our model), and (2) an extra level of privacy determined by a weight, a (given in our model), of his mean absolute deviation standardized by volatility, $\frac{|x_i - \mu_X|}{\sigma_X}$. This specification is motivated by [2] in which the authors show that an individual in a group would demand a higher value of private information if his trait deviates from that of average population segment in the group, and the further his trait is away from that of average population segment, the higher the value of private information is demanded. We set $\alpha_P = 0$ if the firm does not invest the privacy-preserving technology; therefore, the firm would definitely lose all customers and make no money if not investing technology P at a cost K.[4]

---

[3] Again, we assume complete information; thus all parameter values are publicly available information.

[4] Alternatively, denoting α_0 as a level of privacy preservation provided by a firm without investing technology P, we can set $\alpha_0 = \alpha_P - \epsilon > b$, $\epsilon > 0$. It would guarantee that a larger portion of customers is obtained if investing technology P, a property we'd like to retain.

2.3 Firm's (Leader's) Valuation Function

Once customers' optimal decisions have been solved, the firm solves the investment problem by integrating customers' decisions into their valuation. Therefore,
the firm's expected profits from investing technology P is the expected profits
generated from customers using firm's services given:

$$V(x, y, K, D(x)) = \int_{X \times Y} yD(x)dF_{X,Y}(x, y) - K. \tag{1}$$

From (1), a clearly defined joint distribution function is required for valuation. To arrive a richer class of joint distribution functions, we employ copula
functions, which describe the interrelation of several random variables separated
from the marginals to arrive at the joint distribution function.

3 Copulas

We first present a generalized version of copula definition along with important
theorems when exploiting copulas, i.e., Sklar's Theorem and Invariant Theorem.

Definition 1. *Copula, expressed as C, is a multi-dimensional function having
uniform marginal distribution that satisfies the following three conditions:*

1. $C : [0, 1]^n \rightarrow [0, 1]$;
2. C *is a grounded and n-increasing function;*
3. C *has margins C_i that satisfy $C_i(u) = C(1, ..., 1, u, 1, ...1) = u$, $u \in [0, 1]$.*

We next state the important Sklar's Theorem (1959) which shows the existence
of the copula function and the relation between the univariate margins and the
multivariate distribution function.

Theorem 1. *For any multivariate distribution function $F(x_1, ..., x_n) = P(X_1 \leq
x_1, ..., X_n \leq x_n)$ with continuous marginal functions $F_i(x_i) = P(X_i \leq x_i)$ for
$i \leq i \leq n$, there exists a unique n-dimensional copula function, $C(u_1, ..., u_n)$
such that*

$$F(x_1, x_2, ..., x_n) = C(F_1(x_1), F_2(x_2), ..., F_n(x_n)). \tag{2}$$

Corollary 1. *From (2), under the assumption that F_i and C are differentiable,
the following canonical representation holds:*

$$f(x_1, x_2, ..., x_n) = c(u_1, u_2, ..., u_n) \times \prod_{i=1}^{n} f_i(x_i) \tag{3}$$

where

- $f(x_1, x_2, ..., x_n)$ *is the density corresponding to $F(x_1, x_2, ..., x_n)$ and $f_i(x_i)$
  is the density corresponding to $F_i(x_i)$.*
- $c(u_1, u_2, ..., u_n) = \frac{\partial^n C(u_1, ..., u_n)}{\partial u_1 ... \partial u_n}$ *is called the copula density.*

Equation (3) states that, under appropriate conditions, the joint density can be written as a product of marginal densities and the copula density. It is clear that the copula density c encodes information about the dependence among X_i's; thus, c is sometimes called a dependence function. We will rely much on this formulation in our valuation. The Invariant Theorem (stated below) shows that the dependence between random variables is completely captured by the copula, independent of the shape of the marginals.

Theorem 2. *Consider n random variables $X_1, ..., X_n$ with a copula C. Then, if $g_1(X_1), ..., g_n(X_n)$ are continuously strictly increasing on the ranges of $X_1, ..., X_n$, then the random variables $Y_1 = g_1(X_1), ..., Y_n = g_n(X_n)$ have exactly the same copula C.*

4 Valuation under Various Copulas

We are interested in understanding how the two underlying univariate marginals, $F_X(x)$ and $F_Y(y)$, and their dependence structures would affect our valuation. We accomplish this goal by making use of copulas. We explore the impacts of different copulas (equivalently dependence structure) and marginals. We start first with the independent copula to capture the independent case. We then explore the elliptical copulas derived from elliptical distributions. We focus on the Gaussian copula, a member of elliptical copula families, which bears the Gaussian distribution type dependence structure.

4.1 Independent Copula

We first consider the case that the random variables, customers' valuation of private information, X, and customers' profitability to a firm, Y, are independent. As a consequence of Sklar's theorem, random variables are independent if and only if their copula is the independent copula given as: $C(u_1, u_2) = u_1 u_2$. By using the independent copula, the joint distribution function $F_{XY}(x, y)$ is given: $F_{XY}(x, y) = C(F_X(x), F_Y(y)) = F_X(x)F_Y(y)$, and, assuming F_X and F_Y are differentiable, we have: $f_{XY}(x, y) = f_X(x)f_Y(y)$, where f_X and f_Y are density functions corresponding to F_X and F_Y respectively.

Proposition 1. *1. The firm's revenue is positively related to the mean of customers' profitability to the firm, μ_Y, and independent of the shape of distribution of Y.*

2. If $X \in \mathbb{R}$ and is a symmetric probability distribution (ex. Gaussian), the firm's revenue is positively related to the excess level of privacy protection, $\alpha_P - b$, and is negatively related to the "weight", a. The valuation is independent of mean and standard deviation of customers' valuation of private information, μ_X and σ_X.

3. If $X \in \mathbb{R}^+$, the firm's revenue is positively related to $\triangle F$, where $\triangle F = F_X(\frac{(\alpha_P - b) \times \sigma_X}{a} + \mu_X) - F_X((\frac{-(\alpha_P - b) \times \sigma_X}{a} + \mu_X)^+)$.

Proof. See Appendix A.   □

Proposition 1 indicates that if two distributions are independent, the firm's investment tendency in privacy-preserving technologies increases with increases in the mean of customers' profitability to a firm. In addition, the smaller "adjusted" weight a and the basic level of privacy, b, the greater the probability that firms will make the investment. Moreover, if the distribution of customers' valuation of private information is symmetric with domain $\mathbb{R}$, the customer base is implicitly given by the adjusted excess level of privacy protection from adopting technology, i.e., $\frac{\alpha_P - b}{a}$. This in turn suggests implicitly that firms would likely invest in privacy-preserving technologies requiring significant investment costs if government takes intervention.

Corollary 2. *The break-even investment amount, which defines the threshold of firms' investment cost in privacy-preserving technologies*[5]*, is:*

$$
K^{\mathrm{brk}} =
\begin{cases}
\left(F_X\!\left(\frac{(\alpha_P - b) \times \sigma_X}{a} + \mu_X \right) - F_X\!\left(\left(\frac{-(\alpha_P - b) \times \sigma_X}{a} + \mu_X \right)^+ \right) \right)\mu_Y, & \text{if } X \in \mathbb{R}^+ \\[2ex]
\left(F_X\!\left(\frac{(\alpha_P - b) \times \sigma_X}{a} + \mu_X \right) - F_X\!\left(\frac{-(\alpha_P - b) \times \sigma_X}{a} + \mu_X \right) \right)\mu_Y, & \text{if } X \in \mathbb{R}
\end{cases}
.
$$

4.2   Elliptical Copula

Elliptical copulas are the distribution functions of componentwise transformed elliptically distributed random vectors. We choose to work on the Gaussian copula for two reasons: (1) Bivariate normal distributions are widely used in modeling works due to its convenient available formulation of distribution function with linear correlation structures. (2) We can contrast how the results may be misleading if the underlying marginals are not normal but only correlated with Gaussian distribution type correlation structures. For this latter purpose, we work on underlying marginals from either exponential or pareto distributions. The results are different and we caution that management should exercise careful attention to estimating or justifying the underlying marginals.

Gaussian Copula. By $C(u_1, u_2) = F(F_1^{-1}(u_1), F_2^{-1}(u_2))$, we obtain the two dimensional Gausian copula:

$$
\begin{aligned}
C^{\mathrm{Ga}}(u_1, u_2; \rho) &= \Phi_\Sigma\big(\Phi_1^{-1}(u_1), \Phi_2^{-1}(u_2)\big) \\
&= \int_{-\infty}^{\phi_1^{-1}(u_1)} \int_{-\infty}^{\phi_2^{-1}(u_2)} \frac{1}{2\pi\sqrt{1 - \rho^2}} \exp\left(- \frac{s_1^2 - 2\rho s_1 s_2 + s_2^2}{2(1 - \rho^2)} \right) ds_1 ds_2,
\end{aligned}
\tag{4}
$$

where Σ is the 2×2 matrix with 1 on the diagonal and ρ otherwise. Φ is the cumulative distribution function (cdf) of a standard normal distribution while

---

[5] This is the largest investment cost that a firm would be willing to tolerate for undertaking the investment.

Φ_Σ is the cdf for a bivariate normal distribution with zero mean and correlation matrix Σ. Since $F_X(X) \sim U_1$ and $F_Y(Y) \sim U_2$, we again can write

$$F_{XY}(x, y) = C^{\mathrm{Ga}}(u_1, u_2; \rho) = C^{\mathrm{Ga}}(F_X(x), F_Y(y); \rho),$$

and assuming that $F_X(x)$ and $F_Y(y)$ are differentiable, we have:

$$f_{XY}(x, y) = c^{\mathrm{Ga}}(F_X(x), F_Y(y)) f_X(x) f_Y(y), \tag{5}$$

where c^{Ga} is the Gaussian copula density, and by (4), we have:

$$c^{\mathrm{Ga}}(u_1, u_2) = \frac{1}{\sqrt{1-\rho^2}} e^{-\frac{\xi_1^2 + \xi_2^2}{2} + \frac{2\rho\xi_1\xi_2 - \xi_1^2 - \xi_2^2}{2(1-\rho^2)}} \tag{6}$$

where $\xi_1 = \phi^{-1}(u_1)$ and $\xi_2 = \phi^{-1}(u_2)$.

A. Both Random Variables are from Normal Distributions.

The normal distribution is often used to describe any variable that tends to cluster around the mean, and this may well be the case in our two variables. Therefore, we first suppose that both variables X and Y are from normal distributions. That is, $X \sim N(\mu_X, \sigma_X)$ and $Y \sim N(\mu_Y, \sigma_Y)$, thus we have: $F_X(x) = \frac{1}{2}\left[1 + \mathrm{erf}(\frac{x-\mu_X}{\sqrt{2\sigma_X^2}})\right]$, $f_X(x) = \frac{1}{\sqrt{2\pi\sigma_X^2}} e^{-\frac{(x-\mu_X)^2}{2\sigma_X^2}}$, $F_Y(y) = \frac{1}{2}\left[1 + \mathrm{erf}(\frac{y-\mu_Y}{\sqrt{2\sigma_Y^2}})\right]$, and $f_Y(y) = \frac{1}{\sqrt{2\pi\sigma_Y^2}} e^{-\frac{(y-\mu_Y)^2}{2\sigma_Y^2}}$, where $\mathrm{erf}(x) = \int_0^x \frac{2}{\sqrt{\pi}} e^{-t^2} dt$, an error function. Then, by (5) and (6) with $\xi_1 = \phi^{-1}(F_X(x)) = \frac{x-\mu_X}{\sqrt{\sigma_X^2}}$ and $\xi_2 == \phi^{-1}(F_Y(y)) = \frac{y-\mu_Y}{\sqrt{\sigma_Y^2}}$ where we use the fact that $\phi^{-1}(z) = \sqrt{2}\mathrm{erf}^{-1}(2z-1)$, $z \in (0,1)$, we obtain:

$$f_{XY}(x, y) = \frac{1}{2\pi\sigma_X\sigma_Y\sqrt{1-\rho^2}} e^{-\frac{\frac{(x-\mu_X)^2}{\sigma_X^2} - 2\rho(\frac{x-\mu_X}{\sigma_X})(\frac{y-\mu_Y}{\sigma_Y}) + \frac{(y-\mu_Y)^2}{\sigma_Y^2}}{2(1-\rho^2)}} \tag{7}$$

which arrives the bivariate normal distribution with the correlation structure ρ.

Proposition 2. *1. The firm's revenue is positively related to the mean of customers' profitability to the firm, μ_Y, and is independent of the standard deviation.*

2. The firm's revenue is positively related to the excess level of privacy protection, $\alpha_P - b$ and is negatively related to the weight, a. The firm's revenue is independent of mean and standard deviation of customers' valuation of private information, μ_Y and σ_Y.

3. The Pearson correlation structure, ρ, does not have impacts on the valuation.

Proof. See Appendix B. □

Proposition 2 yields a surprising result that when two underlying distributions are from normal, the correlation structure does not affect valuation. The investment rule prediction is exactly the same as that proposed in the independent case.

Corollary 3. *The break-even investment cost is:* $K^{\text{brk}} = \text{erf}\left[\frac{\alpha_P - b)}{\sqrt{2}a}\right] \times \mu_Y$.

B. Both Random Variables are from Exponential Distributions.
By (5) and (6) with $\xi_1 = \phi^{-1}(F_X(x)) = \sqrt{2}\text{erf}^{-1}(2F_X(x) - 1)$ and $\xi_2 = \sqrt{2}\text{erf}^{-1}(2F_Y(y) - 1)$, we obtain the general form of a bivariate joint distribution density under Gaussian correlated structure:

$$f_{XY}(x,y) = \frac{1}{\sqrt{1-\rho^2}} e^{\frac{(\sqrt{2}\text{erf}^{-1}(2F_X(x)-1))^2 + (\sqrt{2}\text{erf}^{-1}(2F_Y(y)-1))^2}{2}}$$

$$\times e^{\frac{2\rho(\sqrt{2}\text{erf}^{-1}(2F_X(x)-1))(\sqrt{2}\text{erf}^{-1}(2F_Y(y)-1)) - (\sqrt{2}\text{erf}^{-1}(2F_X(x)-1))^2 - (\sqrt{2}\text{erf}^{-1}(2F_Y(y)-1))^2}{2(1-\rho^2)}}$$

$$\times f_X(x) \times f_Y(y) \ . \tag{8}$$

Though the exponential distribution is applied mainly in reliability modeling, due to its mathematical simplicity, it has been applied in various other situations such as the product demand distribution, the distribution of individual income...etc. We now assume that the random variables, X and Y, are from exponential distributions. That is, $X \sim \text{Exp}(\lambda_X)$ and $Y \sim \text{Exp}(\lambda_Y)$, thus we have: $F_X(x) = 1 - e^{-\lambda_X x}$, $f_X(x) = \lambda_X e^{-\lambda_X x}$, $F_Y(y) = 1 - e^{-\lambda_Y y}$, and $f_Y(y) = \lambda_Y e^{-\lambda_Y y}$. Moreover, they are jointly distributed with a Gaussian distribution type correlation structure with the correlation ρ. We can easily obtain $f_{XY}(x,y)$ by plugging F_X, f_x, F_Y, and f_Y into (8).

Proposition 3. *1. The firm's revenue is independent of the mean and standard deviation of X, but is positively related to the mean and standard deviation of customers' profitability to the firm.*

2. The firm's revenue is positively related to the excess level of privacy protection, $\alpha_P - b$, and is negatively related to the adjusted weight, a.

3. The firm's revenue is negatively related to the correlation ρ.

Proof. See Appendix C. ☐

Proposition 3 exhibits that although the mean and the standard deviation of customers' valuation of private information are still irrelevant to the valuation, the correlation affects the valuation. The valuation results are negatively related to ρ. In this case, the necessity of government intervention is positively related to the correlation of the underlying two distributions.

C. Both Random Variables are from Pareto Distributions.
The Pareto distribution shows rather well in describing the allocation of wealth among individuals. In some cases, customers profits to the firm may be well correlated to individual wealth, for example the usage of some banking services. In view of this, we assume that both random variables X and Y are from Pareto distributions. That is, $X \sim \text{Pareto}(x_m, \alpha)$, $x_m > 0$, $\alpha > 0$, $x \in [x_m, \infty)$ and $Y \sim \text{Pareto}(y_m, \beta)$, $y_m > 0$, $\beta > 0$, $y \in [y_m, \infty)$, thus we have: $F_X(x) = 1 - \left(\frac{x_m}{x}\right)^\alpha$, $f_X(x) = \frac{\alpha x_m^\alpha}{x^{\alpha+1}}$, $F_Y(y) = 1 - \left(\frac{y_m}{x}\right)^\beta$, and $f_Y(y) = \frac{\beta y_m^\beta}{y^{\beta+1}}$. In addition, we assume

$\alpha > 2$, and $\beta > 2$ to guarantee that the second moment of the distribution is defined. Moreover, they are jointly distributed with a Gaussian distribution type correlation structure with the correlation ρ. We can easily obtain $f_{XY}(x, y)$ by plugging F_X, f_x, F_Y, and f_Y into (8).

Proposition 4. *1. The firm's revenue is negatively related to ρ, and given ρ, the positivity yields better valuation results.*
  2. The firm's revenue is positively related to the mean and standard deviation of customers' profitability to the firm.
  3. The firm's revenue is negatively related to the mean and standard deviation of customers' valuation of private information.

Proposition 4 is obtained through numerical integration with convergence guaranteed. The impacts of correlation structure are similar to those of Proposition 3 with an additional property that, given ρ, the positivity results in better valuation. Unlike Proposition 2 and 1, firms have stronger tendency to invest in privacy-preserving technologies if the distribution of customers' valuation of private information is less volatile, even when it links to the smaller mean.

5 Conclusion

We study firms' optimal investment decisions on privacy-preserving technology adoption in a Stackelberg leader-follower game. We solve the problem backwards. The market mechanism ensured that it is benefitial for firms to undertake the investment as long as it yields non-negative profits. We arrive at the explicit formula for the threshold of a firm's investment cost in privacy-preserving technologies, which is the largest investment cost that a firm would accept for undertaking the investment. By means of copulas, we are able to explore values of privacy-preserving technologies with a richer class of joint distribution functions. We find that dependence structures and underlying distributions affect valuation significantly. Our results identify several cases where the government intervention may be required to have firms invest in privacy-preserving technologies.

For all cases under the independent copula and the Gaussian copula, the mean of distribution of customers' profitability to firms is positively related to firms' revenues. That is, the higher the mean, the stronger is the motivation that firms have to adopt privacy-preserving technologies. However, the impact of its volatility is inconclusive. We caution that this does not mean "volatility" doesn't really matter if for instance the confidence interval of expected profits is discussed; we leave this for our future work. The impact of the distribution of customers' valuation of private information varies with the two underlying marginals. For the independent case, if customers' valuation of private information is a symmetric distribution with domain $\mathbb{R}$, the investment evaluation is irrelevant to its mean and standard deviation. The bi-normal and the bi-exponential distributions yield the same conclusion. In all these three cases, the customer base is implicitly given by the adjusted excess level of privacy protection from technology adoption (i.e., $\frac{\alpha_p - b}{a}$). This in turn implicitly suggests that given these situations, in majority cases, firms would likely invest in privacy-preserving technologies requiring significant costs if the government regulation provides additional motivation. For the bi-pareto distribution, both mean and volatility affect valuation negatively; that is, in this scenario, firms may be motivated to invest in privacy-preserving technologies if its mean and volatility are small. Finally, the dependence structure under Gaussian copulas exhibits different effect for different underlying univariate marginals. Surprisingly, for a bivariate normal distribution, Pearson correlation, ρ, does not affect firms' valuation. This would yield the same investment rule as the independent case. For a bi-exponential distribution and a bi-pareto distribution, the possibility of firms' investment in privacy-preserving technologies is negatively related to ρ. For the bi-exponential distribution, the necessity of government intervention is positively related to the correlation of the two underlying marginals. In addition, for a bi-pareto distribution, given ρ, the positivity yields better valuation results. Thus, for appropriate investment decision making, firms should be cautious about estimating underlying univariate distributions and their dependence structures. If distribution validation is not empirically possible, firms should proceed with distributions and dependence structures which are practically justifiable for their market segments/industries.

Since the Gaussian copula function is radial symmetric and thus does not have tail dependence[6], it may give misleading results if used in the model when in fact the joint distribution has the asymmetric tail dependence property. We suspect that in some market segments/industry, the asymmetric tail dependence property may actually exist and may affect valuation significantly. Therefore, we will study Archimedean copula families which allow for capturing the tail dependence in future work. In addition, the dependence structure may vary as time progresses. Thus, working with dynamic copulas should be an interesting extension. Moreover, the extension to integrating Bayesian learning and competition among firms into our valuation model is left for our future study.

Acknowledgments. This work was partially supported by Air Force Office of Scientific Research MURI Grant FA9550-08-1-0265, National Institutes of Health Grant 1R01LM009989, National Science Foundation Grants Career-0845803, CNS-0964350, and CNS-1016343.

References

1. Acquisiti, A., Grossklags, J.: Losses, Gains, and Hyperbolic Discounting: An Experimental Approach to Information Security Attitudes and Behavior. In: Proc. 2nd Int'l. Workshop Economics and Info. Security (2003)
2. Huberman, B.A., Adar, E., Fine, L.R.: Valuating Privacy. IEEE Security & Privacy 3(5), 22–25 (2005)

---

[6] It measures the amount of dependence in the upper-right-quadrant or lower-left-quadrant tail of the bivariate distribution. In other words, it describes the limiting proportion that one margin exceeds a certain threshold given that the other margin has already exceeded that threshold.

3. Olson, G., Olson, J.: Human-computer Interaction: Psychological Aspects of the Human Use of Computing. Annual Review of Psychology 54, 491 (2003)
4. Hann, I.H., Hui, K.L., Lee, S.-Y.T., Png, I.P.L.: Overcoming Online Information Privacy Concerns: An Information-Processing Theory Approach. Journal of Management Information Systems 24(2), 13–42 (2007)
5. Kleinberg, J., Papadimitriou, C.H., Raghavan, P.: On the Value of Private Information. In: van Benthem, J. (ed.) Proc. 8th Conf. Theoretical Aspects of Rationality and Knowledge (TARK-2001), pp. 249–257. Morgan Kaufmann, San Francisco (2001)
6. Hunker, J.: A Privacy Expectations and Security Assurance Offer System. In: NSPW 2007, North Conway, NH, USA, September 18-21 (2007)
7. Consumers and Health Information Technology: A National Survey, California Health Foundation (April 2010)

A   Proof of Proposition 1

Proof. The firm's expected profit from investing technology P is given:

$$V(x, y, K) = \int_Y \int_X \mathbb{1}_{a \times \frac{|x - \mu_X|}{\sigma_X} + b < \alpha_P} f_X(x) y f_Y(y) dx dy - K \ . \tag{9}$$

If $X \in \mathbb{R}$, from (9), we have:

$$V(x, y, K) = \int_Y \int_{\frac{-(\alpha_P - b) \times \sigma_X}{a} + \mu_X}^{\frac{(\alpha_P - b) \times \sigma_X}{a} + \mu_X} f_X(x) y f_Y(y) dx dy - K$$

$$= \left(F_X \left(\frac{(\alpha_P - b) \times \sigma_X}{a} + \mu_X \right) - F_X \left(\frac{-(\alpha_P - b) \times \sigma_X}{a} + \mu_X \right) \right) \mu_Y$$

$$- K \ , \tag{10}$$

and if $X \in \mathbb{R}^+$, we have:

$$V(x, y, K) = \left(F_X \left(\frac{(\alpha_P - b) \times \sigma_X}{a} + \mu_X \right) - F_X \left(\left(\frac{-(\alpha_P - b) \times \sigma_X}{a} + \mu_X \right)^+ \right) \right) \mu_Y$$

$$- K \ . \tag{11}$$

All propositions directly follow from (10) and (11).   □

B   Proof of Proposition 2

Proof. Using (7), the firm's expected profit from investing technology P is given:

$$V(x, y, K) = \int_{\mu_X - \frac{\sigma_X}{a}(\alpha_P - b)}^{\mu_X + \frac{\sigma_X}{a}(\alpha_P - b)} \int_Y y$$

$$\times \frac{1}{2\pi \sigma_X \sigma_Y \sqrt{1 - \rho^2}} e^{-\frac{\frac{(x - \mu_X)^2}{\sigma_X^2} - 2\rho(\frac{x - \mu_X}{\sigma_X})(\frac{y - \mu_Y}{\sigma_Y}) + \frac{(y - \mu_Y)^2}{\sigma_Y^2}}{2(1 - \rho^2)}} dy dx - K$$

$$= \text{erf}\left[\frac{\alpha_P - b}{\sqrt{2}a} \right] \times \mu_Y - K \ . \tag{12}$$

All results follow directly.   □

C   Proof of Proposition 3

Proof. The firm's expected profit from investing technology P is given:

$$V(x, y, K) = \int_{\frac{1}{\lambda_X}(1-\alpha_P - b)+}^{\frac{1}{\lambda_X}(1+\alpha_P - b)} c^{\mathrm{Ga}}(F_X(x), F_Y(y)) \times \lambda_X e^{-\lambda_X x} dx \int_Y y \lambda_Y e^{-\lambda_Y y} dy$$
$$- K .$$
(13)

Using (6) in the above equation with $F_X(x) = 1 - e^{-\lambda_X x}$, $F_Y(y) = 1 - e^{-\lambda_Y y}$ and performing change of variables in x, it follows immediately that λ_X does not impact $V(x, y, K)$, but $V(x, y, K)$ is positively related to the term $\frac{\alpha_P - b}{a}$. In addition, $\frac{1}{\lambda_Y}$ is positively related to $V(x, y, K)$. We perform numerical integration for the effect of ρ. □

Adversarial Control in a Delay Tolerant Network[*]

Eitan Altman[1], Tamer Başar[2], and Veeraruna Kavitha[1,3]

[1] INRIA, BP93, 06902 Sophia Antipolis, France
[2] University of Illinois, 1308 West Main Street, Urbana, IL 61801-2307, USA
[3] LIA, Avignon University, Agroparc, France

Abstract. We consider a multi-criteria control problem that arises in a delay tolerant network with two adversarial controllers: the source and the jammer. The source's objective is to choose transmission probabilities so as to maximize the probability of successful delivery of some content to the destination within a deadline. These transmissions are subject to interference from a jammer who is a second, adversarial type controller, We solve three variants of this problem: (1) the static one, where the actions of both players, u and w, are constant in time; (2) the dynamic open loop problem in which all policies may be time varying, but independent of state, the number of already infected mobiles; and (3) the dynamic closed-loop feedback policies where actions may change in time and may be specified as functions of the current value of the state (in which case we look for feedback Nash equilibrium). We obtain some explicit expressions for the solution of the first game, and some structural results as well as explicit expressions for the others. An interesting outcome of the analysis is that the latter two games exhibit switching times for the two players, where they switch from pure to mixed strategies and *vice versa*. Some numerical examples included in the paper illustrate the nature of the solutions.

Keywords: Delay-tolerant networks, nonzero-sum game, switching strategies.

1   Introduction

We consider in this paper a delay tolerant network, i.e. a sparse network of mobile relay nodes, where connectivity is very low. There is some source that transmits a file to mobiles that are in the communication range. Each mobile is assumed to be in range with the source at some instants that form a Poisson process. A node that receives a copy of the file stores it so that it may transmit it to some potential destinations that may search for a copy of the file. We consider two controllers whose goals are not aligned: the source and the jammer. They both determine at

---

[*] The work of the second author is supported by an INRIA-UIUC research collaboration grant, as well as by an AFOSR Grant. The other authors were supported by the Indo-French Centre for the Promotion of Advanced Research (IFCPAR), project 4000-IT-1 and by the INRIA association program DAWN.

each time the probability of transmission. Transmission at a time t is successful if and only if the source attempts transmission while the jammer is silent.

We consider three frameworks, which lead to three different games:

(1) The static one, where the actions of both players, that is the probabilities of transmission and of jamming, u and w, respectively, are considered to be constant in time;

(2) The dynamic open-loop problem. Here, all policies may be time varying, but dependent only on the initial state. In solving the open-loop problem, we first show that the game is equivalent (strategically) to a zero-sum differential game, and then seek the saddle-point solution of that game.

(3) The dynamic closed-loop framework, where actions that may change in time are allowed to depend on the current value of the state (the number of mobiles with a copy of the file). In this case the underlying game is a genuine nonzero-sum differential game, where the solution sought is the feedback Nash equilibrium.

This work is another step in our effort of developing a control methodology for delay tolerant networks, which we initiated with our paper [1]. In contradistinction with the simple threshold structure of [1], we obtain here a much richer set of possible structures for the equilibrium policies, exhibiting in some cases multiple switching times between pure and proper mixed strategies.

The use of game theory for jamming problems. Jamming problems are among the first capturing conflicts in networks that have been modeled and solved using tools and the conceptual framework of game theory. The first publications on these games go back almost thirty years with the pioneering work [6] The question of the capacity achievable in channels prone to jamming was one of the main concerns, and was thus naturally studied within the information theory community, as for example in [7,9]. For a recent survey on wireless games that includes jamming games, see [12]. Jamming of specific wireless local area networks were investigated in [11] who study the jamming of IEEE802.11 and [10] who study the jamming of slotted ALOHA. Our current paper falls in this category of papers by specializing to the context of DTNs.

The paper is organized as follows. The next section (Section 2) provides a precise formulation of the problem, which is followed by Sections 3 and 4 which discuss the static and dynamic cases, respectively. These are followed by Section 5 which includes a number of numerical examples, and the concluding remarks of Section 6 concludes the paper.

2   Model and Problem Formulation

2.1   Model

In the model adopted in this paper, there are n relay mobile nodes, a source, and a destination which is assumed to be static. The network serves as a channel that enables the information to reach the destination. Whenever a relay mobile meets the source, the source may forward a packet to it. We consider the two-hop routing scheme [4] in which a mobile that receives a copy of the packet from the source can only forward it if it meets the destination. It cannot copy it into the memory of another mobile. The details of the basic model are as follows:

The source meets each relay node according to a Poisson process with a parameter λ. Each relay node meets the destination according to another Poisson process, with parameter ν. The source attempts to maximize the probability that a packet arrives successfully at a given destination by time ρ. A second transmitter (jammer), however, tries to jam the transmission, and hence attempts to minimize this probability. The jammer is assumed to be located close to the source. Jamming relay nodes is a separate problem that will be considered later. Note that we consider only two hop routing. Therefore jamming at the relays means jamming when transmitting to the destination.

Let $X(t), u_t, w_t$ denote, respectively, the fraction of mobiles with the message, the source's control, and the jammer's control. Here u_t is the probability to transmit at time t if at that time the source meets a relay and w_t is the probability of jamming at time t. We assume that if jamming and transmission occur simultaneously, then the transmitted packet is lost.

Let $x_t = E[X(t)]$ be the expected value of $X(t)$. Then x_t is generated by

$$\dot{x}_t = u_t(1 - w_t)\lambda(n - x_t), \tag{1}$$

with known initial condition x_0 at $t = 0$, and this constitutes the system dynamics.

2.2 Performance Measure: Successful Delivery Probability

During the incremental time interval $[t, t + dt)$, the number of copies of the packet in the network is $X(t)dt$. Then the number of packets that the destination receives during this time interval is a Poisson random variable with parameter $\nu X(t)dt$. In particular, the probability of not receiving any copy of the packet during $[0, \rho]$, conditioned on $X(t)$, is given by

$$P(T > \rho | X(t), 0 \leq t \leq \rho) = \exp\left(-\int_0^\rho \nu X(t)dt\right)$$

where T is the random variable describing the instant when the packet first reaches the destination. Its expectation (over $X(t)$) gives the failure probability, i.e. the complementary of the probability of successful delivery.

Instead of minimizing $P(T > \rho)$, we will minimize a bound on that quantity:

$$P(T > \rho) = E\left[\exp\left(-\int_0^\rho \nu X(t)dt\right)\right] \leq \exp\left(-E\int_0^\rho \nu X(t)dt\right) \tag{2}$$

where the inequality is obtained by applying Jensen's inequality to the concave function $\exp(-x)$. Minimizing the latter (and hence the upper bound on $P(T > \rho)$) is equivalent to maximizing the quantity

$$J(u, w) := \int_0^\rho \nu x_t dt, \tag{3}$$

which we will take as the utility function of the source.

We consider the mean field limit (when we have large number of nodes), in which the randomness in the number of mobiles that have a copy of the nodes as a function of time disappears (we obtain a deterministic time varying limit). In this regime, the difference between the objective function (the delivery failure probability) and the bound (2) vanishes. Indeed, the bound was obtained by exchanging the order of expectation and exponent (using Jensen's inequality), but in the mean-field regime, Jensen's inequality is obtained with equality since the randomness vanishes.

2.3 Related Game Theory Concepts and Some Properties

Saddle-point, maximin and minimax policies: Let $J(u, w)$ be the utility function of the source, as introduced earlier by (3). We assume that the jammer wishes to minimize this quantity and the source wishes to maximize it.

Let Π_c be a set of policies for the controller (both source and relay mobiles) and let Π_j be a set of policies for the jammer. (We will introduce later specific classes of policies.)

We say that $u^* \in \Pi_c$ and $w^* \in \Pi_j$ are saddle-point policies for the game (J, Π_c, Π_j) if for every $u \in \Pi_c$ and $w \in \Pi_j$ we have[1]

$$J(u, w^*) \leq J(u^*, w^*) \leq J(u^*, w)$$

$J(u^*, w^*)$ is then called the value of the game.

In a general zero-sum game saddle-point need not exist. In that case, we are interested in the upper and lower values ($\overline{V}$ and $\underline{V}$) which are always well defined:

$$\overline{V} = \inf_{w \in \Pi_j} \sup_{u \in \Pi_c} J(u, w), \quad \underline{V} = \sup_{u \in \Pi_c} \inf_{w \in \Pi_j} J(u, w),$$

w^* is optimal for the minimax problem if $\overline{V} = \sup_{u \in \Pi_c} J(u, w^*)$. Given such a w^*, the controller u^* is a best response policy if $\overline{V} = J(u^*, w^*)$. Likewise, u^* is optimal for the maximin problem if $\underline{V} = \inf_{w \in \Pi_j} J(u^*, w)$, and given such a u^*, w^* is a best response policy if $\underline{V} = J(u^*, w^*)$.

A policy is said to be *open loop* if it does not depend on the state of the system. It is said to be *Markov* (or a *feedback* policy) if it takes at time t an action that is allowed to depend not only on t but also on the state at time t. A *pure policy* is one for which the actions at all times are deterministic. For example, a pure policy u for the source is a mixed strategy that takes as values only 0 or 1, with a possibility of switching between the two values, depending on t and possibly also the state.[2]

---

[1] By some abuse of notation, we will be using u and w both as policies as well as the realized values of these policies under the adopted information structures which also characterize the sets of policies for the two players (controller and jammer).

[2] Note that this definition is somewhat unconventional, and is made to capture the realization that the 'actions' of the players here are actually probabilities, and hence if these probabilities take the extreme values, 0 or 1, and if this is true for all t, then we call the underlying policies *pure*.

A multiple-criteria game: We next introduce a multiple-criteria problem (game) as follows. The source wishes to maximize with respect to u the function $L^u(x_0, u, w)$, where

$$L^u(x_0, u, w) = J(x, u, w) - \mu \int_0^\rho u_t dt,$$

and the jammer wishes to minimize with respect to w the function

$$L^w(x_0, u, w) = J(x, \rho, u, w) + \theta \int_0^\rho w_t dt,$$

where we have included x_0 in the set of arguments of J (defined earlier by (3)) to emphasize the dependence on the initial state. The pair (u^*, w^*) is a Nash equilibrium for this multiple-criteria problem (nonzero-sum game) if u^* maximizes $L^u(x, \rho, u, w^*)$ over $u \in \Pi_c$ and w^* minimizes $L^w(x, \rho, u^*, w)$ over $w \in \Pi_j$.

Note that in the multi-criteria game, there is antagonism between the two players (related to success probability), but yet it is not a zero-sum game because each player has in addition a second term in his objective function, its own energy cost. However, we can show that this nonzero-sum game is strategically equivalent to a zero-sum game [3], as long as the underlying information structure is open loop; hence every open-loop Nash equilibrium of the multi-criteria game is a saddle-point equilibrium for that particular zero-sum game and *vice versa*.

A strategically equivalent zero-sum game: Let the information structure be open loop for both players, and introduce the objective function

$$L(x, u, w) := J(x, u, w) - \mu \int_0^\rho u_t dt + \theta \int_0^\rho w_t dt,$$

which is obtained by adding $\theta \int_0^\rho w_t dt$ to L^u or equivalently by subtracting $\mu \int_0^\rho u_t dt$ from L^w. Let G_{zs} be the zero-sum game in which the source maximizes $L(x, \rho, u, w)$ and the jammer minimizes it. Note that the addition and subtraction of these additional terms have not changed the Nash equilibrium of the multi-criteria game, because the first term does not depend on the control of the source and the second term does not depend on the control of the jammer, that is[3]

$$\max_u L(x, u, w) = \max_u L^u(x, u, w) + \theta \int_0^\rho w_t dt = [\max_u L^u(x, u, w)] + \theta \int_0^\rho w_t dt,$$

$$\min_w L(x, u, w) = \min_w L^w(x, u, w) - \mu \int_0^\rho u_t dt = [\min_w L^w(x, u, w)] - \mu \int_0^\rho u_t dt,$$

where the first one holds for all open-loop w and the second one for all open-loop u. Then clearly if (u^*, w^*) is an open-loop Nash equilibrium for $(L^u, -L^w)$

---

[3] This argument is not valid if the control policies depend on the state, that is if they are for example feedback policies.

where both players are maximizers, it is also an open-loop Nash equilibrium for $(L, -L)$, and hence an open-loop saddle-point of L (that is game G_{zs}). Likewise, any open-loop saddle-point solution of the zero-sum game G_{zs} is also an open-loop Nash equilibrium of $(L, -L)$, and hence of $(L^u, -L^w)$.

2.4   The Constrained Problem: Energy Constraints

We introduce the constrained game as finding the saddle-point of $J(x, \rho, u, w)$ subject to the following constraints on the source and the jammer controls

$$\int_0^\rho u_t dt \le D_s, \quad \text{and} \quad \int_0^\rho w_t dt \le D_j, \quad \text{respectively.}$$

This constrained problem turns out to be related to the open-loop zero-sum game in the following sense: (u^*, w^*) is a saddle-point if and only if u^* is optimal against w^* and *vice versa*. By the Karush-Kuhn-Tucker (KKT) conditions, there exists $\mu \ge 0$ such that u^* is optimal against w^* if u^* achieves the maximum of $L^\mu(x, u, w^*)$, where

$$L^\mu(x, u, w) = J(x, \ u, w) - \mu \Big(\int_0^\rho u_t dt - D_s \Big).$$

Similarly, there exists $\theta \ge 0$ such that w^* is optimal against u^* if w^* achieves the minimum of $L^\theta(x, u^*, w)$, where

$$L^\theta(x, u, w) = J(x, u, w) + \theta \Big(\int_0^\rho w_t dt - D_j \Big)$$

Hence (u^*, w^*) is an equilibrium in the constrained problem if it is a saddle-point in the zero-sum game

$$L(x, u, w) = J(x, u, w) - \mu \int_0^\rho u_t dt + \theta \int_0^\rho w_t dt \quad \text{for some } \mu \text{ and } \theta.$$

As indicated earlier, we will take the success delivery probability as a performance measure, so that

$$J(x, u, w) = \int_0^\rho \nu x_t dt,$$

where x_t is generated by (1), with initial state x_0.

3   The Static Game

We first restrict the analysis to u and w that are constants in time, in which case we have the unique solution of (1), with initial state x_0, given by

$$x_t = n + (x_0 - n) \exp(-\lambda \kappa t) \tag{4}$$

where $\kappa := u(1 - w)$. Then the objective function of the equivalent zero-sum game can be expressed as:

$$L(x_0, u, w) = \nu \int_0^\rho x_t dt - \rho(\mu u - \theta w) = -\nu(n - x_0)F(\kappa) + \nu n\rho - \rho(\mu u - \theta w),$$

where

$$F(\kappa) := \frac{1 - \exp(-\kappa\lambda\rho)}{\lambda\kappa}.$$

With F' denoting the first derivative of $F(\kappa)$ with respect to κ, and F'' its second derivative, we readily have, for $\kappa \in (0, 1]$:

$$F'(\kappa) = \frac{-1 + (1 + \kappa\lambda\rho)\exp(-\kappa\lambda\rho)}{\lambda\kappa^2}$$

$$F''(\kappa) = \frac{-\kappa^3\lambda^2\rho^2\exp(-\kappa\lambda\rho) + 2\kappa - 2\kappa(1 + \kappa\lambda\rho)\exp(-\kappa\lambda\rho)}{\lambda\kappa^4}$$

$$= \frac{2 - (2 + 2\kappa\lambda\rho + \kappa^2\lambda^2\rho^2)\exp(-\kappa\lambda\rho)}{\lambda\kappa^3}$$

$$> \frac{2 - 2\exp(\kappa\lambda\rho)\exp(-\kappa\lambda\rho)}{\lambda\kappa^3} = 0$$

and for $\kappa = 0$,

$$F'(0) = -\frac{\lambda\rho^2}{2}, \quad F''(0) = \frac{\lambda^2\rho^3}{3}.$$

Hence $F(\kappa)$ is strictly convex in κ, on $[0, 1]$, which implies that $L(x_0, u, w)$ is strictly convex in $\kappa = u(1 - w)$ as long as $x_0 < n$. Since the additional terms in L that depend on u and w are linear, this readily implies that for each $x_0 < n$ $L(x_0, u, w)$ is strictly concave-convex in the pair (u, w) on $(0, 1] \times [0, 1)$, and concave-convex on the closed square $[0, 1] \times [0, 1]$. Hence, we have a concave-convex game defined on a closed and bounded subset of a finite-dimensional space, which is known to admit a saddle-point solution [3]. This result is now captured in the following theorem, which also addresses the uniqueness and characterization:

Theorem 1. *Assume throughout that $x_0 < n$. Then:*

(i) *The static zero-sum game has a saddle-point on $[0, 1] \times [0, 1]$, and it is unique.*

(ii) *If $\nu(n - x_0)\rho\lambda \le 2\mu$, $(u^* = 0, w^* = 0)$ is the unique saddle-point.*

(iii) *The game cannot have a saddle-point with $w = 1$.*

(iv) *If $\nu(n - x_0)\rho\lambda > 2\mu$, the unique saddle-point is in $(0, 1] \times (0, 1)$.*

Proof:

i) As stated prior to the statement of the theorem, existence follows from a standard result in game theory. since we have a concave-convex game. Uniqueness will follow from the proofs of parts (ii) and (iv) below, carried out separately in two regions of the parameter space.

ii) Let $M(u) := -\nu(n - x_0)F(u) - \rho\mu u$, and note that $M'(u) = -\nu(n - x_0)F'(u) - \mu\rho$. Using the earlier expression for $F'(0)$, $M'(0) = \nu(n - x_0)\rho^2\lambda/2 - \mu\rho$ and thus $M'(0) \leq 0$ under the given condition. Further, since $F''(u) > 0$ for all u, $F'(u)$ is an increasing function of u and hence $M'(u)$ is decreasing for all u, which means that

$$M'(0) < 0 \text{ implies } M'(u) < 0 \text{ for all } u > 0. ,$$

and hence that $M(u)$ attains its maximum uniquely at $u = 0$. This means that $u = 0$ is the unique best response to $w = 0$. Further, since $L(x_0, 0, w) = \nu x_0 \rho + \rho\theta w$, the unique minimizing response to $u = 0$ on $[0, 1]$ is $w = 0$. Hence, $(0, 0)$ is a saddle-point solution, and by the ordered interchangeability property of multiple saddle-points and the uniqueness of responses in this case, there can be no other saddle-point.

iii) This readily follows from the observation that the unique maximizing response to $w = 1$ is $u = 0$ while the unique minimizing response to $u = 0$ is $w = 0$. Hence $w = 1$ cannot be part of a saddle-point.

iv) From part (i), we already know that there exists a saddle-point under this condition. Suppose that the saddle-point is not unique, and let (u^*, w^*) and $(\tilde{u}, \tilde{w})$ be two such solutions. By ordered interchangeability of multiple saddle-points, $(\tilde{u}, w^*)$ and $(u^*, \tilde{w})$ are also saddle-point solutions. We know from part (iii) that $w^* \neq 1$, $\tilde{w} \neq 1$, and hence under each of them the objective function is strictly concave in u, which implies that the only way for both u^* and $\tilde{u}$ to be optimal responses to w^* (as well as $\tilde{w}$)) is if they are equal. Hence, u^* has to be unique. Now, if $u^* \neq 0$, then $L(x_0, u^*, w)$ is strictly convex in w, and hence the optimal response by the jammer is unique; hence $w^* = \tilde{w}$ if $u^* \neq 0$. This then leaves out only the case $u^* = 0$ not covered. We already know from the proof of part (ii) that the unique minimizing response to $u = 0$ on $[0, 1]$ is $w = 0$, and under the given condition $u = 0$ is not a maximizing response to $w = 0$ since $M'(0) > 0$. Hence, $u^* = 0$ is ruled out. What we then have is that the saddle-point solution (u^*, w^*) is unique, and necessarily $u^* \in (0, 1]$ and $w^* \in (0, 1)$. $\square$

We now further elaborate on the case when the saddle-point is inside the square, which we know from part (iv) of the Theorem that it happens only when the condition $\nu(n - x_0)\rho\lambda > 2\mu$ holds. We also know that for an inner saddle-point solution, since the game kernel is strictly concave-convex, and jointly continuously differentiable, a necessary and sufficient condition is satisfaction of the stationarity conditions. Toward this end, let

$$K(\kappa) := -\nu(n - x_0)\frac{dF(\kappa)}{d\kappa} = -\nu(n - x_0) \times \frac{-1 + (1 + \kappa\lambda\rho)\exp(-\kappa\lambda\rho)}{\lambda\kappa^2}$$

Then, the inner saddle-point solution (u^*, w^*) uniquely solves

$$\frac{dL(x, u^*, w^*)}{du} = 0, \quad \frac{dL(x, u^*, w^*)}{dw} = 0,$$

which can be written as

$$K(\kappa)(1 - w) - \rho\mu = 0, \quad -K(\kappa)u + \rho\theta = 0.$$

We thus conclude that $\theta(1 - w^*) = \mu u^*$, which leads to

$$\kappa^* = u^*(1 - w^*) = (u^*)^2\mu/\theta \quad \text{or} \quad u^* = \sqrt{\theta\kappa^*/\mu}$$

Finally, substituting this into the second stationarity condition above leads to a single equation for κ^* as below: $K(\kappa^*)\sqrt{\kappa^*} = \rho\sqrt{\theta\mu}$, which we know admits a unique solution in $(0, 1)$. u^* and w^* are then obtained from

$$u^* = \sqrt{\theta\kappa^*/\mu} \quad \text{and} \quad w^* = 1 - (\mu/\theta)u^*.$$

4   The Dynamic Game

We now return to the original dynamic game, and discuss derivation of the equilibrium solution, first for the case of open-loop information and following that for the closed-loop feedback case.

4.1   Open-Loop Information

As discussed earlier, in the open-loop case, every Nash equilibrium of the original differential game is also saddle-point equilibrium of a related strategically equivalent zero-sum differential game. Following the standard derivation of open-loop saddle-point solution [3], we have the single Hamiltonian

$$H(u, w; x, p) = -\mu u + \theta w + \nu x + pu(1 - w)\lambda(n - x), \tag{5}$$

which will be maximized over $u \in [0, 1]$ and minimized over $w \in [0, 1]$. Here p is the co-state variable, which satisfies the associated co-state equation:

$$\dot{p} = -\frac{\partial H}{\partial x} = pu(1 - w)\lambda - \nu, \quad p(\rho) = 0, \tag{6}$$

which constitutes a two-point boundary value problem along with the original state equation

$$\dot{x} = u(1 - w)\lambda(n - x). \tag{7}$$

The source will be maximizing H, and the jammer will be minimizing the same, and if exists we seek a saddle-point solution (u^*, w^*) for the game, which necessarily will also be a saddle-point solution for the Hamiltonian for each t, that is

$$\max_{u \in [0,1]} H(u, w^*; x, p) = \min_{w \in [0,1]} H(u^*, w; x, p) = H(u^*, w^*; x, p).$$

Now, maximizing $H(u, w; x, p)$ over $u \in [0, 1]$ for each $w \in [0, 1]$, and minimizing the same over $w \in [0, 1]$ for each $u \in [0, 1]$ we obtain the complete set of solutions:

$$\arg \max_{u \in [0,1]} H(u, w; x, p) = \begin{cases} 1 & \text{if } p(1 - w)\lambda(n - x) > \mu \\ 0 & \text{if } p(1 - w)\lambda(n - x) < \mu \\ [0, 1] & \text{if } p(1 - w)\lambda(n - x) = \mu \end{cases} \quad (8)$$

$$\arg \min_{w \in [0,1]} H(u, w; x, p) = \begin{cases} 1 & \text{if } pu\lambda(n - x) > \theta \\ 0 & \text{if } pu\lambda(n - x) < \theta \\ [0, 1] & \text{if } pu\lambda(n - x) = \theta \end{cases} \quad (9)$$

Since $p(\rho) = 0$, the unique saddle-point of the Hamiltonian at the terminal time $t = \rho$ is clearly $u^* = w^* = 0$. And clearly, by continuity, the same holds in some left neighborhood of ρ. Integrating the co-state equation backwards from $t = \rho$ with $u = w = 0$, we obtain $p(t) = \nu(\rho - t)$. Note that $u^*(t) = w^*(t) = 0$ is a valid solution as long as

$$p(t)\lambda(n - x_t) < \mu, \quad (10)$$

and the first time (in retrograde time) this is violated will determine the switch time from $u^* = 0$ to some other action for the source. Further note that it is the inequality associated with the source and not the one associated with the jammer that will determine the switching time (in retrograde time) because the LHS of the inequality associated with the jammer, (9), is *zero* as long as $u = 0$. We denote this switching time by $\bar{t}_s$,

$$\bar{t}_s := \sup\{t \leq \rho : \nu(\rho - t)\lambda(n - x_t) < \mu\}$$

When $\theta > \mu$, there exists another threshold t_s such that during the interval $[t_s, \bar{t}_s]$,

$$\mu \leq p(t)\lambda(n - x_t) < \theta$$

and hence from (8) and (9) $u^* = 1$ while $w^* = 0$ during $[t_s, \bar{t}_s]$.

The above two switch times also depend on x_t and $p(t)$, which in turn are generated under the players' actions in the earlier stages of the game. Another observation worth pointing out is that it is not possible for $w^*(t) = 1$ for any t, because this would imply that $u^*(t) = 0$, which in turn implies that $w^*(t) = 0$, a contradiction.

All this reasoning leads to the following theorem which captures the saddle-point solution to the open-loop differential game.

Theorem 2. *(i) If $\theta > \mu$, there exist two switch times $t_s, \bar{t}_s$ with $t_s < \bar{t}_s \leq \rho$ and there exists a saddle-point solution given by*

$$u(t) = \begin{cases} \frac{\theta}{m(t)} & \text{when } t < t_s \\ 1 & \text{when } t_s < t < \bar{t}_s \\ 0 & \text{when } t \geq \bar{t}_s \end{cases} \quad \text{and } w(t) = \begin{cases} 1 - \frac{\mu}{m(t)} & \text{when } t < t_s \\ 0 & \text{when } t_s < t < \bar{t}_s \\ 0 & \text{when } t \geq \bar{t}_s. \end{cases}$$

(ii) When $\theta \leq \mu$, there exists a single switch time t_s such that for $t > t_s$, the saddle-point solution dictates both players to play $u^*(t) = w^*(t) = 0$, and for $t < t_s$

$$u^*(t) = \frac{\theta}{m(t)}, \qquad w^*(t) = 1 - \frac{\mu}{m(t)} \quad where$$
$$m(t) := p(t)\lambda(n - \xi(t)),$$

with p and ξ solving the coupled set of mixed boundary differential equations:

$$\dot{\xi} = \frac{\theta\mu}{p^2\lambda(n - \xi)}, \quad \xi(0) = x_0 \; ; \quad \dot{p} = \frac{\theta\mu}{p\lambda(n - \xi)^2} - \nu, \quad p(t_s) = \nu(\rho - t_s)$$

and t_s is solved from $m(t_s) = \mu$.

Proof: Please see the Appendix, where also the computation of the two switch times, $t_s, \bar{t}_s$, are discussed. $\square$

Remarks: The following are some observations on the saddle-point solution (equivalently Nash solution) obtained in Theorem 2:

- It is an open-loop Nash equilibrium, i.e., the policies obtained depend only upon the time t elapsed from the birth of the message and not on the state x, the number of already infected messages.
- When $\mu > \theta$, i.e., when the power constraint on the source is higher than that on the jammer, the jammer and source are active during the same period and switch off at the same time threshold (t_s of Theorem 2). In a way the jammer is dominating here as it has bigger power resources and hence keeps jamming whenever the source is active.
- When $\theta > \mu$, i.e., when the power constraint on the jammer is high, the jammer is forced to switch off even when the source is active (at time threshold t_s of Theorem 2). The source continues being active for a longer time, until time threshold $\bar{t}_s$. In fact after t_s, the policy is similar to situation without jammer ([1]): the source always infects the contacted mobiles till the threshold $\bar{t}_s$ after which it never infects any further mobiles.
- During the initial time interval, i.e., in the interval $[0, t_s]$ (when the policies are equalizing in nature), the source's probability of transmitting is high whenever the jammer's probability of jamming is low and *vice versa*.

4.2   Closed-Loop Feedback Information

Here we have to stay with the non-cooperative game framework, and seek for Nash equilibria (NE). Let V^u and V^w be the value functions for the two players, where again player u is maximizer and Player w is minimizer. Assuming that these value functions are continuously differentiable jointly in (x, t) (they can even be piecewise continuously differentiable solutions with possibly a finite number of discontinuities in the derivative), the associated HJB equations are ([3]):

$$\frac{\partial V^u}{\partial t} + \max_{u \in [0,1]} \left[\frac{\partial V^u}{\partial x} u(1 - w^*)\lambda(n - x) + \nu x - \mu u \right] = 0 \qquad (11)$$

$$\frac{\partial V^w}{\partial t} + \min_{w \in [0,1]} \left[\frac{\partial V^w}{\partial x} u^*(1 - w)\lambda(n - x) + \nu x + \theta w \right] = 0 \qquad (12)$$

with boundary conditions $V^u(\rho, x) \equiv V^w(\rho, x) \equiv 0$ where (u^*, w^*) is a NE. The corresponding feedback policies are:

$$u^*(x, t) = \arg \max_{u \in [0,1]} \left[\frac{\partial V^u}{\partial x} \lambda u(1 - w^*)(n - x) - \mu u \right] \qquad (13)$$

$$w^*(x, t) = \arg \min_{w \in [0,1]} \left[\frac{\partial V^w}{\partial x} \lambda u^*(1 - w)(n - x) + \theta w \right] \qquad (14)$$

Using these two dynamic programming equations, one can easily establish the following two lemmas.

Lemma 1. *Any feedback Nash equilibrium (NE) will feature a jammer policy taking values only in the semi-open interval $[0, 1)$.*

Proof: If w^* was 1, at some (t, x), then, from equation (13) the corresponding optimal controller would be $u^* = 0$. This in turn implies from equation (14) that $w^* = 0$, which is a contradiction. □

Lemma 2. *If $\nu\lambda(n - x_0)\rho < \mu$ then at NE, $u^* = w^* \equiv 0$, i.e., the optimal policies of both the jammer and the source are to never jam/transmit.*

Proof: From the pair of HJB equations (11) and (12), if it is possible to make the point-wise optimizers in both the Hamiltonians equal to zero, the solution of both PDEs would have been $V^u(x, t) = V^w(x, t) = \nu x(\rho - t)$ for all x, t. And this is exactly the case under the given hypothesis as for any $x \in [x_0, n]$, $t \in [0, \rho]$ and for any $w \in [0, 1]$,

$$\frac{\partial V^u}{\partial x} u(1 - w)\lambda(n - x) = \nu(\rho - t)(1 - w)\lambda(n - x) < \nu\rho\lambda(n - x_0) < \mu$$

and hence $u^* \equiv 0$, and thus from equation (14) $w^* \equiv 0$ □

The first lemma rules out the possibility of pure-strategy NE with nonzero jammer policy.[4] What this leaves as possibility is a NE which is 1) completely inner (or completely mixed NE, i.e., where both players' policies take values in the open interval $(0, 1)$) for some states and time and 2) with $w^* = 0$ for the rest of the states and time. Lemma 2 gives the condition under which the second

---

[4] Again, by *pure strategy* here we mean one that does not take the extreme values 0 or 1, for both players. A *mixed-strategy* NE in this context is one where at least one player's policy takes values in the open interval $(0, 1)$ for some time and state.

situation always (for all states and time) happens. We now consider the case in which this condition is negated, i.e., henceforth we assume that $\nu\lambda(n-x_0)\rho > \mu$. We show the existence of a switching time until which the first possibility occurs and beyond which the second scenario (that of $w^* = 0$) occurs.

Let us consider the first possibility. This would happen if the policies would actually be *equalizer rules*, with $u^* \in (0,1)$ making the expression to be minimized on the right-hand-side of (14) independent of w, and simultaneously $w^* \in (0,1)$ making the expression to be maximized on the right-hand-side of (13) independent of u. Such a (u^*, w^*) would be the solution of the fixed point equations:

$$\frac{\partial V^u}{\partial x}\lambda(1 - w^*)(n - x(t)) = \mu \tag{15}$$

and

$$\frac{\partial V^w}{\partial x}\lambda u^*(n - x(t)) = \theta. \tag{16}$$

If there exist such solutions, then the HJB equations will be simplified to

$$\frac{\partial V^u}{\partial t} + \nu x = 0 , \quad \frac{\partial V^w}{\partial t} + \nu x + \theta 1_{\{u^*>0\}} = 0; \tag{17}$$
$$V^u(\rho, x) = 0 = V^w(\rho, x) \quad \text{for all } x.$$

The simplification in the second PDE is obtained using (16). For future reference we note that we would have arrived at these PDEs if both u and w were taken to be identically zero–a property we will have occasion to utilize shortly.

One can easily solve and obtain the solution $V^u(t, x) = \nu x(\rho - t)$ and hence that $\partial V^u/\partial x = \nu(\rho - t)$. Hence the objective function in (13) is non-positive for all $t > t_c(x)$, where

$$t_c(x) := \frac{\rho\nu\lambda(n - x) - \mu}{\nu\lambda(n - x)} \quad \text{and hence} \quad u^*(x, t) = 0 \text{ for all } t \geq t_c(x).$$

This in turn yields from equation (14) that $w^*(x, t) = 0$ for all $t \geq t_c(x)$. Now the second PDE in (17) can be solved:

$$V^w(t, x) = \theta(t_c(x) - t)1_{\{t<t_c(x)\}} + \nu x(\rho - t).$$

Both PDEs can be brought to the above simplified form and hence the simplified solutions of the fixed point equations (15) and (16) can be obtained for all $t \leq t_c(x)$. By definition of t_c, whenever $t < t_c(x)$, the fixed point equation (15) can be satisfied with a $w^* \in (0, 1)$ if we assume $\mu/\theta - \lambda(\rho - t_c(x_0)) > 1$ as then

$$\frac{\partial V^w(t, x)}{\partial x}\lambda(n - x) = \left(\nu(\rho - t) - \frac{\mu\theta}{\nu\lambda(n - x)^2}\right)\lambda(n - x)$$
$$> \mu - \frac{\mu\theta}{\nu(n - x)} = \theta\left(\frac{\mu}{\theta} - \lambda(\rho - t_c)\right) > \theta$$

for all (t, x) with $t < t_c(x)$. Under this assumption, the fixed point equation (16) can also be satisfied with $0 < u^* < 1$. Thus we have

Theorem 3. *Under the assumption* $\mu/\theta - \lambda(\rho - t_c(x_0)) > 1$, *the closed-loop mixed strategy NE exists with the optimal state trajectory given as the solution of the following ODE:*

$$\dot{x} = f(x,t) \text{ with } f(x,t) := \frac{\mu\theta(n-x)}{(\rho-t)(\nu^2\lambda(\rho-t)(n-x)^2 - \mu\theta)} 1_{\{t \le t_c(x)\}}.$$

and the optimal controls are given by,

$$u^*(t) = \frac{\theta 1_{\{t \le t_c(x_t)\}}}{\lambda(n-x_t)\left(\nu(\rho-t) - \frac{\mu\theta}{\nu\lambda(n-x_t)^2}\right)}$$

$$w^*(t) = \left(1 - \frac{\mu}{\nu(\rho-t)\lambda(n-x_t)}\right)1_{\{t \le t_c(x_t)\}}. \qquad \square$$

Optimal controls for larger values of θ. Now we consider the cases that may not satisfy $\mu/\theta - \lambda(\rho - t_c(x_0)) > 1$. We may not find a $w^* \le 1$ that satisfies the fixed point equation (16) for all $t \le t_c$. Let us start with the extreme case: assume θ is very large ($\theta \gg \mu$) such that the fixed point equation can not be satisfied for all (x,t) with $t \le t_c(x)$. In this case, one can easily verify that the jammer's optimal strategy is to never jam (i.e., $w^* \equiv 0$) and the source's optimal policy is

$$\bar{u}(t,x) := 1_{\left\{\lambda(n-x)\frac{d\bar{V}^u}{dx} > \mu\right\}}$$

where $\bar{V}^u$ is the solution of the PDE

$$\frac{d\bar{V}^u}{dt} + \nu x + \left(\lambda(n-x)\frac{d\bar{V}^u}{dx} - \mu\right)\bar{u}(t,x); \qquad \bar{V}^u(\rho,x) = 0 \qquad (18)$$

and the optimal state trajectory x^* is the solution of $\dot{x}(t) = \lambda(n-x)\bar{u}(t,x)$. The corresponding Hamiltonian PDE for the jammer will be

$$\frac{d\bar{V}^w}{dt} + \nu x + \bar{u}(t,x)\lambda(n-x)\frac{d\bar{V}^w}{dx} = 0. \qquad (19)$$

Thus, a sufficient condition for the optimal jammer policy to be *zero* is that

$$\frac{d\bar{V}^w}{dt}\lambda(n-x)\bar{u}(t,x) < \theta \text{ for all } (x,t). \qquad (20)$$

This condition can only be verified on numerical examples.

Remark: The PDE solutions $\bar{V}^u$, $\bar{V}^w$ are both equal to $\nu x(\rho - t)$ for all (x,t) with $t > t_c(x)$ (the boundary condition is at left boundary $t = \rho$). Thus, $\bar{u}(t,x) = \bar{w}(t,x) = 0$ for all (x,t) with $t \le t_c(x)$. $\qquad \square$

Continuing further, consider now the case when (20) is not true for some (x,t). Then there exists an $0 \le \bar{t}_c(x) \le t_c(x)$ such that,

$$\bar{t}_c(x) = \inf_{t < t_c(x)}\left\{\frac{d\bar{V}^w}{dx}\bar{u}(t,x)\lambda(n-x) > \theta\right\}. \qquad (21)$$

Let $\tilde{V}^u$, $\tilde{V}^w$ represent the solutions of the PDEs,

$$\frac{d\tilde{V}^u}{dt} + \nu x + \left(\lambda(n-x)\frac{d\tilde{V}^u}{dx} - \mu\right)\bar{u}(t,x)1_{\{\bar{t}_c(x) \leq t < t_c(x)\}} = 0; \tilde{V}^u(\rho,x) = 0$$

$$\frac{d\tilde{V}^w}{dt} + \nu x + \bar{u}(t,x)\lambda(n-x)\frac{d\tilde{V}^w}{dx}1_{\{\bar{t}_c(x) \leq t < t_c(x)\}} + \theta 1_{\{t < \bar{t}_c(x)\}} = 0; \tilde{V}^w(\rho,x) = 0$$

Then the optimal controls will be given by,

$$u_t^* := \tilde{u}(t,x_t^*) \text{ with } \tilde{u}(t,x) := \bar{u}(t,x)1_{\{t \geq \bar{t}_c(x)\}} + 1_{\{t < \bar{t}_c(x)\}}\frac{\theta}{\frac{d\tilde{V}^w}{dx}\lambda(n-x)} \quad (22)$$

$$w_t^* := \tilde{w}(t,x_t^*) \text{ with } \tilde{w}(t,x) := 1_{\{t < \bar{t}_c(x)\}}\left(1 - \frac{\mu}{\frac{d\tilde{V}^u}{dx}\lambda(n-x)}\right) \quad (23)$$

where x_t^* is now the solution of the ODE,

$$\dot{x} = \lambda(n-x)\tilde{u}(t,x)(1 - \tilde{w}(t,x)).$$

Remark. The PDE solutions $(\tilde{V}^u, \tilde{V}^w)$ equal to $(\bar{V}^u, \bar{V}^w)$ for all (x,t) with $t > \bar{t}_c(x)$. Further, the solution can be obtained numerically.   □

Remarks: The following are some observations on the closed-loop feedback NE policies:

- The solution is a genuine closed-loop feedback NE, i.e., the policies depend both upon the time t elapsed from the birth of the message and upon the state x, the number of already infected messages.
- The nature of these controls is exactly the same as that in the case of open loop controls.
  - When $\mu/\theta - \lambda(\rho - t_c(x_0)) > 1$ (the case of Theorem 3), i.e., when the power constraint on the source is higher than that on the jammer, the jammer and source are active during the same period and switch off at the same time threshold $(t_c(x),)$. The jammer is dominating even in closed-loop strategies, as it has bigger power resources and hence keeps jamming whenever the source is active. The switch off threshold t_c, unlike in the case of open loop strategies, also depends upon the number of infected mobiles, x.
  - When the power constraint on the jammer is high, the jammer is forced to switch off even when the source is active (at time threshold $\bar{t}_c(x)$ given by (21)). The source continues being active for a longer time, till time threshold $t_c(x)$. In fact after $\bar{t}_c(x)$, the policy is similar to situation without jammer ([1]): the source always infects the contacted mobiles till the threshold $\bar{t}_s$ after which it never infects any further mobiles.

5   Numerical Examples

We now compute the optimal policies obtained in the previous section for the closed-loop case for some numerical examples and verify the same using HJB equations. For example, to verify that (u^*, w^*) is a NE, we obtain a second set of PDEs by replacing the optimal value in the Hamiltonians of (11), (12) with the values evaluated at u^*, w^* respectively. We compare the solutions of this new set of PDEs with that of the HJB solutions.

The two sets of PDE solutions are compared in Figure 1. In Figure 1 the thick lines represent the solution of the simplified HJB equations (17) while thin dotted curves represent the corresponding ones of the PDEs with the optimal policies. We note that the two trajectories almost match, thereby reinforcing the existence of Mixed strategy NE. We also plot the optimal policies as a function of time in Figure 2. It is interesting to observe that the jammer jams with higher probability in the beginning while the probability, with which the source transmits, increases with time till it reaches the switching threshold. This behavior could be because, whenever the jammer jams with large probability, the source better attempt with smaller probability and use the resources at some other time point. Further both the jammer and the source do not transmit after the switching threshold. Note that this threshold is given by $\inf_t\{t > t_c(x_t^*)\}$ where x_t^* is the optimal state trajectory.

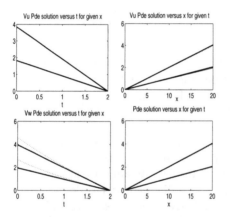

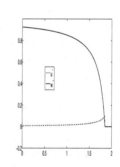

Fig. 1. HJB solutions versus PDE Solutions at the computed NE

Fig. 2. Optimal Controls

We conclude this section with an example which considers two values of θ in Figures 3 4, 5 and 6. For $\theta = 9$, the condition of Theorem 2 is satisfied and hence the optimal control is given by Theorem 2. For $\theta = 200$, we compute the optimal policies using the procedure explained and verify the same by showing that the optimal policies satisfy the HJB equations given earlier in this section. We plot the optimal policies for both cases in Figure 4 while both optimal state trajectories are plotted in Figure 3. In this example, the switching period $t_c(x_0) \approx 1981$ is very close to $\rho = 2000$ and hence in both cases the source is active for almost all the time. We notice that with large θ, there exists another switching time $\bar{t}_c \approx 1698$ beyond which the source is completely active, while the jammer is completely inactive. For small θ $\bar{t}_c$ coincides with t_c. And before this switching time the optimal policies are always mixed in nature (*equalizer rules*) for all the cases. We finally verify that the optimal policies satisfy the HJB equations and the corresponding PDE solution is plotted in Figure 5 for large θ. Both the switching periods $t_c, \bar{t}_c$, when $\theta = 200$, are plotted as functions of x in Figure 6. We also plot the two optimal state trajectories in Figure 3. For larger values of θ the jammer is constrained more and hence the infected population size at any point in time is bigger with larger θ.

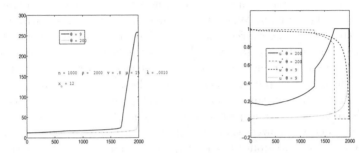

Fig. 3. Example 2: Optimal state trajectory   **Fig. 4.** Example 2: Optimal Controls

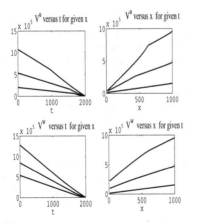

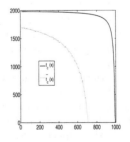

Fig. 5. PDE Solutions: When $\mu/\theta > 1 + \lambda(\rho - t_c(x_0))$ **Fig. 6.** With θ large, $t_c(x)$, $\bar{t}_c(x)$ versus x

6 Conclusions

We have considered a multi-criteria control problem that arises in a delay tolerant network with two adversarial controllers: the source and the jammer. The source's objective was to choose transmission probabilities so as to maximize the probability of successful delivery of some content to the destination within a given time interval, and the jammer's objective was to cause collisions. We considered two types of information structures; the closed loop and the open loop. In the closed loop structure we assume both the jammer and the source have the knowledge of the current number of mobiles with a copy of the message and in this case the game is a genuine nonzero-sum differential game. In the open loop structure they do not have such knowledge and the game becomes strategically equivalent to a zero-sum differential game. The structure of the policies are similar for both types of information structures. In both cases, the optimal policies have two or one switch time(s) depending upon the energy constraints of the source and the jammer. When the jammer has a tighter constraint on its energy resources than the source, the policies have two switch times. Before the first switch time, both the source and jammer policies are inner (i.e., the transmission probabilities are not one of the extreme cases, 0 or 1) and are given by equalizer policies. After the first switch time, the jammer switches off and the source continues transmitting at maximum probability and after the second switch time, both the source and jammer are off. When the source has a tighter constraint on its energy resources than the jammer, there exists only one switch time before which both use inner equalizer policies and after which both are switched off.

References

1. Altman, E., Başar, T., De Pellegrini, F.: Optimal monotone forwarding policies in delay tolerant mobile Ad-Hoc networks, InterPerf, Athens, Greece (October 2008)
2. Altman, E., Azad, P., Başar, T., De Pellegrini, F.: Optimal Activation and Transmission Control in Delay Tolerant Networks. In: Proceedings of IEEE Infocom, San Diego, USA (March 2010)
3. Başar, T., Olsder, G.J.: Dynamic Noncooperative Game Theory, Philadelphia. SIAM Series in Classics in Applied Mathematics (January 1999)
4. Hanbali, A.A., Nain, P., Altman, E.: Performance of ad hoc networks with two-hop relay routing and limited packet lifetime. In: Proc. of Valuetools, p. 49. ACM, New York (2006)
5. Guo, X., Hernandez-Lerma, O.: Zero-sum continuous-time Markov games with unbounded transition and discounted payoff rates. Bernoulli 11(6), 1009–1029 (2005)
6. Başar, T.: The Gaussian test channel with an intelligent jammer. IEEE Trans. Inform. Theory 29(1), 152–157 (1983)
7. Medard, M.: Capacity of correlated jamming channels. In: Allerton Annual Conf. on Comm., Control and Computing (1997)
8. Shafiee, S., Ulukus, S.: Capacity of multiple access channels with correlated jamming. MILCOM 1, 218–224 (2005)
9. Kashyap, A., Başar, T., Srikant, R.: Correlated jamming on MIMO Gaussian fading channels. IEEE Trans. Inform. Theory 50(9), 2119–2123 (2004)
10. Zander, J.: Jamming games in slotted Aloha packet radio networks. In: IEEE MILCOM 1990, Monterey, CA, USA, September 30-October 03, vol. 2, pp. 830–834 (1990)
11. Bayraktaroglu, E., Kingy, C., Liu, X., Noubir, G., Rajaraman, R., Thapa, B.: On the Performance of IEEE 802.11 under Jamming. In: Proceedings of Infocom (2008)

12. Felegyhazi, M., Hubaux, J.-P.: Game Theory in Wireless Networks: A Tutorial, EPFL Technical report: LCA-REPORT-2006-002
13. Love, J.: Cell Phone Jammers, methodshop.com, site created at 09/01/2005 and updated at 04/24/2009,
    http://www.methodshop.com/gadgets/reviews/celljammers
14. Shah, S.W., Babar, M.I., Arbab, M.N., Yahya, K.M., Ahmad, G., Adnan, T., Masood, A.: Cell phone jammer. In: Multi-topic Conference, INMIC 2008. IEEE International Issue Date: December 23-24 (2008)

Appendix: Proof of Theorem 2

Case 1: $\theta > \mu$**:** In the interval $[0, t_s]$, both u^* and w^* are simultaneously inner (i.e., have values in $(0,1)$). For this to happen, we need equalizer policies, i.e., u^* should make Hamiltonian (5) independent of w and w^* should make the same Hamiltonian independent of u simultaneously. Thus,

$$u(t) = \frac{\theta}{m(t)} \text{ and } w(t) = 1 - \frac{\mu}{m(t)} \text{ where } m(t) := p(t)\lambda(n - x(t)).$$

In the above, $p(t)$ and $x(t)$ are the co-state and state trajectories for the saddle-point that we are constructing and these are obtained in retrograde while constructing the saddle-point policy. The equalizer policies must be in open interval $(0, 1)$ and hence for all $t < t_s$,

$$m(t) > \max\{\theta, \mu\} = \theta.$$

Thus t_s is given by

$$t_s = \inf\{t : m(t) \leq \theta\}$$

or by continuity it satisfies $m(t_s) = \theta$. Substituting the policies back in the state and co-state equations, the state and co-state trajectories in the interval $[0, t_s]$ are obtained by solving the ODE's:

$$\dot{x} = \frac{\mu\theta}{\lambda p^2(n - x)} \qquad \text{with } x(0) = x_0 \qquad (24)$$

$$\dot{p} = \frac{\mu\theta}{p\lambda(n - x)^2} - \nu \qquad \text{with } p(t_s) = p_{t_s}. \qquad (25)$$

where the expression for p_{t_s} will be given shortly.

In the interval $[t_s, \bar{t}_s]$, $u^* = 1$ and $w^* = 0$ and $\mu < m(t) \leq \theta$ and hence the co-state trajectory can be solved for this interval as

$$p(t) = c(\bar{t}_s)e^t + \nu \text{ with } c(t) := e^{-t}(p_{\bar{t}_s} - \nu) \text{ for all } t \in [t_s, \bar{t}_s], \qquad (26)$$

where the expression for $p_{\bar{t}_s}$ will also be given shortly. Now p_{t_s} is calculated in terms of $p_{\bar{t}_s}$ as:

$$p_{t_s} = c(\bar{t}_s)e^{t_s} + \nu = e^{t_s - \bar{t}_s}(p_{\bar{t}_s} - \nu) + \nu.$$

The state trajectory in this interval would be,

$$x(t) = n - (n - x(t_s))e^{-\lambda(t-t_s)} \text{ for all } t \in [t_s, \bar{t}_s]. \tag{27}$$

For all $t > \bar{t}_s$ $u^*(t) = 0 = w^*(t)$. Thus solving backwards,

$$p(t) = \nu(\rho - t) \text{ and } x(t) = x(\bar{t}_s) \text{ for all } t \in [\bar{t}_s, \rho].$$

Thus,

$$p_{\bar{t}_s} = \nu(\rho - \bar{t}_s) \text{ and hence } p_{t_s} = e^{t_s - \bar{t}_s}\left(\nu(\rho - \bar{t}_s) - \nu\right) + \nu. \tag{28}$$

Further,

$$m(t) = \lambda\nu(n - x(\bar{t}_s))(\rho - t) \text{ for all } t \in [\bar{t}_s, \rho]$$

and hence $m(t)$ is strictly decreasing for t beyond $\bar{t}_s$, i.e., in the interval $[\bar{t}_s, \rho]$. It is possible that there can be no $t \le \rho$ for which $m(t) = \mu$ and in this case we define $\bar{t}_s = \rho$. In the other case we define $\bar{t}_s$ as the time which satisfies the equation $m(\bar{t}_s) = \mu$, i.e.,

$$\lambda\nu(n - x(\bar{t}_s))(\rho - \bar{t}_s) = \mu.$$

From (27),

$$x(\bar{t}_s) = n - (n - x(t_s))e^{-\lambda(\bar{t}_s - t_s)}$$

and hence

$$\lambda\nu(n - x(t_s))e^{-\lambda(\bar{t}_s - t_s)}(\rho - \bar{t}_s) = \mu. \tag{29}$$

From (26) and (27) for all $t \in [t_s, \bar{t}_s]$,

$$\begin{aligned} m(t) &= \lambda c(\bar{t}_s)e^{(1-\lambda)t}(n - (x_{t_s}))e^{\lambda t_s} + \lambda\nu(n - x(t_s))e^{-\lambda(t-t_s)} \\ &= \lambda\nu(n - x(t_s))\left(e^{-(\bar{t}_s - t)}(\rho - \bar{t}_s - 1) + 1\right)e^{-\lambda(t-t_s)} \end{aligned} \tag{30}$$

Further at t_s, $m(t_s) = \theta$ and hence we get the second equation in terms of t_s and $\bar{t}_s$:

$$m(t_s) = \lambda\nu(n - x(t_s))\left((\rho - \bar{t}_s - 1)e^{t_s - \bar{t}_s} + 1\right) = \theta \tag{31}$$

The thresholds t_s and $\bar{t}_s$ are obtained by solving (29) and (31), and by further using the solutions of the ODEs (24) and (25) with boundary condition (28).

Case 2: $\mu \ge \theta$: The solution can be obtained as in the previous case but now with $\bar{t}_s = t_s$. With $\mu \ge \theta$, at t_s $m(t) = \max\{\mu, \theta\} = \mu$ and hence it is not possible for u^* to be 1. Thus the solutions are obtained by solving the joint ODEs (24) and (25) where t_s is obtained from (29) after replacing $\bar{t}_s = t_s$.   □

Security Interdependencies for Networked Control Systems with Identical Agents

Saurabh Amin[1], Galina A. Schwartz[2], and S. Shankar Sastry[2]

[1] Department of CEE, University of California, at Berkeley - Berkeley, CA, USA
amins@berkeley.edu
[2] Department of EECS, University of California, at Berkeley - Berkeley, CA, USA
{schwartz,sastry}@eecs.berkeley.edu

Abstract. This paper studies the security choices of identical plant- controller systems, when their security is interdependent due the exposure to network induced risks. Each plant is modeled by a discrete-time stochastic linear system, which is sensed and controlled over a communication network. We model security decisions of the individual systems (also called players) as a game. We consider a two-stage game, in which first, the players choose whether to invest in security or not; and thereafter, choose control inputs to minimize the average operational costs. We fully characterize equilibria of the game, which give us the individually optimal security choices. We also find the socially optimal choices. The presence of security interdependence creates a negative externality, and results in a gap between the individual and the socially optimal security choices for a wide range of security costs. Due to the negative externality, the individual players tend to under invest in security.

1 Introduction

In this article, we investigate security choices of individual operators of networked controlled systems (NCS) when security interdependencies are present due to network induced risks. Today, NCS already exhibit substantial interdependence. An imminent wider deployment of smart devices is only likely to result in a higher degree of interdependence [26],[5].

The current state-of-the-art literature on NCS assumes independent and identically distributed (IID) packet losses for systems, even when the systems use the same communication network for their operation [16],[18],[23],[12],[2]. In such settings, attacks on the availability of sensor and control data packets for one system do not affect the availability of data packets for other systems.

The analysis based on the IID packet loss models does not capture the environments in which an attack on the availability of data packets of one system can propagate to other systems due to fact that they share the same communication network. An important example of such attacks are the so-called distributed-denial-of-service (DDOS) attacks [26],[9],[2]. Since the DDOS attacks affect the availability of sensor and control data packets of multiple systems, any security

T. Alpcan, L. Buttyan, and J. Baras (Eds.): GameSec 2010, LNCS 6442, pp. 107–122, 2010.
© Springer-Verlag Berlin Heidelberg 2010

choice of one system is also likely to influence the security of other systems. This paper contributes to the existing literature by considering a setting where security of one system affects security of other systems.

Several factors exacerbate the severity of the losses which may be caused by security interdependencies. First, only a small number of vendors provide embedded controller devices [26], causing a danger of highly correlated software–hardware malfunctions. Due to the prevalence of identical devices, a single glitch could bring major disruption of NCS functioning. Second, since the NCS will soon govern the operation of critical infrastructure systems, the NCS interdependencies could be exploited by nation states. So far, no such occurrences have been recorded, but presence of aforementioned cyber attack capabilities is well documented [25], and cannot be ignored [26]. The risks of such rare (but extremely disruptive) events are similar to risks of terrorist attacks [7], and it is established that private mitigation of such risks fails, thus likely requiring governments to step in [15].

We model the problem of operator's security choice as a non-cooperative two-stage game between M plant-controller systems (or players). Each of these players is modeled in a standard NCS setting (see for e.g., [18],[23]). In the first stage, each player has a binary choice of investing versus non-investing into enhanced security measures at his plant. In the second stage, players choose optimal control inputs for their respective plants. Each players' objective is to minimize the average long-term cost, which is comprised of the plant operating costs and the cost of security measures. We compare the individually optimal choices with that of the social planner, whose objective is to minimize the sum of aggregate operating costs of all the players (which include costs of security measures). The approach in this paper compliments the existing and growing literature on investment efficient security strategies for critical systems systems [10],[11],[1],[6]. By imposing penalties on the players not investing in security, we induce individually optimal player choices that coincide with the socially optimal ones. Such correction of individual incentives is frequently referred as internalizing the externalities [1].

The importance of network externalities for incentives to invest in security have been noted and modeled by numerous researches (see for e.g., [3],[8] and the references therein). The relevance of these effects for critical infrastructures, and in particular, the provision of electricity was raised in [4],[5], but to the best of our knowledge, so far nobody attempted formal modeling of security interdependencies in NCS. The closest models to ours are the application of security interdependencies to Internet security such as [21], where the authors apply [15], and present an analytical model, which permits them to study the deployment of security features and protocols in the sub-nets with different network topologies. Also, [20] expands on [15] to study economics of malware (propagation of viruses and worms).

Our modeling of security choices builds on the Heal and Kunreuther's interdependent security model (see [15],[14],[19]). We refer the reader to [22], [13],[17] for similar approaches.

This paper is organized as follows: In Section 2, we formulate the game between NCS when interdependencies are present. In Sections 3 and 4, we present the analysis of the game of two and M players respectively. Concluding remarks are drawn in Section 5.

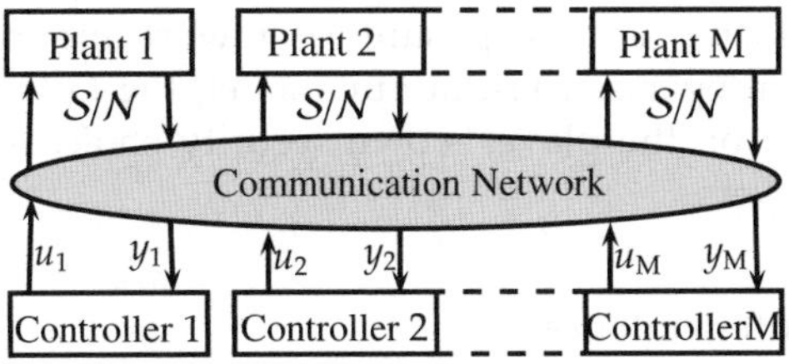

Fig. 1. Networked Control System (NCS)

2 Problem Setup and Preliminaries

We consider M identical NCS; each of these systems consist of a plant and a controller communicating over a network [18] as shown in Fig. 1. We model the interaction of these systems, henceforth referred as *players*, as a two-stage game. In the first stage, each player chooses to make a security investment (S) or not (N). Once player security choices are made, they are irreversible and observable by all the players. In the second stage, each player chooses a control input sequence $u_{0,i}, u_{1,i}, \ldots$. Each player objective in the game is to minimize an infinite-horizon average expected cost criterion, as defined in Section 2.3.

Let L_i denote the security choice of player i:

$$L_i := \begin{cases} S, & \text{player } i \text{ invests in security,} \\ N, & \text{player } i \text{ does not invest in security} \end{cases}$$

and let L denote the vector of security choices of all players:

$$L := \left(L_1, \ldots, L_M\right).$$

We also define the indicator function:

$$I_i := \begin{cases} 0, & L_i = S, \\ 1, & L_i = N. \end{cases}$$

For each plant, the sensor measurements $y_{t,i}$ are sent to the respective controller over a communication network. The controller sends the computed control inputs $u_{t,i}$ to the actuators over the network. The sensor-controller and controller-actuator communication links are subject to losses induced by the network. Following [18],[23], we model the loss processes of the sensor and control communication channel of the i-th system at time $t \in \mathbb{N}$ as Bernoulli random processes $\gamma_{t,i}$ and $\nu_{t,i}$ with the respective failure probabilities $\bar{\gamma}_i$ and $\bar{\nu}_i$:

$$P[\gamma_{t,i} = 0] = \bar{\gamma}_i, \quad P[\nu_{t,i} = 0] = \bar{\nu}_i.$$

In contrast to [18],[23], we assume that the failure probabilities $\bar{\gamma}_i$ and $\bar{\nu}_i$ are interdependent due to the exposure to network induced insecurities. To reflect security interdependencies, in our model, the failure probabilities for each player depend on the player's own security choice *and* on the security choices of other players.

2.1 Security Interdependence

To start, let us consider a two player game ($M = 2$), where in the first stage the players choose S or N. We assume the following failure probabilities for player i:

$$\bar{\gamma}_i(L) = I_i\bar{\gamma} + (1 - I_i\bar{\gamma})\alpha(I_i, I_{-i}),$$
$$\bar{\nu}_i(L) = I_i\bar{\nu} + (1 - I_i\bar{\nu})\alpha(I_i, I_{-i}),$$
$$(1)$$

where the subscript $-i$ denotes the other player. In (1), the first term reflects the probability of a *direct* failure, and the second term reflect an *indirect* failure probability. This second term in (1) reflects player interdependence due to being networked and subjected to external attacks, for e.g., distributed denial-of-service (DDOS) attacks. We define the interdependence term $\alpha(\cdot, \cdot) : \{0, 1\}^2 \to$ $]0, 1[$ as follows

$$0 =: \alpha(0,0) = \alpha(1,0) < \alpha(0,1) := \underline{\alpha} < \alpha(1,1) := \bar{\alpha} < 1,$$

where $\bar{\alpha}$ is such that $\bar{\gamma} + (1 - \bar{\gamma})\bar{\alpha} < 1$, and $\bar{\nu} + (1 - \bar{\nu})\bar{\alpha} < 1$. Thus, we assume that the second term becomes higher when more players are insecure. Notice that $\bar{\gamma}$ (resp. $\bar{\nu}$) respectively denote the failure probabilities of the sensor (resp. control) communication links (identical for both players) if $\alpha(I_i, I_{-i})$ would have been zero, i.e., no interdependence. Then, the failure probabilities in our model coincide with the existing literature [18],[23].

From (1), when player i invests in security, he reduces his direct probability of failure (the first term in (1)) to 0. However, the players are unable to affect (and thus reduce) the indirect probability of failure resulting from network interdependence (the second term in (1)).

We now generalize (1) to any $M \in \mathbb{N}$ players by assuming:

$$\bar{\gamma}_i(L) = I_i\bar{\gamma} + (1 - I_i\bar{\gamma})\alpha(\eta_{-i}),$$
$$\bar{\nu}_i(L) = I_i\bar{\nu} + (1 - I_i\bar{\nu})\alpha(\eta_{-i}),$$
$$(2)$$

where $\eta_{-i} := \frac{\sum_{-i} I_i}{M-1}$ denotes the fraction of players (excluding player i) who have chosen N. As in the two-player case, the failure probabilities should be higher when a higher fraction of players is insecure. To reflect this, we assume that the interdependence term $\alpha(\eta_{-i})$ increases with η_{-i}. With a slight abuse of notation, we define $\alpha(\cdot) : (0,1) \longrightarrow (0,1)$ as follows

$$0 =: \alpha(0) < \cdots < \alpha(\eta) < \cdots < \alpha(1) := \bar{\alpha} < 1, \tag{3}$$

where $\bar{\gamma} + (1 - \bar{\gamma})\bar{\alpha} < 1$, and $\bar{v} + (1 - \bar{v})\bar{\alpha} < 1$.

In contrast to (1), the interdependence for the i–th system as defined in (2) does not depend its own choice of security investment. Notice that although we do not specifically model the interdependencies *between* the failure probabilities of sensor and control links, these links are also interdependent due to the term α in (1) and (2).

2.2 LQG Problem

Consider the following discrete-time stochastic linear system:

$$\begin{aligned} x_{t+1,i} &= Ax_{t,i} + Bv_{t,i}u_{t,i} + w_{t,i}, \\ y_{t,i} &= \gamma_{t,i}Cx_{t,i} + v_{t,i}, \end{aligned} \quad t \in \mathbb{N}_0, \tag{4}$$

where $x_{t,i} \in \mathbb{R}^d$ is the state of the i–th player's system at time t, $u_{t,i} \in \mathbb{R}^m$ is the control input, and $y_{t,i} \in \mathbb{R}^p$ is the measured output. The matrix $A \in \mathbb{R}^{d \times d}$ is the dynamics matrix, $B \in \mathbb{R}^{d \times m}$ is the input matrix, $C \in \mathbb{R}^{p \times d}$ is the observation matrix. The $\mathbb{R}^d$-valued system noise process $w_{t,i}$ (resp. the $\mathbb{R}^p$-valued measurement noise process $v_{t,i}$) is a Gaussian processes with mean 0 (resp. 0) and covariance Q (resp. R). The initial condition $x_{0,i}$ is also Gaussian with mean $\bar{x}_0$ and covariance $\bar{P}_0$. We assume uncorrelated $x_{0,i}$, $w_{t,i}$, and $v_{t,i}$.

Let us define the information vector $X_{t,i}$ as follows[1]:

$$X_{0,i} := \left(L, y_{0,i}, \gamma_{0,i}\right),$$

and for $t = 1, 2, 3, \ldots,$

$$X_{t,i} := \left(L, y_{0,i}, \ldots, y_{t,i}, v_{0,i}, \ldots, v_{t-1,i}, \gamma_{0,i}, \ldots, \gamma_{t,i}\right).$$

We consider the class of control policies consisting of the sequence of measurable functions $\pi_i = \{\mu_{0,i}, \mu_{1,i}, \ldots\}$, where each function maps the information set available to player i at time $t \in \mathbb{N}_0$, denoted $X_{t,i}$ into the control space (here $\mathbb{R}^m$):

$$u_{t,i} = \mu_{t,i}(X_{t,i}), \quad t \in \mathbb{N}_0, \quad i = 1 \ldots M.$$

Let

$$U := \{u_{t,1}, \ldots, u_{t,M} | t \in \mathbb{N}_0\}.$$

---

[1] This information set corresponds to the packet acknowledgment behavior of TCP-like protocols (see [18]).

For given choices L and U, the LQG infinite-horizon average cost for the system (4) can be defined as follows:

$$J_i(L, U) = \limsup_{T \to \infty} \frac{1}{T} \mathsf{E} \left[\sum_{t=0}^{T-1} x_{t,i}^{\top} G x_{t,i} + v_{t,i} u_{t,i}^{\top} H u_{t,i} \right], \tag{5}$$

where $G > 0$ and $H \geqslant 0$ are the matrices of appropriate dimensions.

The above infinite horizon LQG problem is addressed in [23], and we will draw on these results[2]. Following Theorem 5.6 of [23], we will assume that the maximum failure probabilities are below certain thresholds $\bar{\gamma}_i^c$ and $\bar{v}_i^c$, i.e., $\bar{\gamma} + (1 - \bar{\gamma})\bar{a} < \bar{\gamma}_i^c$ and $\bar{v} + (1 - \bar{v})\bar{a} < \bar{v}_i^c$.

For a given choice of security levels L, let the minimum average LQG cost be denoted as $J_i^*(L)$. The corresponding failure probabilities $\bar{\gamma}_i(L)$ and $\bar{v}_i(L)$ are given by definitions (1) and (2) for the case of 2 and M players respectively. In general, an analytical expression of $J_i^*(L)$ can not be obtained; however, [23] provide analytical expressions for the upper and lower bounds of this cost (see Appendix A for these expressions).

To simplify the exposition, we will restrict our attention to the special case of invertible observation matrix C and the measurement noise covariance matrix $R = 0$[3]. Then, the upper and lower bounds on the minimum expected average cost coincide. This cost can be computed as:

$$J_i^*(L) = \bar{\gamma}_i(L) \operatorname{tr} \left((A^{\top} S_{\infty,i} A + G - S_{\infty,i}) \underline{P}_{\infty,i} \right) + \operatorname{tr}(S_{\infty,i} Q) \tag{6}$$

Here $S_{\infty,i}$ and $\underline{P}_{\infty,i}$ are the positive definite solutions of the following algebraic matrix equations:

$$S_{\infty,i} = A^{\top} S_{\infty,i} A + G$$
$$\qquad - (1 - \bar{v}_i(L)) A^{\top} \bar{S}_{\infty,i} B (B^{\top} S_{\infty,i} B + H)^{-1} B^{\top} S_{\infty,i} A, \tag{7}$$
$$\underline{P}_{\infty,i} = \bar{\gamma}_i(L) A \underline{P}_{\infty,i} A^{\top} + Q.$$

We will make use of the following lemma.

Lemma 1. *For any $\bar{\gamma}_i(L^1) < \bar{\gamma}_i(L^2)$ and $\bar{v}_i(L^1) < \bar{v}_i(L^2)$,*

$$J_i^*(L^1) < J_i^*(L^2). \tag{8}$$

Proof. From (7) $S_{\infty,i}$ and $\underline{P}_{\infty,i}$ are increasing with $\bar{v}_i$ and $\bar{\gamma}_i$ respectively. The proof follows from (6).   □

Intuitively, Lemma 1 provides that expected minimum cost decreases in failure probabilities.

---

[2] In [23], these expressions are given for the arrival probabilities $1 - \bar{\gamma}_i$ and $1 - \bar{v}_i$, while we work with failure probabilities here.

[3] One could always deal with a general case by using empirically computed minimum average cost obtained via Monte-Carlo simulations.

Example 1. Consider (4) for the scalar setting with $d = 1$, $B = 1$, $C = 1$. Then Q, R, G, H are scalars. For $|A| > 1$, the critical failure probability is A^{-2} for both sensor and control links. Let $\tilde{\gamma} < A^{-2}$ and $\tilde{\nu} < A^{-2}$. Using Appendix A, the upper and lower bounds for the optimal cost J^* simplify to:

$$J_\infty^{\max} = QS + \bar{P}\left(\frac{\bar{P}\tilde{\gamma} + R}{\bar{P} + R}\right)\left((A^2 - 1)S + G\right),$$

$$J_\infty^{\min} = QS + \tilde{\gamma}\underline{P}\left((A^2 - 1)S + G\right),$$

$$\tag{9}$$

where

$$\bar{P} = \frac{(A^2 R + Q - R) + \sqrt{(A^2 R + Q - R)^2 + 4QR(1 - A^2\tilde{\gamma})}}{2(1 - A^2\tilde{\gamma})}, \quad \underline{P} = \frac{Q}{1 - A^2\tilde{\gamma}},$$

and

$$S = \frac{(A^2 H + G - H) + \sqrt{(A^2 H + G - H)^2 + 4GH(1 - A^2\tilde{\nu})}}{2(1 - A^2\tilde{\nu})}.$$

Notice that $J_\infty^{\max} = J_\infty^{\min}$ if $R = 0$.   □

2.3 The Game

The objective of player i in our game M players is to minimize the following average expected cost:

$$F_i(L, U) = J_i(L, U) + (1 - I_i)\ell, \tag{10}$$

where $\ell \in [0, \infty)$ is the per period security cost (incurred only if in first stage, the player i has chosen S). In the first stage, each player (say i) chooses the security level $I_i \in \{0, 1\}$. In the subgame that starts after the first stage, i.e., once the security choices I_i are made, each player i chooses the control input sequence $u_{0,i}, u_{1,i}, \ldots$ to minimize the LQG cost (5). The solution concept for our two-stage game is subgame perfect Nash equilibrium.

Finally, we also introduce the baseline case: the game of a social planner, whose objective is to minimize the aggregate average costs given by

$$F^{\mathrm{SO}} = \sum_{i=1}^{M} F_i(L, U). \tag{11}$$

3 Equilibria for Two Player Game

Consider a game with 2 players and interdependent failure probabilities specified by (1). For any fixed player security choices, the minimum expected costs are obtained in Section 2.2, and player objectives (10) become as shown in Fig. 2(top). Minimum expected costs for the social planner are shown in Fig. 2(bottom).

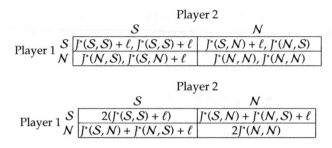

Fig. 2. Costs: 2–player game (top) & social planner (bottom)

Next, we derive player optimal actions in the first stage. We will distinguish the following two cases:

$$J^*(N,N) - J^*(S,N) \leqslant J^*(N,S) - J^*(S,S), \tag{12}$$

$$J^*(N,S) - J^*(S,S) \leqslant J^*(N,N) - J^*(S,N). \tag{13}$$

If (12) holds and a player invests in security, then other player gain from investing in security *increases*. On the other hand, if (13) holds, then each player decision to secure *decreases* the other player gain from investing in security.

Next, in Sections 3.1 and 3.2, we present equilibria for different ℓ, and compare with social optima.

3.1 Increasing Incentives [(12) Imposed]

Let (12) hold, and let us define

$$\underline{\ell}_1 := J^*(N,N) - J^*(S,N), \quad \bar{\ell}_1 := J^*(N,S) - J^*(S,S).$$

Using Fig. 2(top), we infer that if $\ell < \underline{\ell}_1$ (resp. $\ell > \bar{\ell}_1$) then (S,S) (resp. (N,N)) is unique Nash equilibrium. Thus, $\underline{\ell}_1$ (resp. $\bar{\ell}_1$) is the cut-off cost below (resp. above) which both players invest (resp. neither player invests) in security. However, if

$$\underline{\ell}_1 \leqslant \ell \leqslant \bar{\ell}_1, \tag{14}$$

then both, (S,S) and (N,N) are individually optimal.

For the social planner, from Fig. 2(bottom), the optimum is (S,S) if $\ell < \ell_1^{SO}$, with

$$\ell_1^{SO} := \min\left[J^*(N,N) - J^*(S,S), J^*(N,S) + J^*(S,N) - 2J^*(S,S)\right]$$
$$= J^*(N,N) - J^*(S,S),$$

and the second equality is due to (12). From (8), if (14) holds, for ℓ in the entire range $\bar{\ell}_1 < \ell \leqslant \ell_1^{SO}$, individually optimal decision is (N,N), while the socially optimal decision is still (S,S). Finally, if $\ell > \ell_1^{SO}$, the individual and socially optimum choices coincide at (N,N). Fig. 3, Case 1, summarizes the equilibria for different ℓ.

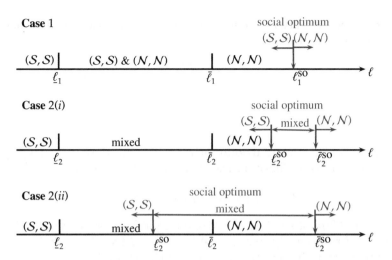

Fig. 3. Nash equilibria and social optima for different ℓ

3.2 Decreasing Incentives [(13) Imposed]

Let (13) hold, and let us define

$$\underline{\ell}_2 := J^*(N, S) - J^*(S, S), \quad \bar{\ell}_2 := J^*(N, N) - J^*(S, N).$$

Using Fig. 2(top), we infer that if $\ell < \underline{\ell}_2$ (resp. $\ell > \bar{\ell}_2$) then (S, S) (resp. (N, N)) is unique Nash equilibrium. Thus, $\underline{\ell}_2$ (resp. $\bar{\ell}_2$) is the cut-off cost below (resp. above) which both players invest (resp. neither player invests) in security. However, if ℓ is in the intermediate range, i.e.,

$$\underline{\ell}_2 < \ell < \bar{\ell}_2, \tag{15}$$

then no pure strategy equilibrium exists. In this case, equilibrium will be mixed. Let θ_i (resp. $(1 - \theta_i)$) denote the mixing probability with which a player chooses S (resp. N). Then, the probability θ_i, for which the player i expected cost, independent of the realizations S or N, is obtained as the positive solution of the following quadratic equation:

$$\theta_i \theta_{-i}(J^*(S, S) + \ell) + \theta_i(1 - \theta_{-i})(J^*(S, N) + \ell)$$
$$= (1 - \theta_i)\theta_{-i}(J^*(N, S)) + (1 - \theta_i)(1 - \theta_{-i})J^*(N, N),$$

Then, by writing a similar equation for player $-i$, it is easy to check that $\theta_i = \theta_{-i}$. Thus, mixed equilibrium is symmetric.

From Fig. 2(bottom), the social optimum is (S, S) if $\ell \leqslant \ell_2^{SO}$, with

$$\ell_2^{SO} := \min [J^*(N, N) - J^*(S, S), J^*(N, S) + J^*(S, N) - 2J^*(S, S)]$$
$$= J^*(N, S) + J^*(S, N) - 2J^*(S, S),$$

and the second equality is due to (13). Notice that ℓ_2^{SO} can be either above or below the cut-off cost $\bar{\ell}_2$, i.e.,

$$\bar{\ell}_2 < \ell_2^{SO}, \tag{16}$$

or

$$\bar{\ell}_2 > \ell_2^{SO}. \tag{17}$$

If (15) and (16) hold, the social optimum is (S, S), while the individually optimal outcome is a mixed equilibrium. The social optimum is (N, N) if $\ell > \bar{\ell}_2^{SO}$, where

$$\bar{\ell}_2^{SO} := 2J^*(N, N) - J^*(S, N) - J^*(N, S).$$

Cases 2(*i*) and 2(*ii*) of Fig. 3 summarize the equilibria for different ℓ when (16) and (17) hold respectively. (Notice that $\ell_2 < \ell_2^{SO} < J^*(N, N) - J^*(S, S) < \bar{\ell}_2^{SO}$).

We now provide an example system for each case of Fig 3.

Example 2. **Case 1.** Let $A = 0.80, G = Q = H = R = 1$, and $\bar{\gamma} = \bar{v} = \underline{\alpha} = \bar{\alpha} = 0.1$. From (9), this system satisfies (12). **Case 2(*i*).** Let $A = 1.2, G = H = Q = R = 1$, $\bar{\gamma} = \bar{v} = 0.1, \underline{\alpha} = \bar{\alpha} = 0.25$. This system satisfies (13) and (16). **Case 2(*ii*).** Let $\bar{\gamma} = \bar{v} = 0.25$ and all other parameters be as in Case 2(*i*). This system satisfies (13) and (17).   □

4   Equilibria for M Player Game

We now extend the analysis of Section 3 to the case of M identical players. Here we will focus on two instructive cases, which generalize the increasing and decreasing incentive cases for the two-player game (Sections 3.1 and 3.2 respectively). Recall from Section 2.1 that η_{-i} denotes the fraction of insecure players (excluding player i). To simplify the notation, we will henceforth omit the subscript $-i$. The player failure probabilities are interdependent as specified by (2), and the interdependence term $\alpha(\eta)$ satisfies (3) for $\eta = 0, \frac{1}{M-1}, \frac{2}{M-1}, \ldots, 1$.

Consider the scenario when the i−th player chooses S or N, and the security choices of all other players are fixed. Let the fraction η of other players be insecure. Without loss of generality, we can assume that the players $1, \ldots, i - 1$ are secure and the players $i + 1, \ldots, M$ are insecure, where $\frac{M-i-1}{M-1} = \eta$. We use the following simplifying notation:

$$L = \begin{cases} (S, \eta), & L_1 = \cdots = L_i = S, L_{i+1} = \cdots = L_M = N, \\ (N, \eta), & L_1 = \cdots = L_{i-1} = S, L_i = \cdots = L_M = N. \end{cases} \tag{18}$$

4.1   Increasing Incentives

Let the following M − 1 conditions hold:

$$J^*(N, \eta) - J^*(S, \eta) < J^*(N, \eta^-) - J^*(S, \eta^-)), \tag{19}$$

where $\eta = \frac{1}{M-1}, \frac{2}{M-1}, \ldots, 1$, and $\eta^- := \left(\eta - \frac{1}{M-1}\right)$ corresponds to the fraction for which one more player invests in security than η. In (19), the left-hand-side term $J^*(N, \eta) - J^*(S, \eta)$ is a player's gain from investing in security when the fraction η of other players are insecure. Similarly, the right-hand-side term $J^*(N, \eta^-) - J^*(S, \eta^-))$ is a player gain from securing when one more player invests in security (relative to the fraction η).

Thus, similar to (12), (19) corresponds to the case when the decision of an extra player to invest in security *increases* other players' gains from investing in security. Analogous to Section 3.1, we have:

Theorem 1. *In the game of* M *players with* (19) *imposed, a pure strategy equilibrium exists, and is symmetric.*

Proof. First, with (19) imposed, the existence of symmetric pure strategy Nash equilibrium is straightforward from adopting the construction of Section 3.1. Depending on the magnitude of $\ell \in [0, \infty[$, the equilibrium is

$$(S, \ldots, S) \quad \text{if } \ell < J^*(N, 1) - J^*(S, 1), \tag{20}$$

$$(N, \ldots, N) \quad \text{if } \ell > J^*(N, 0) - J^*(S, 0), \tag{21}$$

$$(S, \ldots, S) \text{ or } (N, \ldots, N) \quad \text{if } \ell \text{ is such that} \tag{22}$$
$$J^*(N, 1) - J^*(S, 1) \leqslant \ell \leqslant J^*(N, 0) - J^*(S, 0).$$

Second, we show that no asymmetric equilibrium exists. Assume on the contrary that $(S, \ldots, S, N)$ is an equilibrium, i.e., the first M − 1 players invest in security and the M-th player does not. Since the M-th player chooses N over S and $\eta = 0$ because all other players are secure, we have

$$J^*(N, 0) - J^*(S, 0) < \ell. \tag{23}$$

Since the (M − 1)-th player chooses S, when M − 2 players choose S and the M-th player chooses N, we have

$$\ell < J^*(N, \eta) - J^*(S, \eta), \tag{24}$$

where $\eta = \frac{1}{M-1}$. From (23) and (24):

$$J^*(0, N) - J^*(0, S) < J^*(N, \eta) - J^*(S, \eta),$$

which contradicts (19) for $\eta = \frac{1}{M-1}$. Thus, $(S, \ldots, S, N)$ is not an equilibrium.

The same contradiction could be demonstrated for any other asymmetric equilibrium. Indeed, let $(S, \ldots, S, \underbrace{N \ldots, N})_{M_1 \text{ players}}$ be an equilibrium, i.e., when the first M − M_1 players invest in security and the last M_1 players do not. For the $(M - M_1 + 1)$-th player,

$$J^*(N, \eta^-) - J^*(S, \eta^-)) < \ell,$$

where $\eta^- = \frac{M_1-1}{M-1}$, and for the $(M - M_1)$-th player,

$$\ell < J^*(N,\eta) - J^*(S,\eta),$$

where $\eta = \frac{M_1}{M-1}$. Combining these two inequalities contradicts (19) for $\eta = \frac{M_1}{M-1}$. Thus, we have proven that no asymmetric equilibrium exists, and the theorem is proven.   □

From the proof of Theorem 1, in equilibrium, if ℓ is below a critical minimum value (given by (20)) all players invest in security $(S,\dots,S)$. If ℓ exceeds a critical maximum value (given by (21)) no player invests in security $(N,\dots,N)$; for ℓ between these critical values (see (22)), both outcomes $(S,\dots,S)$ and $(N,\dots,N)$ are equilibria.

4.2 Decreasing Incentives

Now, let the following $M - 1$ conditions hold:

$$J^*(N,\eta^-) - J^*(S,\eta^-) < J^*(N,\eta) - J^*(S,\eta), \tag{25}$$

where $\eta = \frac{1}{M-1}, \frac{2}{M-1} \dots 1$. Notice that analogous to (13), from (25), each player gain from security investment *decreases* as more players invest in security. Analogous to Section 3.2, we have:

Theorem 2. *In the game of* M *players with* (25) *imposed, equilibrium exists; it is unique and symmetric.*

Proof. First, the existence of equilibrium follows by adopting the construction of Section 3.2. Depending on the magnitude of ℓ, the equilibrium is:

$$(S,\dots,S) \quad \text{if } \ell < J^*(N,0) - J^*(S,0), \tag{26}$$

$$(N,\dots,N) \quad \text{if } \ell > J^*(N,1) - J^*(S,1), \tag{27}$$

Mixed if ℓ is such that

$$J^*(N,0) - J^*(S,0) < \ell < J^*(N,1) - J^*(S,1). \tag{28}$$

Next, we show that the mixed equilibrium is symmetric. Assume on the contrary, i.e, without loss of generality, let the first two players mix with different probabilities $\theta_1 \neq \theta_2$, and other players' mixing probabilities be fixed. Then, as in Section 3.2, we have $\theta_1 = \theta_2$. Thus, only a symmetric mixed equilibrium exists.

Lastly, uniqueness is obvious from construction.   □

From the computation of mixed strategies, we have:

Remark 1. When the equilibrium is mixed, the probability of investment in security decreases with ℓ.   □

5   Discussion and Concluding Remarks

In this paper, we investigated the incentives to invest in security for players which operate interdependent and identical NCS. We presented a new model of interdependendent NCS, where the players' failure probabilities are dependent on the security investments of other players.

In our setting, player actions differ from social optimum ones; this reflects the presence of externalities. Indeed, in general, when player costs are affected by other player's choices, players impose externalities on each other. The externalities manifest by the gap between the individually and socially optimal security choices [1]. In the case of negative externalities, players tend to under-invest in security. This is commonly referred to free-riding in economics.

We hope that our findings are relevant for analyzing the effects of DDoS attacks on NCS governing the critical infrastructures, for e.g., the next generation electric power grid. It is well accepted that in the future grid, a large number of commodity IT solutions will be deployed [26],[24]. A wider deployment of smart devices is likely to result in a higher number of players (higher M), a higher degree of interdependence between the players (a higher second terms in (3)), and also a higher security cost ℓ due to the increased configuration (and overall system) complexity. Thus, we expect that with the NCS becoming increasingly "smarter", the magnitude of negative externalities, and therefore the gap between the individually and socially optimal outcomes will only widen.

Such underinvestment in the presence of interdependencies raises the possibility of major breakdowns, see [26],[7], which would create losses (due to higher costs) far beyond the NCS losses considered in this paper. Our model does not incorporate these extra loses, which makes our estimates of security investments, including the socially optimal ones, rather conservative.

Acknowledgments. The authors are grateful to Professor Pravin Varaiya for useful discussions. This work was supported by TRUST (Team for Research in Ubiquitous Secure Technology), which receives support from the National Science Foundation (#CCF-0424422) and the following organizations: AFOSR (#FA9550-06-1-0244), BT, Cisco, DoCoMo USA Labs, EADS, ESCHER, HP, IBM, iCAST, Intel, Microsoft, ORNL, Pirelli, Qualcomm, Sun, Symantec,TCS, Telecom Italia, and United Technologies.

References

1. Alpcan, T., Başar, T.: Network Security: A Decision and Game Theoretic Approach. Cambridge University Press, Philadelphia (2011)
2. Amin, S., Cárdenas, A.A., Sastry, S.: Safe and secure networked control systems under denial-of-service attacks. In: Majumdar, R., Tabuada, P. (eds.) HSCC 2009. LNCS, vol. 5469, pp. 31–45. Springer, Heidelberg (2009)
3. Anderson, R., Böhme, R., Clayton, R., Moore, T.: Security economics and European policy. In: Proceedings of the Workshop on the Economics of Information Security WEIS, Hanover, USA (June 2008)

4. Anderson, R., Fuloria, S.: Security economics and critical national infrastructure. In: The Eighth Workshop on the Economics of Information Security (2009)
5. Anderson, R., Fuloria, S.: On the security economics of electricity metering. In: The Ninth Workshop on the Economics of Information Security (2010)
6. Başar, T., Olsder, G.J.: Dynamic Noncooperative Game Theory, 2nd edn., Philadelphia. SIAM Series in Classics in Applied Mathematics (1999)
7. Bier, V., Oliveros, S., Samuelson, L.: Choosing what to protect: Strategic defensive allocation against an unknown attacker. Journal of Public Economic Theory 9(4), 563–587 (2007)
8. Böhme, R., Schwartz, G.A.: Modeling cyber-insurance: Towards a unifying framework. In: Proceedings of the Workshop on the Economics of Information Security WEIS, Harvard University, Cambridge (June 2010)
9. Cárdenas, A.A., Amin, S., Sastry, S.S.: Research challenges for the security of control systems. In: Provos, N. (ed.) HotSec. USENIX Association (2008)
10. Carin, L., Cybenko, G., Hughes, J.: Cybersecurity strategies: The QuERIES methodology. Computer 41
11. Cavusoglu, H., Mishra, B., Raghunathan, S.: The value of intrusion detection systems in information technology security architecture. Info. Sys. Research 16(1), 28–46 (2005)
12. Garone, E., Sinopoli, B., Casavola, A.: LQG control over lossy TCP-like networks with probabilistic packet acknowledgements. International Journal of Systems, Control and Communications 2(1/2/3), 55–81 (2010)
13. Grossklags, J., Christin, N., Chuang, J. (eds.): Secure or Insure? A Game-Theoretic Analysis of Information Security Games. In: Proceedings of the 17th International World Wide Web Conference (April 2008)
14. Heal, G., Kunreuther, H.: Interdependent security. Journal of Risk and Uncertainty 26(2-3), 231–249 (2003)
15. Heal, G., Kunreuther, H.: Interdependent security: A general model. NBER Working Papers 10706, National Bureau of Economic Research, Inc. (August 2004)
16. Hespanha, J.P., Naghshtabrizi, P., Xu, Y.: A survey of recent results in networked control systems. Proceedings of the IEEE 95(1), 138–162 (2007)
17. Hofmann, A.: Internalizing externalities of loss prevention through insurance monopoly: an analysis of interdependent risks. The GENEVA Risk and Insurance Review 32(1), 91–111 (2007)
18. Imer, O.C., Yüksel, S., Başar, T.: Optimal control of LTI systems over unreliable communication links. Automatica 42(9), 1429–1439 (2006)
19. Kunreuther, H., Heal, G.: Interdependent security: The case of identical agents. Working Paper 8871, National Bureau of Economic Research (April 2002)
20. Lelarge, M.: Economics of malware: epidemic risks model, network externalities and incentives. In: Allerton 2009: Proceedings of the 47th Annual Allerton Conference on Communication, Control, and Computing, Piscataway, NJ, USA, pp. 1353–1360. IEEE Press, Los Alamitos (2009)
21. Lelarge, M., Bolot, J.: Network externalities and the deployment of security features and protocols in the internet. SIGMETRICS Perform. Eval. Rev. 36(1), 37–48 (2008)
22. Mounzer, J., Alpcan, T., Bambos, N.: Dynamic control and mitigation of interdependent IT security risks. In: Proceedings of the IEEE Conference on Communication (ICC), IEEE Communications Society (May 2010)
23. Schenato, L., Sinopoli, B., Franceschetti, M., Poolla, K., Sastry, S.S.: Foundations of control and estimation over lossy networks. Proceedings of the IEEE 95, 163–187 (2007)

24. Tabors, R.D., Parker, G., Caramanis, M.C.: Development of the smart grid: Missing elements in the policy process. In: Proceedings of the Hawaii International Conference on System Sciences, Los Alamitos, CA, USA, pp. 1–7 (2010)
25. Dam, K.W., Owens, W.A., Lin, H.S.: Technology, Policy, Law, and Ethics Regarding U.S. Acquisition and Use of Cyberattack Capabilities. Committee on Offensive Information Warfare, National Research Council, Philadelphia (2009)
26. Weiss, J.: Protecting Industrial Control Systems from Electronic Threats. Momentum Press (2010)

A   LQG Cost for Optimization General Case

We describe the computation of minimum expected average cost for the general case with non-invertible C and $R \neq 0$. Following [23], we assume that $\tilde{\gamma}_i < \tilde{\gamma}_i^c$, $\tilde{v}_i < \tilde{v}_i^c$, (A,B) and $(A,Q^{1/2})$ are controllable, and (A,C) and $(A,G^{1/2})$ are observable. Then $J_i^*(L)$ can be computed by taking the limit $T \longrightarrow \infty$ of $\frac{1}{T}J_{i,T}^*(L)$, where $J_{i,T}^*(L)$ is the finite-horizon optimal LQG cost.

The state estimate $\hat{x}_{t,i} = \mathsf{E}[x_{t,i}|X_{t,i}]$, and error covariance matrix

$$P_{t,i} = [(x_{t,i} - \hat{x}_{t,i})(x_{t,i} - \hat{x}_{t,i})^\top | X_{t,i}]$$

are given by the following update equations starting with $\hat{x}_0 = \bar{x}_0$ and $\bar{P}_0 = \bar{P}_0$:

$$\tilde{x}_{t,i} = A\hat{x}_{t-1,i} + (1 - v_{t,i})Bu_{t-1,i},$$
$$\tilde{P}_{t,i} = AP_{t-1,i}A^\top + Q,$$

and

$$\hat{x}_{t,i} = \tilde{x}_{t,i} + (1 - \gamma_{t,i})\tilde{P}_{t,i}C^\top \left(C\tilde{P}_{t,i}C^\top + R\right)^{-1}(y_{t,i} - C\tilde{x}_{t,i}),$$
$$P_{t,i} = \tilde{P}_{t,i} - (1 - \gamma_{t,i}) \times \tilde{P}_{t,i}C^\top \left(C\tilde{P}_{t,i}C^\top + R\right)^{-1}C\tilde{P}_{t,i}.$$

The optimal control input $u_{t,i}$ is given by

$$u_{t,i} = -(B^\top S_{t+1,i}B + H)^{-1}B^\top S_{t+1,i}A\hat{x}_{t,i},$$

where matrices $S_{t,i}$ are computed using the following (backward) recursion starting with $S_T = G$:

$$S_{t,i} = A^\top S_{t+1,i}A + G$$
$$- (1 - \tilde{v}_i)A^\top \bar{S}_{t+1,i}B(B^\top S_{t+1,i}B + H)^{-1}B^\top S_{t+1,i}A.$$

The finite-horizon optimal LQG cost $J_{i,T}^*(L)$ is:

$$J_{i,T}^*(L) = \bar{x}S_{0,i}\bar{x} + \mathrm{tr}(S_0\bar{P}_0) + \sum_{t=0}^{T-1} \mathrm{tr}(S_{t,i}Q)$$

$$+ \sum_{t=0}^{T-1} \mathrm{tr}\left((A^\top S_{t+1,i}A + G - S_{t,i})\mathsf{E}_{\gamma_i}[P_{t,i}]\right), \tag{29}$$

where the notation $\mathsf{E}_{\gamma_i}[\cdot]$ indicates that the expectation is taken w.r.t. $\{\gamma_{k,i}\}_{k=0}^t$. Following [23], we note that due to the nonlinear dependence of $P_{t,i}$ on $\{\gamma_{k,i}\}_{k=0}^t$, the expected error covariance matrix $\mathsf{E}_{\gamma_i}[P_{t,i}]$ (and hence $J_{i,T}^*(L)$) cannot be computed analytically. However, one can find upper and lower bounds of $J_{i,T}^*(L)$ by two deterministic sequences

$$J_{i,T}^{\min}(L) \leqslant J_{i,T}^*(L) \leqslant J_{i,T}^{\max}(L).$$

Under the assumption of $\tilde{\gamma}_i < \tilde{\gamma}_i^c$ and $\tilde{v}_i < \tilde{v}_i^c$, the infinite-horizon optimal cost $J_i^*(L)$ is bounded from below and above by $J_{i,\infty}^{\min}(L)$ and $J_{i,\infty}^{\max}(L)$ respectively, where $J_{i,\infty}^{\min}(L) := \lim_{T \to \infty} \frac{1}{T} J_{i,T}^{\min}(L)$ is same as (6), and

$$
\begin{aligned}
J_{i,\infty}^{\max}(L) &:= \lim_{T \to \infty} \frac{1}{T} J_{i,T}^{\max}(L) \\
&= \text{tr} \left\{ (A^\top S_{\infty,i} A + G - S_{\infty,i}) \times (\bar{P}_{\infty,i} \right. \\
&\quad \left. - (1 - \tilde{\gamma}_i(L)) \bar{P}_{\infty,i} C^\top (C P_{\infty,i} C^\top + R)^{-1} C \bar{P}_{\infty,i}) \right\} \\
&\quad + \text{tr}(S_{\infty,i} Q).
\end{aligned}
$$

Here $S_{\infty,i}$ and $\underline{P}_{\infty,i}$ are the positive definite solutions of the system of equations (7), and $\bar{P}_{\infty,i}$ is the positive definite solution of:

$$
\begin{aligned}
\bar{P}_{\infty,i} &= A \bar{P}_{\infty,i} A^\top + Q \\
&\quad - (1 - \tilde{\gamma}_i(L)) \times A \bar{P}_{\infty,i} C^\top (C \bar{P}_{\infty,i} C^\top + R)^{-1} C \bar{P}_{\infty,i} A^\top.
\end{aligned}
$$

Notice, that for the special case of invertible C and $R = 0$, the upper and lower bounds for the infinite-horizon optimal cost coincide.

Robust Control in Sparse Mobile Ad-Hoc Networks

Eitan Altman[1], Alireza Aram[3], Tamer Başar[2],
Corinne Touati[4], and Saswati Sarkar[5]

[1] INRIA, BP93, 06902 Sophia Antipolis, France
[2] University of Illinois, 1308 West Main Street, Urbana, IL 61801-2307, USA
[3] Finance Department, The Wharton School, Univ of Pennsylvania, USA
[4] Mescal/LIG project, ENSIMAG, 38330 Montbonnot du St Martiin, France
[5] Dept. of Electrical and Systems Eng., University of Pennsylvania

Abstract. We consider a two-hop routing delay-tolerant network. When the source encounters a mobile then it transmits, with some probability, a file to that mobile, with the probability itself being a decision variable. The number of mobiles is not fixed, with new mobiles arriving at some constant rate. The file corresponds to some software that is needed for offering some service to some clients, which themselves may be mobile or fixed. We assume that mobiles have finite life time due to limited energy, but that the rate at which they die is unknown. We use an H^∞ approach which transforms the problem into a worst case analysis, where the objective is to find a policy for the transmitter which guarantees the best performance under worst case conditions of the unknown rate. This problem is formulated as a zero-sum differential game, for which we obtain the value as well as the saddle-point policies for both players.

1 Introduction

We consider in this paper a delay tolerant network, i.e. a sparse network of mobile relay nodes, where connectivity is very low. There is some source that transmits a file to mobiles that are in the communication range. Each mobile is assumed to be in range with the source at some instants that form a Poisson process. A node that receives a copy of the file stores it so that it may transmit it to some potential destinations that may search for a copy of the file.

We assume that it is desirable that the number of mobiles that have a copy of the file be close to some fixed threshold. Distributing the file to a number of mobiles larger than this threshold is not desirable since transmitting and storing the file costs resources (such as memory).

We further assume that the life time of the mobiles is finite due to the finite energy stored in their battery. There is some rate at which the battery empties, which is assumed to be unknown and to change in time in an unpredictable way. We call this the departure rate.

T. Alpcan, L. Buttyan, and J. Baras (Eds.): GameSec 2010, LNCS 6442, pp. 123–134, 2010.

In order to compensate for the mobiles that become not operational due to battery energy limitation, other mobiles are added to the network (these could be viewed as mobiles that managed to recharge their battery). The new mobiles join at some instants, described by a Poisson process. The rate of this process is assumed to be controlled by the network.

We describe the random evolution of this system and obtain an ODE (ordinary differential equation) for the evolution of the expected number of mobiles that have a copy of the file and of those that do not. We assume that the departure rate od mobiles is unknown and may change in time in an unpredictable way. We then formulate a robust control problem, that of obtaining a policy for the network control that guarantees the best performance under the worst (time varying) departure rate. We formulate the problem as a H$^\infty$ control problem, which we transform into an equivalent zero-sum differential game. We provide an explicit solution to this game using the theory of linear-quadratic zero-sum differential games [5,11]. This problem can thus be seen as one with an additional malicious player that has an opposite performance objective.

Various adversarial approaches have been often used in networking to solve problems in which some parameter may vary in time in an unknown way. A paradigm that has been used frequently in scheduling and routing problems which falls into this category is the competitive online algorithmic paradigm [1]. The H$^\infty$ control theory [5] that we use here is yet another such paradigm that has used in the context of flow control [3,6,16]. We are not aware of previous applications of adversarial type approachess for the control of epidemic type models.

The two-hop routing protocol considered here was first introduced by Gröss-glauser and Tse in [9]; the main goal there was to characterize the capacity of mobile ad-hoc networks and the two-hop protocol was meant to overcome severe limitations of static networks capacity obtained in [10]. Two-hop routing, in particular, provides a convenient compromise of energy versus delay compared to epidemic routing; the standard reference work for the analysis of the two-hop relaying protocol is [8]. Fluid approximations and infection spreading models similar to those we use here are described extensively in [13].

Algorithms to control forwarding in DTNs have been proposed in the recent literature, e.g., [12], [7]. In [12], the authors describe an epidemic forwarding protocol based on the *susceptible-infected-removed* (SIR) model [13]. They show that it is possible to increase the message delivery probability by tuning the parameters of the underlying SIR model. In [7] a detailed general framework is proposed in order to capture the relative performances of different self-limiting strategies. Finally, under a fluid model approximation, the work in [4] provides a general framework for the optimal control of the broad class of monotone relay strategies, i.e., policies where the number of copies do not decrease over time. It is proved there that optimal forwarding policies are of threshold type.

The present paper is the first one we know of that utilizes tools and framework of robust control and linear-quadratic zero-sum differential games in the context of controlling DTNs, or more generally, controlling propagations of epidemics.

2   The Model

Each mobile of the DTN is assumed to be in the communication range of the
source at some time instant governed by a Poisson process with rate η_t. Further:

- x is the **expected number of nodes** that have the file and y is the expected
  number of nodes that do not, with x_t and y_t denoting their values at time t.
- The source is in contact with each mobile without file at a rate η_t. Thus at
  a rate of $\eta_t y_t$, mobiles without a file transform into type x_t mobiles.
- There is a stream of new mobiles (without files) that join the system at a
  rate λ_t
- Mobiles that have the file die at a rate $\nu_t x$, where ν_t is unknown to the
  source and may change in time.

Introduce the following pair of coupled ordinary differential equations (ODEs),
for $t \geq 0$:

$$\begin{cases} \dot{x}_t = \eta_t y_t - \nu_t x_t\,, & x_0 = 0\,, \\ \dot{y}_t = -\eta_t y_t + \lambda_t\,, & y_0 = 0\,. \end{cases} \tag{1}$$

These equations can be shown to correspond to the mean field limit of the actual number of mobiles with and without the file, with some appropriate scalings
when scaled properly [15,17]. Interestingly, however, these are not just approximations; they in fact describe the precise dynamics of the **expected** numbers
of mobiles with and without a copy of the file.

We will now let $u_t := \eta_t y_t$ and $\mu_t := \nu_t x_t$, in view of which (1) is written as:

$$\begin{cases} \dot{x}_t = u_t - \mu_t\,, & x_0 = 0\,, \\ \dot{y}_t = -u_t + \lambda_t\,, & y_0 = 0\,, \end{cases} \tag{2}$$

which is a linear, controlled dynamics.

Next we introduce the cost structure. We assume that it is desired for some
target number $\bar{x}$ of nodes to have a copy of the file. Due to energy and memory
constraints, it is desirable not to exceed this number. The energy for the source
to transmit at a rate u_t increases with u_t, so the corresponding cost should be
an increasing function of u_t.

The reasoning above as well as practical implementations lead to the following
instantaneous cost:

$$c(t) = (x_t - \bar{x})^2 + u_t^2 \tag{3}$$

The controller, u, will be picked to minimize C given by

$$C(u, \mu) = \int_0^{t_f} c(t) dt\,, \tag{4}$$

where $[0, t_f]$ is the horizon of interest.

Remark 1. Since, as it will turn out, u_t is positive and $x_t < \bar{x}$, the instantaneous
cost is indeed an increasing function of u_t, and it forces x_t to stay close to $\bar{x}$.

Remark 2. As another motivation or justification for the quadratic cost on the control, we offer the following explanation. Signaling is needed in order for the source to be able to know whether a mobile that does not have the file is within its transmission range. Assume that this signaling is done by each such mobile by periodical sending of beacons. The control variable η_t can then be interpreted as the frequency of beaconing per mobile, and u_t is the total beaconing rate of all mobiles. Thus u_t can be interpreted as the signaling energy expended by all terminals.

Because of the way the control u was introduced, as a control policy we have to restrict it to depend on the current value of y and not on the current value of x, and hence with respect to state x it is an open-loop policy. Note also that since the cost does not depend on y, the value of y (and hence the dynamics that generate y) does not enter into the optimization problem. Hence we can essentially work with a scalar state equation, and once we obtain the optimal u (say u_t^* at time t) the optimal value of the original decision variable η can be obtained by dividing u_t^* by y_t, assuming that the latter is nonzero.

Now, since μ_t drives the state equation (and hence affects the value of the cost function), optimal choice of the control cannot be obtained independently of μ. Further, since the control does not know μ_t, it is reasonable to adopt a worst-case approach, where we see μ as controlled by an adversary. This is the framework of robust control, or H^∞ control [5], where the goal is to minimize the effect of μ on C, by a proper choice of u. We therefore seek to solve the inf sup problem:

$$\inf_u \sup_\mu \frac{C(u,\mu)^{\frac{1}{2}}}{\|\mu\|} =: \gamma^*$$

where $\|\mu\|^2 := \int_0^{t_f} \mu_t^2 dt$, and are interested in the corresponding minimizing control u^*. Since, it is not always possible to achieve γ^*, it would be sufficient to find a control that would achieve a value of γ slightly higher than γ^*, say by an $\epsilon > 0$. Then what we are looking for is a control u^γ achieving[1]

$$\sup_\mu \frac{C(u^\gamma,\mu)^{\frac{1}{2}}}{\|\mu\|} \leq \gamma^* + \epsilon =: \gamma(\epsilon)$$

Now following [5], this is equivalent to finding u^γ that guarantees that

$$L_\gamma(u^\gamma,\mu) := C(u^\gamma,\mu) - \gamma^2 \|\mu\|^2 \leq 0$$

for all μ, and doing this for "smallest" possible γ. What we have here is a zero-sum differential game with kernel $L_\gamma(u,\mu)$, parametrized by γ. It turns out [5]

---

[1] Since the goal is to drive x to $\bar{x}$ and not to *zero* as in standard H^∞ control, the formulation here does not fit the standard H^∞ optimal control formulation, but after the problem is brought into the linear-quadratic differential game format, the standard theory becomes applicable as to be seen shortly.

that for each $\gamma > \gamma^*$ this differential game admits a saddle-point solution, that is (u^γ, μ^γ) such that for every u and μ,

$$L_\gamma(u^\gamma, \mu) \le L_\gamma(u^\gamma, \mu^\gamma) \le L_\gamma(u, \mu^\gamma).$$

Here $L_\gamma(u^\gamma, \mu^\gamma)$ is the *value* of the game with objective function

$$L_\gamma(u, \mu) = \int_0^{t_f} ((x - \overline{x})^2 + u^2 - \gamma^2 \mu^2) dt$$

An interpretation of the zero-sum differential game

We may view the above zero-sum differential game as arising in the context of a transmission problem with a malicious player. Instead of assuming that each mobile leaves at an unknown rate of ν_t, we assume that it would stay in the system in the absence of the malicious player. However, the latter transmits to the mobiles some virus at a rate of $\mu_t = \nu_t x_t$, and as a result, these mobiles go out of operation at the rate μ_t. And the malicious player attempts to maximize the quantity the control is trying to minimize, by also respecting some cost on energy

$$D = \int_0^{t_f} \mu_t^2 dt$$

which enters the objective function as a *soft* constraint. Note that, because of the relationship $\mu_t = \nu_t x_t$, we are looking for a maximizing policy that is a function of the state x, that is a closed-loop policy. As well known from theory of H$^\infty$ control, however, whether the maximizing player uses closed-loop or open-loop policy is inconsequential, and the optimum value of γ, γ^*, does not depend on it [5]. Accordingly, in the presentation of the solution in the next section, we will occasionally also use open-loop policies for the maximizer.

3   The Saddle-Point Solution

Our goal is to find a saddle point of L_γ. Introducing the shifted state variable,

$$\tilde{x}_t := x_t - \overline{x},$$

we have:

$$\dot{\tilde{x}} = u_t - \mu_t, \ \tilde{x}_0 = -\overline{x}, \tag{5}$$

$$L_\gamma(u, \mu) = \int_0^{t_f} (\tilde{x}_t^2 + u_t^2 - \gamma^2 \mu_t^2) dt. \tag{6}$$

This is a standard linear quadratic zero-sum differential game (the dynamics are linear and the cost is quadratic in both the state and the control variables (of both players)), and recall also that as discussed in the previous section the minimizing player has access to open-loop information (that is u will be only a function of t and $\overline{x}$), whereas the maximizing player has access to closed-loop information but could as well be taken (initially) also to be open loop without any loss (or gain) in performance. The theory of chapter 4 of [5] directly applies, and we have the solution presented in stages in the subsections below.

3.1 Computing the Value γ^*

Before solving for the saddle point, we first need to determine the values of $\gamma > 0$ for which such a saddle point exists, or in other words the upper value of the game is bounded. The answer lies in the solution of the following Riccati differential equation

$$\dot{s} + 1 + \frac{1}{\gamma^2} s^2 = 0, \ s(t_f) = 0, \tag{7}$$

which has the following general solution:

$$s = \gamma \frac{a \cos \frac{t}{\gamma} - b \sin \frac{t}{\gamma}}{a \sin \frac{t}{\gamma} + b \cos \frac{t}{\gamma}} \tag{8}$$

Using the terminal condition $s(t_f) = 0$, we get

$$\tan \frac{t_f}{\gamma} = \frac{a}{b} \tag{9}$$

A feasible γ is one for which s is finite in the interval $[0, t_f]$, that is there is no finite escape. S goes to infinity only if the term in the denominator goes to zero, which will happen for all times t that satisfy $\tan(t/\gamma) = -b/a = -1/\tan(t_f/\gamma)$. Hence γ is feasible if the t that satisfies this condition falls outside the interval $[0, t_f]$, which happens if $t_f/\gamma < \pi$, or in other words, $\gamma > t_f/\pi$. Therefore, the open-loop zero sum differential game has a unique open-loop saddle point for all $\gamma > \frac{t_f}{\pi}$, and the γ^* introduced earlier is equal to this value, that is $\gamma^* = \frac{t_f}{\pi}$. For $\gamma < \gamma^*$, the upper value of the game is unbounded when the minimizing player is restricted to open-loop policies.

3.2 The Solution

We now proceed to solve for the saddle point. The relevant Riccati differential equation in this case is (see, [5], p. 135):

$$\dot{z} = -1 + qz^2; \ z(t_f) = 0, \quad q := (1 - \frac{1}{\gamma^2}) \tag{10}$$

The general solution for z is $-t + k$ if $\gamma = 1$. Else, $z = -\dot{w}/qw$, where w solves

$$\ddot{w} - qw = 0, \tag{11}$$

which admits the general solution

$$w = \begin{cases} a \exp +\sqrt{q}t + b \exp -\sqrt{q}t, & q > 0 \\ a \sin \sqrt{-q}t + b \cos \sqrt{-q}t, & q < 0 \end{cases} \tag{12}$$

where a and b are parameters to be determined from the terminal condition (actually, the only relevant quantity is a/b, and thus we henceforth let $b = 1$).

Note also that we are interested in this solution to the extent that w remains bounded away from *zero*, because at $w(t) = 0$, z will be unbounded. As we will see next, this will place a restriction on the range of values q and thus γ can take. What we know (without doing any computation), however, is that the the Riccati differential equation (10) will admit a unique solution whenever (7) does, and hence if $(\overline{\gamma}, \infty)$ is the range of feasible values of γ for (10), then $\overline{\gamma} < \gamma^*$, and further that $\overline{\gamma} < 1$; see [5].

Therefore, we have (as the solution to the original Riccati differential equation (10)):

$$z = \begin{cases} \frac{-1}{\sqrt{q}} \frac{a \exp \sqrt{q}t - \exp(-\sqrt{q}t)}{a \exp \sqrt{q}t + \exp(-\sqrt{q}t)} & \gamma > 1 \\ -t + k & \gamma = 1 \\ \frac{1}{\sqrt{-q}} \frac{a \cos \sqrt{-q}t - \sin \sqrt{-q}t}{a \sin \sqrt{-q}t + \cos \sqrt{-q}t} & \overline{\gamma} < \gamma < 1 \end{cases} \tag{13}$$

where the parameters a and k are determined by the condition $z(t_f) = 0$, and $\overline{\gamma}$ is the value of γ which makes the denominator of the third expression in (13) *zero*. Carrying this out, we obtain: $k = t_f$,

$$a = \begin{cases} \exp(-2\sqrt{q}t_f), & \gamma > 1 \\ \tan \sqrt{-q}t_f, & \overline{\gamma} < \gamma < 1 \end{cases} \tag{14}$$

and $\overline{\gamma} = 2t_f / \sqrt{\pi^2 + 4t_f^2}$. In view of these, (13) becomes (parameterized by γ)

$$z_\gamma = \begin{cases} \frac{1}{\sqrt{q}} \tanh(\sqrt{q}(t_f - t)) & \gamma > 1 \\ t_f - t & \gamma = 1 \\ \frac{1}{\sqrt{-q}} \tan(\sqrt{-q}(t_f - t)) & \overline{\gamma} < \gamma < 1 \end{cases} \tag{15}$$

The open-loop saddle-point solution is given by

$$u_t^* = -z_\gamma(t)\tilde{x}_t^*, \quad \mu_t^* = -\gamma^{-2}z_\gamma(t)\tilde{x}_t^*,$$

where γ has to satisfy $\gamma > \gamma^* = t_f/\pi$, and $\tilde{x}_t^*$, $t \geq 0$, is the corresponding trajectory of the shifted state, obtained from

$$\dot{\tilde{x}}_t^* = -(1 - \gamma^{-2})z_\gamma(t)\tilde{x}_t^*, \quad \tilde{x}_0^* = -\bar{x}. \tag{16}$$

The corresponding trajectory for y can be obtained by substituting u back into the second ODE of (1):

$$y_t^* = \int_0^t [z_\gamma(s)\tilde{x}_s^* + \lambda_s] \, dt.$$

To ensure that y^* stays positive, the input rate should be chosen to satisfy

$$\lambda_t > -z_\gamma(t)\tilde{x}_t^*$$

A few remarks are in order here:

First, for the saddle-point solution obtained above to relate to the original problem posed, we have to have x_t^* positive for all t, which means that $\tilde{x}_t^*$ should be nondecreasing in t. This will be achieved only if $\gamma > 1$, and hence we need (in view of the earlier open-loop condition)

$$\gamma > \max(1, t_f/\pi).$$

Second, to obtain the corresponding contact rate (η_t), we have to divide u_t^* by y_t^*, leading to

$$\eta_t^* = -z_\gamma(t)\tilde{x}_t^*/y_t^*,$$

which is positive. This, however, is well defined as long as we start y^* not at 0 at $t = 0$, but at some positive value (that is, the expected value of mobiles without the file initially, or at the time optimization kicks in, should be positive); otherwise the rate will be infinite.

Third, to obtain the corresponding dying rate, we have to divide μ_t^* by x_t^*, leading to

$$\nu_t^* = -\gamma^{-2}z_\gamma(t)\tilde{x}_t^*/(\tilde{x}_t^* + \overline{x}),$$

which is positive. Again, the denominator of this expression is 0 at $t = 0$, and hence an adjustment has to be made so that $x_0^* > 0$ and not 0, which would be a small perturbation on the initial state.

Note that the condition on the positivity of the initial conditions is not consistent with the assumption that we had made in eq (1) (that the initial conditions are zero). However, since the solutions are continuous in the initial conditions, the departure from optimality will be tolerable if we take them to be positive but close to zero.

One can obtain explicit expressions for the corresponding contact rate and the dying rate by first noting that

$$\frac{d\tilde{x}}{dt} = -qz_\gamma(t)\tilde{x}$$

Hence

$$ln(-\tilde{x}(T)) - ln(-\tilde{x}(0)) = -q\int_0^T \frac{1}{\sqrt{q}}\tanh(\sqrt{q}(t_f - t))dt$$

$$= -\sqrt{q}\int_0^T \tanh(\sqrt{q}(t_f - t))dt$$

$$= -\sqrt{q}\int_0^T \frac{exp(\sqrt{q}(t_f - t)) - exp(-\sqrt{q}(t_f - t))}{exp(\sqrt{q}(t_f - t)) + exp(-\sqrt{q}(t_f - t))}dt$$

$$= \int_{exp(\sqrt{q}t_f)+exp(-\sqrt{q}t_f)}^{exp(\sqrt{q}(t_f-T))+exp(-\sqrt{q}(t_f-T))} \frac{1}{u}du \quad \begin{array}{l}\text{(with the change of variable} \\ u = exp(\sqrt{q}(t_f - t)) \\ + exp(-\sqrt{q}(t_f - t)))\end{array}$$

$$= ln\left(\frac{cosh(\sqrt{q}(t_f - T))}{cosh(\sqrt{q}t_f)}\right)$$

Hence, for any $\gamma > 1$, we have:

$$\tilde{x}_t = -\bar{x}\left(\frac{cosh(\sqrt{q}(t_f - t))}{cosh(\sqrt{q}t_f)}\right)$$

We obtain finally the following expression for the control η_t, which gives the the arrival rate of mobiles with the file:

$$\frac{u_t}{y_t} = \frac{-z_\gamma(t)\tilde{x}_t^*}{\int_0^t [z_\gamma(s)\tilde{x}_s^* + \lambda_s]\,dt} = \frac{\frac{\bar{x}}{\sqrt{q}}\frac{sinh(\sqrt{q}(t_f-t))}{cosh(\sqrt{q}t_f)}}{\frac{-\bar{x}}{\sqrt{q}}\int_0^t \frac{sinh(\sqrt{q}(t_f-t))}{cosh(\sqrt{q}t_f)}\,dt + \int_0^t \lambda_s\,dt + y_0}$$

$$= \frac{\frac{\bar{x}}{\sqrt{q}}\frac{sinh(\sqrt{q}(t_f-t))}{cosh(\sqrt{q}t_f)}}{\frac{\bar{x}}{q}\frac{cosh(\sqrt{q}(t_f-t))-cosh(\sqrt{q}t_f)}{cosh(\sqrt{q}t_f)} + \int_0^t \lambda_s\,dt + y_0}$$

$$= \frac{\bar{x}\sqrt{q}sinh(\sqrt{q}(t_f - t))}{\bar{x}cosh(\sqrt{q}(t_f - t)) - \bar{x}cosh(\sqrt{q}t_f) + qcosh(\sqrt{q}t_f)\left(\int_0^t \lambda_s\,dt + y_0\right)} \quad (17)$$

For ν we get the following expression:

$$\frac{\frac{1-q}{\sqrt{q}}sinh(\sqrt{q}(t_f - t))}{cosh(\sqrt{q}t_f) - cos(\sqrt{q}(t_f - t))}$$

Numerical examples. Below we present the evolution of the states and controllers over the time interval $\le 0 \le 10$. We use the parameters $\lambda = 3$ (constant in time), $y_0 = 4, \gamma = 5, \bar{x} = 40$. The figures display the mean number of infected nodes, the control actions of both controllers and the instantaneous cost; all these are plotted against time.

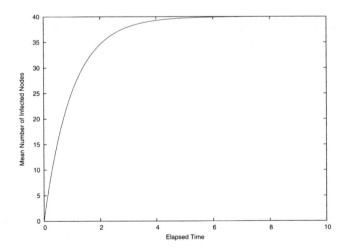

Fig. 1. Mean number of infected nodes (vertical axis) as a function of time (the horizontal axis)

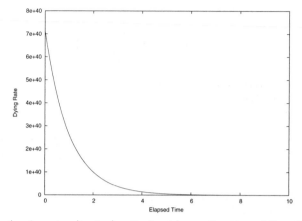

Fig. 2. Dying (or departure) rate (vertical axis) as a function of time (horizontal axis)

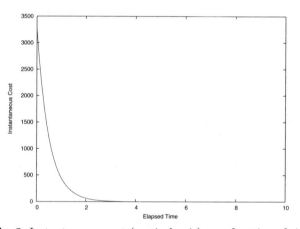

Fig. 3. Instantaneous cost (vertical axis) as a function of time

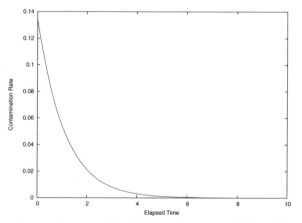

Fig. 4. Contamination rate (controlled by the original controller) at the vertical axis, as a function of time (horizontal axis)

4 Conclusions

DTNs exhibit many special features that make them hard to control: the population of mobiles may be large, relevant information may be unavailable or may take time to arrive. The complexity of these systems often pushes us to come with simplified models (including Poisson contact processes) which allows us to describe some features of the model through exact closed form formulas. The linear quadratic control is one central framework that enables derivation of optimal control through the use of closed form formulae.

In deriving simplified control models, it is desirable to have not only simplicity but also robustness. In particular, the model should not to be too sensitive to the simplifying assumptions that lead to it. The H^∞ control is an appropriate framework that allows one on the one hand to include the simplicity of the linear quadratic framework, and, on the other, to account for some complexities and imprecisions. To our knowledge, this is the first attempt to use the H^∞ paradigm in the context DTNs. The problem presented here is not directly one of security, but the solution approach is well adapted to many potential security problems.

Acknowledgement. The work of the first and third authors has been supported in part by an INRIA-UIUC collaborative research grant. The work of the first author has been partially supported by the European Commission within the framework of the BIONETS project IST-FET-SAC-FP6-027748, see URL:-www.bionets.eu. The work of the second author was supported by the DAWN associated team program between INRIA, UPenn and IISc. The work of the third author was also supported by an AFOSR Grant.

References

1. Albers, S., Leonardi, S.: On-line algorithms. ACM Computing Surveys (CSUR) archive 31(3) (September 1999)
2. Al-Hanbali, A., Nain, P., Altman, E.: Performance of ad hoc networks with two-hop relay routing and limited packet lifetime. In: Proc. of Valuetools, Pisa, Italy, October 11-13 (2006)
3. Altman, E., Başar, T., Hovakimian, N.: Worst-case rate-based flow control with an ARMA model of the available bandwidth. In: Gaitsgory, et al. (eds.) Annals of Dynamic Games, vol. 5, pp. 3–29. Birkhäuser, Basel (2000)
4. Altman, E., Başar, T., De Pellegrini, F.: Optimal monotone forwarding policies in delay tolerant mobile ad-hoc networks. In: Proc. of ACM/ICST Inter-Perf, Athens, Greece, October 24 (2008)
5. Başar, T., Bernhard, P.: H^∞-Optimal Control and Relaxed Minimax Design Problems: A Dynamic Game Approach, 2nd edn. Birkhäuser, Boston (1995)
6. Biberovic, E., Iftar, A., Özbay, H.: A solution to the robust flow control problem for networks with multiple bottlenecks. In: Proceedings of the IEEE Conference on Decision and Control, Orlando, FL, pp. 2303–2308 (December 2001)
7. Fawal, A.E., Boudec, J.-Y.L., Salamatian, K.: Performance analysis of self limiting epidemic forwarding. EPFL, Tech. Rep. LCA-REPORT-2006-127 (2006)

8. Groenevelt, R., Nain, P., Koole, G.: The message delay in mobile ad hoc networks. In Posters ACM SIGMETRICS 2005, Canada (2005)
9. Grossglauser, M., Tse, D.: Mobility increases the capacity of ad hoc wireless networks. IEEE/ACM Trans. on Networking 10(4), 477–486 (2002)
10. Gupta, P., Kumar, P.R.: The capacity of wireless networks. IEEE Trans. on Information Theory 46(2), 388–404 (2000)
11. Başar, T., Olsder, G.J.: Dynamic Noncooperative Game Theory. SIAM Series in Classics in Applied Mathematics. SIAM, Philadelphia (1999)
12. Musolesi, M., Mascolo, C.: Controlled Epidemic-style Dissemination Middleware for Mobile Ad Hoc Networks. In: Proc. of ACM Mobiquitous, San Jose, California, July 17-21 (2006)
13. Zhang, X., Neglia, G., Kurose, J., Towsley, D.: Performance modeling of epidemic routing. Computer Networks 51(10) (July 2007)
14. Altman, E., Neglia, G., De Pellegrini, F., Miorandi, D.: Decentralized stochastic control of delay tolerant networks. In: Proc. of Infocom, Rio de Janeiro, April 15-19 (2009)
15. Kurtz, T.G.: Solutions of Ordinary Differential Equations as Limits of Pure Jump Markov processes. J. Appl. Prob. 7, 49–58 (1970)
16. Quet, P.-F., Ramakrishnan, S., Özbay, H., Kalyanaraman, S.: On the H controller design for congestion control with a capacity predictor. In: Proceedings of the IEEE Conference on Decision and Control, Orlando, FL, pp. 598–603 (December 2001)
17. Weiss, A., Shwartz, A.: Large Deviations for Performance Analysis. Chapman and Hall, Boca Raton (1995)

A Game-Theoretical Approach for Finding Optimal Strategies in a Botnet Defense Model

Alain Bensoussan[1], Murat Kantarcioglu[2], and SingRu(Celine) Hoe[2]

[1] University of Texas at Dallas, USA and The Hong Kong Polytechnic University, HK
[2] University of Texas at Dallas, USA
alain.bensoussan@utdallas.edu, muratk@utdallas.edu, hoceline02@yahoo.com

Abstract. Botnets are networks of computers infected with malicious programs that allow cybercriminals/botnet herders to control the infected machines remotely without the user's knowledge. In many cases, botnet herders are motivated by economic incentives and try to significantly profit from illegal botnet activity while causing significant economic damage to society. To analyze the economic aspects of botnet activity and suggest feasible defensive strategies, we provide a comprehensive game theoretical framework that models the interaction between the botnet herder and the defender group (network/computer users). In our framework, a botnet herder's goal is to intensify his intrusion in a network of computers for pursuing economic profits whereas the defender group's goal is to defend botnet herder's intrusion. The percentage of infected computers in the network evolves according to a modified SIS (susceptible-infectious-susceptible) epidemic model. For a given level of network defense, we define the strategy of the botnet herder as the solution of a control problem and obtain the optimal strategy as a feedback on the rate of infection. In addition, using a differential game model, we obtain two possible closed-loop Nash equilibrium solutions. They depend on the effectiveness of available defense strategies and control/strategy switching thresholds, specified as rates of infection. The two equilibria are either (1) the defender group defends at maximum level while the botnet herder exerts an intermediate constant intensity attack effort or (2) the defender group applies an intermediate constant intensity defense effort while the botnet herder attacks at full power.

Keywords: Botnet Defense, Differential Game, Nash Equilibrium.

1 Introduction

According to recent reports from Russian-based Kaspersky Labs [22] and Symantec [21], botnets (zombie networks)currently pose the biggest threat to the cybersecurity. In fact, Botnets have become a significant source of income for cybercriminals. According to [18], sources of income for the botnet business include distributed denial of service attacks, theft of confidential information,

T. Alpcan, L. Buttyan, and J. Baras (Eds.): GameSec 2010, LNCS 6442, pp. 135–148, 2010.
© Springer-Verlag Berlin Heidelberg 2010

spam, phishing, search engine optimization spam, click fraud, and distribution of adware and malicious programs.

To analyze the economic aspects of botnet activity and suggest feasible defensive strategies, we provide a comprehensive game theoretical framework that models the interaction between the botnet herder and the defender group (network/computer users). In our framework, a botnet herder's goal is to maximize his profits (equivalently minimize his cost) by intensifying his intrusion in a network of computers whereas the defender group's goal is to maximize his profits/benefits (equivalently minimize his cost/loss) by defending the infection of computers. In view of the contagion of malicious programs used to expand botnets among computers, we model the evolution of percentage of infected computers in the network with an SIS epidemic model, in which a computer state may be either susceptible or infectious. The reason we work on an SIS model is due to the fact that a computer may be subject to multiple vulnerabilities and thus a computer is still vulnerable even recovering from one susceptibility. Moreover, [3] has confirmed that an SIS model depicts the dynamics of botnet transmission well in a 6-month experimental study.

Under a fixed level of defense applied by the defender group, we define botnet herder's optimal attack strategy as the solution to a cost minimization control problem. In addition, we solve the simultaneous move differential game between the botnet herder and the defender group. Each player optimizes his objective while considering their opponent's action. For the differential game, under equilibrium, we predict that either one of the players but not both will always play "full effort" strategy. The outcome hinges on the effectiveness of available defense strategies and control/strategy switching thresholds. Switching thresholds are determined by rates of infection, at which point the player alters his action optimally. The existence of playing "full effort" strategy by one party under equilibrium is because the intermediate effort level strategy is not admissible. When the most effective defense strategy available cannot efficiently reduce the spread of malicious program, it drives the botnet activity towards the equilibrium such that the defender group defends at maximum level and the botnet herder exerts an intermediate constant attack effort. One point to mention here is that the defender group may reduce the equilibrium infection size by posing severe penalties to botnet herder's attack effort. The other equilibrium is that the defender group exerts an intermediate constant defense effort and the botnet herder attacks at full power. This equilibrium occurs when the least effective defense strategy cannot efficiently deter the propagation of malicious program but the most effective one can reduce the infection successfully. These results indicate that in some cases trying to defend against botnet activity may not be economically feasible instead the goal should be to limit the damage to an acceptable level. Also when significant resources are allocated to defend against particular botnet activity, botnet herders will choose to reduce their attack effort even if the defensive strategies are not very effective. To our knowledge, none of previous works suggested these two different equilibrium strategies.

2   Related Work

Our framework exploits the epidemic model to characterize the fact that bots (infected machines) may spread malware to other hosts connected to the network. Studies related to epidemic models and computer worms in networks center on two themes. One theme focuses on the study of the epidemic thresholds on the network, for example [7]. The other theme combines the epidemic model with defense measures to study worm propagation or the optimal patch deployment process across networks, for example, [14], [10] and among others.

Recognizing the interdependent security impacts, researchers recently have combined the epidemic model with game theoretical modeling to capture interdependent security decisions, for example [5], [11], [12] and among others. Our paper instead focuses on strategic interactions between the botnet herder and the defender group in which both groups can affect infection evolution with interdependent security problems captured by the contact transmission dynamics.

We develop a differential game model between the botnet herder and the defender group for simultaneous moves. Both players take actions that optimize their objective while also considering the actions of their opponent. The use of game theory in modeling interactions between an attacker and a defender has been adopted widely in the computer security domain recently, for example, [16], [17] and [9] among others. Most of the works focus on the matrix game setting. Our work focuses on continuous time state evolution and control application, solving differential games for optimal policies.

As botnet threats have become an increasing concern, the volume of research papers dealing with this issue increases. [18] discusses the economics of botnet in detail. [15] studies the use of honeypots to deter the development of a botnet. [11] studies the botnet security problem focusing on the interconnected host's security solution to the contagious risk. [19] models botnet related cybercrimes as a result of profit-maximizing decision making from the economic prospectives of both botnet masters and renters/attackers.

We study botnet problems in a game theoretical framework from two different angles. For a given level of network defense, like [19], we consider botnet related cybercrimes as a result of profit-maximizing (equivalently cost minimization) decision making. However, instead of studying uncertainty impacts from honeypots, we concentrate on finding the botnet herder's optimal attack strategy. This provides a useful idea of potential botnet activity equilibrium for a given level of network defense. Unlike [19] but similar to [11], the contagious risk of malware propagation is considered to incorporate interdependent security. Nevertheless, we employ the epidemic evolutionary process directly rather than exogenously assigning the transition probability of states as in [11]. For simultaneous moves, different from [11] which considers the defender group's interdependent security game, we work on the botnet herder's and the defender group's strategic optimization in response to their opponent's actions while incorporating interdependent security problems.

The other article similar to ours is [13] in which they mainly focus on one shot game and analyze botnet herder's attack coordinations as well as defender's interdependent security defense decision. They obtain multiple Nash equilibria under different conditions. We focus on a continuous state evolution model, in which the state evolves according to the botnet herder's and the defender group's actions and contact transmission, so do both players' corresponding payoffs/costs. For a given level of network defense, we obtain botnet herder's optimal attack strategy as a feedback on the rate of infection. The steady state equilibrium exists with an intermediate constant intensity attack effort . This prediction is somewhat similar to that of short discussion of extensive-form games in the appendix of [13]. Surprisingly, for simultaneous moves, we arrive at different equilibrium results in which either one of the players will always play "full effort" strategies.

3   Epidemic Model

In the area of virus and worm modeling, many studies have employed epidemiological models to understand the general characteristic of worm propagation. Depending on the model specifications, the state of a computer at a given time can be infectious, susceptible (vulnerable to a worm) or immune (excluded from further dynamics). In this study, we base our dynamics of percentage of infected computers in a network on a modified SIS model. This allows us to incorporate the attacker's and defender's strategies into the system dynamics.

3.1   Deterministic SIS Model

Because a computer may be subject to multiple vulnerabilities, a computer is still vulnerable even recovering from one susceptibility. Therefore, it is reasonable to work on an SIS model. Moreover, [3] has confirmed that an SIS model depicts the dynamics of botnet transmission well in a 6-month experimental study. In the classical SIS model, such as [8] and [4], a recovered host immediately becomes susceptible again. That is, each host stays in one of two states: susceptible or infectious. For a fixed population system with N hosts, let $S(t)$ and $I(t)$ denote the number of susceptible and infectious hosts at time t respectively; then the dynamic of the system is described by the following set of differential equations:

$$\begin{cases} \frac{dy(t)}{dt} = -\beta x(t)y(t) + \gamma x(t), \ y(0) = 1 - x_0 \\ \frac{dx(t)}{dt} = \beta x(t)y(t) - \gamma x(t), \ x(0) = x_0, \ 0 \le x_0 \le 1 \end{cases} , \qquad (1)$$

where $y(t) = S(t)/N$, the percentage of susceptible nodes at time t, $x(t) = I(t)/N$, the percentage of infectious nodes at time t, $\beta \ge 0$ is the average number of transmissions possible from a given infected host in each period, and $\gamma \ge 0$ is the recovery rate.

3.2   Modified SIS Model − The State Equation

Our model bases on the classical SIS model with some modifications given[1]:

$$\frac{dx(t)}{dt} = cv_{\mathrm{H}}(t)\big(1 - x(t)\big) + \beta x(t)\big(1 - x(t)\big)$$
$$- \big(\gamma_{\min} + v_{\mathrm{D}}(t)(\gamma_{\max} - \gamma_{\min})\big)x(t),\ x(0) = x_0,\ 0 \le x_0 \le 1 \qquad (2)$$

The time argument will be suppressed in future where no confusion arises. In (2),

1. $cv_{\mathrm{H}}(1 - x)$ expresses the increment of percentage of infected computers due to botnet herder's ongoing direct attack effort (not from contagion), where c is the average attack successful rate and $v_{\mathrm{H}} \in [0,1]$ is the attack effort intensity, the botnet herder's control, indicating how aggressively the botnet herder tries to intensify his intrusion[2].
2. $v_{\mathrm{D}} \in [0,1]$ depicts the defender group's defense effort, the defender group's control. In our specification, we assume that there is a set of available defense strategies which can range the effectiveness of defense from minimum level to maximum level. We relate the effectiveness to the recovery rate coefficient[3] γ, and associate defensive strategies with minimum and maximum effectiveness by $\gamma_{\min}$ and $\gamma_{\max}$ respectively. From the term $\gamma_{\min} + v_{\mathrm{D}}(\gamma_{\max} - \gamma_{\min})$, it is clear that, through the defense effort, v_{D}, the defender group is able to attain the defense effectiveness within the range of minimal and maximal effectiveness provided by available defense strategies. That is, v_{D} signals the defense strategy chosen by the defender group. For example, if $v_{\mathrm{D}} = 1$, the defender group exerts a full defense effort to achieve the maximal effectiveness, $\gamma_{\max}$. In other words, the defender group chooses the best defense strategy available.

4   Game between Botnet Herder and Defender Group

We begin with a simplified game, in which the botnet herder solves his optimal intrusion in response to a given level of defense strategies, and continue with a simultaneous move game between the botnet herder and the defender group, in which both parties solve their own optimal strategy in response to their opponent's action.

---

[1] From (1), $dx(t)/dt + dy(t)/dt = 0$ and $x(0) + y(0) = 1$, this implies that $x(t) + y(t) = 1$ for all $t \ge 0$. Now let $y(t) = 1 - x(t)$, we need only use the percentage of infected computers to completely describe the network dynamics.

[2] In [13], they do not consider the evolution explicitly, and they assume that the attacker has the control of successful attack rate; in such a setting, we simply let $c = 1$ and then v_{H} will be equivalent to what they refer as a probability of successful attack, p.

[3] To control the defense effectiveness from recovery coefficient has the similar implications as control from the other coefficients, such as β, c.

4.1   Main Assumption

We first briefly justify the rationality of main assumptions used in the model.

1. Botnet herder's operational cost function $f_H(x)$ satisfies the conditions $f_H(x)'$ < 0 and $f_H(x)'' > 0$. This assumption implies that botnet herder's operational cost decreases at a decreasing rate as the number of infected computer increases. In fact, the botnet herder's operational cost can be defined as the sum of fixed development cost[4] and the loss from mismatching market demand[5]. We next illustrate a possible cost function. Given a fixed network size N, assuming that the development cost of the malicious program is $\overline{C} > 0$ and per unit loss from mismatching marketing demand, $D > 0$, is $b > 0$, the botnet herder's operational cost function can be specified as: $f_H(x) = \overline{C} + b \times e^{D-Nx}$. The first term captures the fact of free duplicates after development, and the second term captures the loss from mismatching market demand.
2. The defender group is assumed to be homogenous; thus we do not consider the externality effect caused by the heterogenous behavior. We leave this for our future work.
3. Defender group's operational cost function (equivalently, loss value function) $f_D(x)$ satisfies $f_D(x)' > 0$ and $f_D(x)'' > 0$. It is expected that the defending cost increases at an increasing rate as the number of infected machine increases because the complexity of workloads and defense software programming escalates and more professionals are needed to fix the problem.
4. The recovery rate provided by the defense strategy is faster than the contact transmission rate, i.e., $\gamma > \beta$ (or $\gamma_{\max} > \beta$). This is a reasonable assumption since otherwise it may imply the case that all computers in the network are compromised.
5. $x_H^* < x_D^*$: The steady-state infection percentage achieved when the botnet herder exerts an intermediate attack effort and the defender group defends at the maximum level is less than steady-state infection percentage reached by the situation where the defender group exerts an intermediate defense effort and the botnet herder attacks at the maximum level. This assumption is intuitively understandable, since, cetris paribus, full attack coupled with an intermediate level of defense shall cause a higher infection rate than the maximal defense coupled with an intermediate level of attack.

4.2   Game under a Fixed Level of Defense

In this game, we solve the botnet herder's best response when facing a fixed level of defense. [6] The problem of interest here is the botnet herder's optimal

---

[4] Once the malicious program is developed, it is free to duplicate.

[5] By [18], the more bots that the botnet herder owns, the more they can charge for their bots.

[6] It is reasonable to assume that defenders' actions are observable to the botnet herder since defenders are "known" subjects to the botnet herder, but not vice versa. In addition, to focus on an interested subject's problem, it is common to fix a counterparty's strategy, for example [6] solves defenders' five different interdependent security games by assuming a certain level of attack.

strategy. We do not need to consider the defender group's various strategies and action (the control); thus, for ease of presentation, we rewrite the state equation (2) as:

$$\frac{dx}{dt} = cv(1 - x) + \beta x(1 - x) - \gamma x, \ x(0) = x_0, \ 0 \le x_0 \le 1 \ . \tag{3}$$

Let $f(x)$ be botnet herder's cost function with $f'(x) < 0$, and $f''(x) > 0$. Next, let $k > 0$, a constant, be the per unit time cost associated with botnet herder's attack effort; thus his total cost of attack effort per unit time is $v \times k$. We may interpret this effort cost as the extra penalty cost from increasing probability of getting caught due to the increasing severity of attack. This interpretation is supported by an observed phenomenon in the real world botnet operation as suggested by [20] and [13]. The botnet herder's objective, subject to the dynamics of (3), is to seek optimal intensity of attack effort v for minimizing discounted total cost (operation cost plus effort cost) with a constant discount rate r over an infinite time horizon:

$$\inf_{v(\cdot)} \left\{ J_x\big(v(\cdot)\big) = \int_0^\infty e^{-rt}\big(f(x) + kv\big)dt \right\}, \qquad 0 \le v \le 1 \ . \tag{4}$$

To solve the minimization problem, we form the current value Hamiltonian associated with (4) given:

$$H(x, v(t), p) = f(x) + kv + p\big(cv(1 - x) + \beta x(1 - x) - \gamma x\big) \ . \tag{5}$$

The first two terms in (5), $f(x) + kv$, represent botnet herder's instantaneous cost, while the third term represents the future cost of percentage change of infected computers. We can interpret $p(t)$ as the botnet herder's marginal cost at time t. The optimal control, $\hat{v}(t)$, is obtained by minimizing the Hamiltonian H. Because the Hamiltonian is linear in $v(t)$, the optimal control, $\hat{v}(t)$ takes the following bang-bang and (a possible) singular form:

$$\hat{v}(t) = \mathbb{1}_{H_v < 0} + u \mathbb{1}_{H_v = 0} \text{ with } 0 < u < 1, \text{ to be determined,} \tag{6}$$

where $H_v = \frac{\partial H}{\partial v(t)} = k + pc(1 - x)$. When $H_v < 0$, the botnet herder exerts full attack effort ($\hat{v}(t) = 1$), and when $H_v > 0$, the botnet herder exerts zero attack effort ($\hat{v}(t) = 0$). When $H_v = 0$ and stays at this value, an intermediate level of effort $0 < u < 1$ is exerted. This phase is referred to as singular, which has the additional steady-state property that the values of the control and the state variables are constant in this region.

The adjoint equation is:

$$\dot{p} = -\frac{\partial H}{\partial x} + rp = -f'(x) + p\big(cv + \beta(2x - 1) + \gamma + r\big) \ . \tag{7}$$

Substituting (3) and (7) into $\dot{H}_v (= \frac{\partial H_v}{\partial t}) = \dot{p}c(1 - x) - pc\dot{x}$, and setting it equal to zero, we obtain:

$$f'(x) = \frac{k}{c(1 - x)}\Big(\beta(1 - x) - \frac{\gamma}{1 - x} - r\Big) \ . \tag{8}$$

We can solve (8) for the steady state percentage of infected computers, x^*, a constant; the optimal feedback $\hat{v}(x)$ in this singular region is a fixed rate and found by solving $\dot{x} = 0$ at x^*:

$$\hat{v}(x^*) = u = -\frac{\beta x^*(1 - x^*) - \gamma x^*}{c(1 - x^*)} . \tag{9}$$

Theorem 1. *The optimal feedback of the botnet herder is given:*

$$\hat{v}(x) = 1 \text{ if } x < x^*; \ \hat{v}(x) = u \text{ if } x = x^*; \ \hat{v}(x) = 0 \text{ if } x > x^*, \tag{10}$$

where $u = -\dfrac{\beta x^*(1 - x^*) - \gamma x^*}{c(1 - x^*)}$, *and* $x^* < \dfrac{\sqrt{(c+\gamma-\beta)^2+4c\beta}-(c+\gamma-\beta)}{2\beta}$.

Proof. See Appendix A in [2]. ☐

Theorem 1 states that if the starting percentage of infected computer $x_0 > x^*$, then it is optimal for the botnet herder to "reduce" his percentage of invasion in the network to x^* by exerting zero attack effort, $\hat{v}(x) = 0$. The reason is that once the percentage of infection passes the steady-state level, x^*, the opportunity cost of getting caught/traced outweighs the size benefits of the operation cost. If $x_0 < x^*$, then the botnet herder aggressively leverages his infection in the network up to x^* by applying full attack effort, $\hat{v}(x) = 1$. If $x = x^*$, then the constant intensity of attack effort, $\hat{v}(x) = u$, would be implemented and stays at the same level afterwards. It may not be optimal for botnet herders to infect too many computers since it would then attract too much attention. Therefore, the possibility of getting caught increases, and in turn the botnet business may eventually get hurt. The prediction of converging to an optimal steady state level is supported by the recent observation of dormant Confiker botnet [20].

4.3   Nash Game between Botnet Herder and Defender Group

We now consider a simultaneous move game in which each player optimizes his objective while considering his opponent's action.

A. Botnet Herder. The botnet herder's problem is similar to Sect. 4.2 except that now he must consider defender group's strategic interaction in solving his optimization problem. Due to introducing another player, we now denote botnet herder's cost function, control, and the cost of effort per unit time by $f_H(\cdot)$, $v_H(x)$ and k_H respectively.

B. Defender Group. As described in Sect. 3.2, there exists a set of available defense strategies which can range the effectiveness of defense from minimal level to maximal level, and we relate defense strategies with minimal and maximal effectiveness by γ_{min}, and γ_{max} respectively. The defender group optimizes their defense strategy by exerting their defense effort $v_D \in [0, 1]$, which can in turn allow them to attain the effectiveness of defense within the range of minimal effectiveness and maximal effectiveness. That is, it signals the defense strategy chosen by the defender group.

Without doubt, the defender group needs to pay costs to defend against infection. We relate the defender group's operational cost with percentage of infected computers denoted by $f_D(x)$ with $f_D'(x) > 0$, and $f_D''(x) > 0$. In addition, the defender group needs to pay extra costs as additional efforts taken to achieve higher levels of defense. Let $k_D > 0$, a constant, be the per unit time cost associated with the defense effort; then the defender group's total defense effort cost per unit time is $v_D \times k_D$. The defender group's goal is to seek optimal intensity of defense effort v_D for minimizing discounted total cost over an infinite time horizon taking into account the botnet herder's action.

C. State Equation Recall. From Sect. 3.2, the dynamics of percentage of infected computers in a network considering both parties' strategies is given:

$$\frac{dx}{dt} = cv_H(1 - x) + \beta x(1 - x) - \left(\gamma_{\min} + v_D(\gamma_{\max} - \gamma_{\max})\right)x, \ 0 \le x_0 \le 1 \ . \quad (11)$$

D. Differential Game. The botnet herder's optimization problem is given:

$$\begin{cases} \phi_H(x) = \inf_{v_H(\cdot)} \left\{ J_x^H\big(v_H(\cdot), v_D(\cdot)\big) = \int_0^\infty e^{-rt}\big(f_H(x) + k_H v_H\big)dt \right\} \\ 0 \le v_H \le 1 \end{cases},$$

and the defender group's optimization problem is given:

$$\begin{cases} \phi_D(x) = \inf_{v_D(\cdot)} \left\{ J_x^D\big(v_H(\cdot), v_D(\cdot)\big) = \int_0^\infty e^{-rt}\big(f_D(x) + k_D v_D\big)dt \right\} \\ 0 \le v_D \le 1 \end{cases};$$

both are subject to the dynamics of (11). The current value Hamiltonian associated with the botnet herder's and defender group's optimization problems are given in (12) and (13) respectively:

$$H^H(x, v_H(t), v_D, p_1) = f_H(x) + v_H k_H + p_1\big(cv_H(1 - x) \\ + \beta x(1 - x) - \big(\gamma_{\min} + v_D(\gamma_{\max} - \gamma_{\min})\big)x\big) \ . \quad (12)$$

$$H^D(x, v_H(t), v_D, p_2) = f_D(x) + v_D k_D + p_2\big(cv_H(1 - x) \\ + \beta x(1 - x) - \big(\gamma_{\min} + v_D(\gamma_{\max} - \gamma_{\min})\big)x\big) \ . \quad (13)$$

The interpretations of (12) and (13) are similar to (5). Since the Hamiltonian is linear in the corresponding controls, v_H and v_D, the optimal controls, $\hat{v}_H$ and $\hat{v}_D$, take the bang-bang-(possible)singular forms, given in (14) and (15) respectively:

$$\hat{v}_H = \mathbb{1}_{H^H_{v_H} < 0} + u_H \mathbb{1}_{H^H_{v_H} = 0}, \text{ where } 0 < u_H < 1 \text{ is to be determined}, \quad (14)$$

$$\hat{v}_D = \mathbb{1}_{H^D_{v_D} < 0} + u_D \mathbb{1}_{H^D_{v_D} = 0}, \text{ where } 0 < u_D < 1 \text{ is to be determined}, \quad (15)$$

where $H^H_{v_H} = \frac{\partial H^H}{\partial v_H} = k_H + p_1 c(1 - x)$ and $H^D_{v_D} = \frac{\partial H^D}{\partial v_D} = k_D - p_2(\gamma_{\max} - \gamma_{\min})$. The interpretations for (14) and (15) are the same as that for (6).

E. Equilibrium Solution. We look for a Nash equilibrium solution such that

$$J_x^H(\hat{v}_H(\cdot), \hat{v}_D(\cdot)) \leq J_x^H(v_H(\cdot), \hat{v}_D(\cdot)), \; v_H \in [0,1] \; . \tag{16}$$

$$J_x^D(\hat{v}_H(\cdot), \hat{v}_D(\cdot)) \leq J_x^D(\hat{v}_H(\cdot), v_D(\cdot)), \; v_D \in [0,1] \; . \tag{17}$$

Employing (14) and (15), we can write the Hamiltonians ,(12) and (13), as:

$$\hat{H}^H(x, \hat{v}_H, \hat{v}_D, p_1, p_2) = f_H(x) - \left(k_H + p_1 c v_H(1-x)\right)^-$$
$$+ p_1\left(\beta x(1-x) - \gamma_{\min}x\right) - p_1(\gamma_{\max} - \gamma_{\min})x\left(\mathbb{1}_{H_{v_D}^D < 0} + u_D \mathbb{1}_{H_{v_D}^D = 0}\right) \; . \tag{18}$$

$$\hat{H}^D(x, \hat{v}_H, \hat{v}_D, p_1, p_2) = f_D(x) - \left(k_D - p_2(\gamma_{\max} - \gamma_{\min})x\right)^-$$
$$+ p_2(\beta x(1-x) - \gamma_{\min}x) + p_2 c(1-x)\left(\mathbb{1}_{H_{v_H}^H < 0} + u_H \mathbb{1}_{H_{v_H}^H = 0}\right) \; . \tag{19}$$

From Dynamic Programming, the Nash equilibrium solution must satisfy the following system of Bellman equations:

$$\begin{cases} r\phi_H(x) = \hat{H}^H(x, \hat{v}_H, \hat{v}_D, \phi_H'(x), \phi_D'(x)) \\ r\phi_D(x) = \hat{H}^D(x, \hat{v}_H, \hat{v}_D, \phi_H'(x), \phi_D'(x)) \end{cases} \; . \tag{20}$$

For facilitating presentation, we define the following notations:

(1) Define θ_H as the switching threshold for the botnet herder such that $\hat{v}(\theta_H) = 1$ if $x < \theta_H$, and $\hat{v}(\theta_H) = 0$ if $x > \theta_H$.
(2) Define θ_D as the switching threshold for the defender group such that $\hat{v}(\theta_D) = 0$ if $x < \theta_D$, and $\hat{v}(\theta_D) = 1$ if $x > \theta_D$.
(3) Set $\gamma = \gamma_{\max} - \gamma_{\min}$.

We are now ready to state solutions related to (20).

Theorem 2. *Assume* $x_H^* < x_D^*$ *where* x_H^* *and* x_D^* *are solutions to* $F_H(x) = f_H'(x)c(1-x) + k_H\left(r - \beta(1-x) + \frac{\gamma_{\max}}{1-x}\right) = 0$ *and* $F_D(x) = f_D'(x)\gamma x - k_D\left(r + \beta x + \frac{c}{x}\right) = 0$ *respectively. There exist two Nash equilibrium solutions:*

1. *For* $\theta_H = x_H^*$ *and* $\theta_D \leq x_H^*$.
   Assume (i) $\beta < \gamma_{\max}$ *and (ii)* $c(1 - x_H^*) + \beta x_H^*(1 - x_H^*) - \gamma_{\max}x_H^* > 0$.
   There exists a equilibrium solution at x_H^* *such that the botnet herder applies optimal feedback policy with* $\hat{v}_H(x_H^*) = u_H = -\frac{1}{c(1-x_H^*)}\left(\beta x_H^*(1 - x_H^*) - \gamma_{\max}x_H^*\right)$ *and the defender group applies the optimal policy* $\hat{v}_D(x_H^*) = 1$. *The optimal feedback for the botnet herder and the defender group can be summarized as:*
   Botnet Herder:

$$\hat{v}_H(x) = 1 \; if \; x < x_H^*; \; \hat{v}_H(x) = u_H \; if \; x = x_H^*; \; \hat{v}_H(x) = 0 \; if \; x > x_H^*. \tag{21}$$

   Defender Group:

$$\hat{v}_D(x) = 1 \; if \; x > \theta_D; \; \hat{v}_D(x) = 0 \; if \; x < \theta_D; \; \hat{v}_D(x_H^*) = 1. \tag{22}$$

2. *For $\theta_D = x_D^*$ and $x_D^* < \theta_H$.*

Assume (i) $c(1-x_D^*)+\beta x_D^*(1-x_D^*)-\gamma_{\max}x_D^* < 0$. and (ii) $c(1-x_D^*)+\beta x_D^*(1-x_D^*)-\gamma_{\min}x_D^* > 0$. There exists a equilibrium solution at x_D^* such that the botnet herder applies optimal feedback policy with $\hat{v}_H(x_D^*) = 1$ and the defender group applies the optimal policy $\hat{v}_D(x_D^*) = u_D = \frac{c(1-x_D^*)+\beta x_D^*(1-x_D^*)-\gamma_{\max}x_D^*}{\gamma x_D^*}$. The optimal feedback for the botnet herder and the defender group can be summarized as:
Botnet Herder:

$$\hat{v}_H(x) = 1 \; if \; x < \theta_H; \; \hat{v}_H(x) = 0 \; if \; x > \theta_H; \; \hat{v}_H(x_D^*) = 1. \tag{23}$$

Defender Group:

$$\hat{v}_D(x) = 1 \; if \; x > x_D^*; \; \hat{v}_D(x) = u_D \; if \; x = x_D^*; \; \hat{v}_D(x) = 0 \; if \; x < x_D^*. \tag{24}$$

In addition, if (1) $c(1 - x_H^*) + \beta x_H^*(1 - x_H^*) - \gamma_{\max}x_H^* > 0$, (2) $c(1 - x_D^*) + \beta x_D^*(1-x_D^*)-\gamma_{\max}x_D^* < 0$, (3) $c(1-x_D^*)+\beta x_D^*(1-x_2^*)-\gamma_{\min}x_D^* > 0$, and $\tilde{x}_H > x_D^*$, $\tilde{x}_D < x_H^*$, we can either take $\theta_H = \theta_D = x_H^*$ or $\theta_H = \theta_D = x_D^*$.

Proof. See Appendix B in [2]. □

The above theorem states that there are two possible Nash equilibria. One equilibrium occurs at the infection rate equal to x_H^*, a steady-state solution to the botnet herder's minimization problem given the full defense effort exerted from the defender group. This equilibrium occurs where the most effective defense strategy available cannot efficiently reduce the spread of malicious program; thus an intermediate defense effort is not an admissible strategy (control) to the defender group. This is an equilibrium such that the infection percentage remains at $x_H^* > \theta_D$. The botnet herder will exert full attack effort if $x < x_H^*$, exert zero attack effort if $x > x_H^*$, and apply the intermediate constant control $\hat{v}_H(x^*) = u_H$ when $x = x_H^*$ and the state and the attack effort will remain at this level. The defender group applies zero defense effort ($\hat{v}_D = 0$) if $x < \theta_D$ and maximum defense effort ($\hat{v}_D = 1$) if $x > \theta_D$. At $x = \theta_D$, the defender group is indifferent in taking either control since the state cannot remain on θ_D. When the state reaches x_H^*, the equilibrium occurs and will stay at this level in which the botnet herder applies the intermediate constant control u_H and the defender group maximizes his defense effort.

The other equilibrium occurs at the infection rate $x_D^* < \theta_H$, a solution to the defender group's steady-state minimization problem given the full attack effort from the botnet herder. This equilibrium occurs where the least effective defense strategy cannot efficiently deter the propagation of malicious program, but the one with maximal effectiveness can reduce the infection successfully. In this environment, an intermediate level of attack effort is not admissible to the botnet herder. This is an equilibrium such that the infection percentage remains at $x_D^* < \theta_H$ with an intermediate constant defense effort from the defender group and maximum attack from the botnet herder. The interpretation of the optimal feedback policy for both parties is similar to the above paragraph.

We find that the equilibrium solution in which the botnet herder exerts a constant attack effort, $u_H \in (0,1)$ and the defender group maximizes his defense level may be supported by the observation of Confiker botnet. It seems that the Confiker botnet herder chooses to stay dormant to reduce its size since the infection size may pass the optimal one, causing severe penalties from high probability of getting caught. In addition, there may not be a powerful defense strategy which could effectively deter this botnet. In the equilibrium, the best that the defender group could do would be to apply maximal defense level available and the Confiker botnet owner may exert an intermediate constant attack effort to stay at the optimal infection size for minimizing costs. We thus expect this botnet may not stay dormant forever. It is to mention that the defender group may reduce the equilibrium infection size by posing severe penalties. The other equilibrium solution where the defender group exerts a constant defense level, $u_D \in (0,1)$ and the botnet herder maximizes his attack may predict a situation such that this equilibrium infection size may yield the most benefits to both parties.

5   Conclusion

We employ a game theoretical framework to analyze the economic aspects of botnet activities and suggest feasible defensive strategies. The dynamics of infected computers evolve according to a modified SIS epidemic model considering both parties' actions. For a given level of network defense, we obtain the botnet herder's optimal strategy as a feedback on the rate of infection. For simultaneous moves, we solve the differential game and obtain two possible closed-loop Nash equilibrium solutions. These predictions provide insights to the network society as to how the equilibrium botnet activity might be given available defense strategies. In addition, the optimal feedback policies may offer guidelines for the network society to respond to the botnet attack strategically optimal. Realizing that the evolution of infected computers may in fact not be fully controlled by both parties' actions alone, we are working on extending this SIS dynamics to incorporating stochastic disturbances [1].

In addition, to implement the optimal control policies suggested by the model, effective parameter values need to be obtained. We suggest to estimate the values from empirical data. If the data is not empirically available, we suggest to implement an experimental project to simulate the possible attack-defense dynamics for obtaining justifiable parameter values.

Acknowledgments. This work is partially funded by Office of Naval Research Grant, N000140910776.

References

1. Bensoussan, A., Kantarcioglu, M., Hoe, C.: Botnet Defense Under Uncertainty: A Stochastic Differential Game Approach, Working Paper, UT Dallas (2010)
2. Bensoussan, A., Kantarcioglu, M., Hoe, C.: A Game-Theoretical Approach for Finding Optimal Strategies in a Botnet Defense Model, Technical Report, UTDCS-14-10, http://www.utdallas.edu/~mxk055100/publications/botnet-defense-game.pdf
3. Dagon, D., Zou, C., Lee, W.: Modeling Botnet Propagation Using Time Zones. In: Proc. of the 13th Network and Distributed System Security Symposium NDSS
4. Cohen, F.: Computer Viruses Theory and Practice. Computer and Security 6, 22–35 (1987)
5. Theodorakopoulos, G., Baras, J.S., Le Boudec, J.-Y.: Dynamic Network Security Deployment under Partial Information. In: Proc. of the 46th Annual Allerton Conference on Communication, Control, and Computing, pp. 261–267 (2008)
6. Grossklags, J., Christin, N., Chuang, J.: Security investment (failures) in five economic environments: A comparison of homogeneous and heterogeneous user agents. In: Proc. of the 7th Workshop on the Economics of Information Security (WEIS 2008) (2008)
7. Liu, J., Tang, Y., Yang, Z.R.: The Spread of Disease with Birth and Death on Networks. Journal of Statistical Mechanics: Theory and Experiment (2004)
8. Kephart, J.O., White, S.R.: Directed-Graph Epidemiological Models of Computer Viruses. In: Proc. of IEEE Symposium on Security and Provacy, pp. 343–361 (1991)
9. Lye, K.W., Wang, J.: Game Strategies in Network Securities. International Journal of Information Security 1(1-2), 71–86 (2005)
10. Bloem, M., Aplcan, T., Basar, T.: Optimal and Robust Epidemic Response for Multiple Networks. IFAC Control Engineering Practice 17(5), 525–533 (2009)
11. Lelarge, M.: Economics of Malware: Epidemic Risks Model, Network Externalities and Incentives. In: The 8th Workshop on the Economics of Information Security
12. Lelarge, M., Bolot, J.: A Local Mean Field Analysis of Security Investments in Networks. In: Proc. of the 3rd International Workshop on Economics of Networked Systems, pp. 25–30 (2008)
13. Fultz, N., Grossklags, J.: Blue versus Red: Towards a model of distributed security attacks. In: Dingledine, R., Golle, P. (eds.) FC 2009. LNCS, vol. 5628, pp. 167–183. Springer, Heidelberg (2009)
14. Toutonji, O., Yoo, S.-M.: An Approach against a Computer Worm Attack. International Journal of Communication Networks and Information Security 1(2), 47–53 (2009)
15. Baucher, P., Holz, T., Kotter, M., Wicherski, G.: Konw your Enemy: Tracking Botnets, http://www.honeynet.org/papers/bots
16. Alpcan, T., Basar, T.: A Game Theoretic Appropach to Decision and Analysis in Network Intrusion Detection. In: Proceeding of the 42nd IEEE Conference on Decision and Control, pp. 2595–2600
17. Alpcan, T., Basar, T.: An Inrtusion Detection Game with Limited Observations. In: The 12th Int. Symp. on Dynamic Games and Applications (2006)
18. Namestnikov, Y.: The Economics of Botnets, http://www.viruslist.com/en/downloads/pdf/ynam_botnets_0907_en.pdf
19. Li, Z., Liao, Q., Striegel, A.: Botnet Economics: Uncertainty Matters. In: The 7th Workshop on the Economics of Information Security (WEIS 2008) (2008)

20. Conficker Botnet 'Dead In the Water', Researcher Says,
    http://darkreading.com/vulnerability_management/security/attacks/
    showArticle.jhtml?articleID=224201115
21. Symantec Global Internet Security Threat Report,
    http://eval.symantec.com/mktginfo/enterprise/white_papers/
    b-whitepaper_internet_security_threat_report_x_04-2010.en-us.pdf
22. Kaspersky Security Bulletin: Malware evolution (2008),
    http://www.securelist.com/en/analysis?pubid=204792051

ISPs and Ad Networks Against Botnet Ad Fraud

Nevena Vratonjic, Mohammad Hossein Manshaei,
Maxim Raya, and Jean-Pierre Hubaux

Laboratory for computer Communications and Applications (LCA), EPFL, Switzerland
{nevena.vratonjic,hossein.manshaei,jean-pierre.hubaux}@epfl.ch,
maxim.raya@gmail.com

Abstract. Botnets are a serious threat on the Internet and require huge resources to be thwarted. ISPs are in the best position to fight botnets and there are a number of recently proposed initiatives that focus on how ISPs should detect and remediate bots. However, it is very expensive for ISPs to do it alone and they would probably welcome some external funding. Among others, botnets severely affect ad networks (ANs), as botnets are increasingly used for ad fraud. Thus, ANs have an economic incentive, but they are not in the best position to fight botnet ad fraud. Consequently, ANs might be willing to subsidize the ISPs to do so. We provide a game-theoretic model to study the strategic behavior of ISPs and ANs and we identify the conditions under which ANs are likely to solve the problem of botnet ad fraud by themselves and those under which the AN will subsidize the ISP to achieve this goal. Our analytical and numerical results show that the optimal strategy depends on the ad revenue loss of the ANs due to ad fraud and the number of bots participating in ad fraud.

Keywords: Ad Fraud, Botnets, ISP, Ad Network, Security, Game Theory.

1 Introduction

Today, botnets are a very popular tool for perpetrating distributed attacks on the Internet. Botnets are a serious threat for a number of entities: end users, enterprises with online businesses, websites, Internet Service Providers (ISPs), advertisers and ad networks (ANs). Botnets usually consist of compromised end users' PCs. Thus, depending on the malware, the consequences for end users can be severe (e.g., stolen credentials). Very often botnets are used for sending spam, which creates problems for ISPs, enterprises and end users. Botnet operators (aka bot masters) also use botnets to extort money from websites' owners under the threat of Distributed Denial of Service Attacks (DDoS). Lately, it is becoming more and more popular to use botnets for ad fraud [4], which creates a loss of ad revenue for advertisers, associated websites and ad networks and security threats for end users (e.g., fraudulent ads that lead to phishing attacks).

Consequently, thwarting botnets would benefit everyone and would reduce the level of online crime on the Internet. However, the problem of botnets in general cannot be solved exclusively by users (lack of know-how), ISPs (too expensive to fight botnets alone), ad networks, advertisers, websites and enterprises (lack of tools and resources).

Recent initiatives propose that ISPs should perform the detection of botnets and remediation of the infected devices [20] [24]. Indeed, it is the ISPs that are in the best

T. Alpcan, L. Buttyan, and J. Baras (Eds.): GameSec 2010, LNCS 6442, pp. 149–167, 2010.

position to detect the presence of a botnet and to take measures against it. Yet, the revenue of ISPs are not (directly) affected by the botnets and ISPs would probably welcome some external funding in the efforts to fight botnets. One possible approach is a government-sponsored program, as in Australia [7] and Germany [10]. In the case governments are unwilling to fund these initiatives, ISPs need to find a way to make them, at the very least, cost neutral if not cost positive.

Over the last decade, online advertising has become a major component of the Web, leading to annual revenues expressed in tens of billions of US Dollars (e.g., $22.4 billion in the US in 2009 [5]). The business model of a fast growing number of online services is based on online advertising and much of the Internet activity depends on that source of revenue. Unsurprisingly, the ad revenue has caught the eye of many ill-intentioned people who have started abusing the advertising system in various ways. In particular, click fraud has become a phenomenon of alarming proportions [4]. Recently, a new type of ad fraud attack has appeared, consisting in the on-the-fly modification of the ads themselves. A prominent example is the *Bahama botnet*, in which malware causes infected systems to display altered ads, as well as altered results for Google or Yahoo searches to the end users [17]. Another example of such a botnet is Gumblar [16]. If the modification of ads is successful, users see ads that are different from what they would otherwise be. Consequently, users' clicks on the altered ads generate a revenue for the bot master instead of the AN. Thus, the modification of the ads negatively affects the revenues of the "legitimate" advertisers and undermines the business model of the ANs.

Considering the increasing trend of botnet ad-fraud attacks and the consequently increasing loss of ad revenue for ad networks, ANs have economic incentives to fight botnets. However, ANs are not in the best position to thwart botnets themselves and thus ANs might be willing to subsidize the ISPs to achieve that goal. In this paper, we investigate whether ad fraud botnets alone are cause enough for ISPs and ANs to cooperate. Such cooperation would help ISPs deploy detection and remediation mechanisms and would be a first step towards fighting all botnets.

The contributions of our paper are threefold. First, we identify two potential countermeasures that ANs could use to address the problem of botnet ad fraud and we propose a cooperation scheme in which ISPs and ANs jointly fight botnets. Second, we provide a game-theoretic model to study the interactions between ISPs and ANs, as well as to identify an optimal countermeasure strategy of ANs and ISPs under different conditions. Finally, we apply the results to a real data set to study the practical impact. To the best of our knowledge, this paper is the first to model the behavior of ISPs and ad networks facing botnet ad fraud.

The rest of the paper is organized as follows. After a brief presentation of the state of the art in Section 2, we describe the impact of botnets on the online advertising business in Section 3. We then address the various threats and countermeasures in Section 4 and provide a case study of a botnet ad fraud in Section 5. In Section 6, we present a game-theoretic model with two players, the ISP and the AN, and identify equilibrium outcomes of that game. We provide a numerical example to study the practical impact of the obtained results in Section 7 and conclude the paper in Section 8.

2   Related Work

There are two main categories of literature that are relevant to our work: research on fraud in online advertising and analyses of security investments on the Internet.

Research on online advertising fraud is mostly focused on click fraud [12] [15] [21]. Many problems that stem from online advertising and security gaps, especially the consequences for the end users, are addressed in [13]. The economics of click fraud are briefly addressed in [21]. In [8], the economic analysis based on a game-theoretic model of the online advertising market, shows that ad networks that deploy effective algorithms for click fraud detection gain a significant competitive advantage. If it is the case that some ad networks do not fight click fraud, mechanisms are proposed in [14] to protect online advertisers from paying for fraudulent clicks. In comparison, our model does not address click fraud detection mechanisms but introduces a new strategic player - the ISP - in addition to the traditional players in online advertising (i.e., ad networks, advertisers and publishers). Our results show that this player can yield significant implications for the security of the Internet.

In [30], the authors investigate novel man-in-the-middle attacks on online advertising systems, which can be perpetrated by access networks (e.g., an ISP) to exploit online advertising systems. The authors propose a collaborative secure scheme that relies on web servers and ad networks to fix the identified vulnerabilities of online advertising systems. This solution also relies on the fact that most of online advertising networks own digital authentication certificates and can become a source of trust. The authors explain why the deployment of this solution would benefit the Web browsing security in general. In this paper we propose new approaches to thwart distributed ad fraud attacks, where we address the possibility of collaboration between ISPs and ad networks.

Related to the second issue - finding the right incentives to increase the security on the Internet - there are several contributions in the literature. The game-theoretic approach of [18] models how users choose between investments in security (e.g., firewalls) or insurance (e.g., backup) mechanisms. The positive effect of cyberinsurance on the investment of agents in self-protection is analyzed using a game-theoretic model in [23]. The main conclusion of this work is that cyberinsurance without regulation does not provide sufficient incentives for self-protection. Another line of work proposes a centralized certification mechanism to encourage ISPs to secure their traffic and analyzes the resulting scheme using game theory [33]. In contrast to these works, our analysis shows that Internet security can be increased, under given conditions, without any central oversight and thanks to self-interested decisions by only a few key players (namely, the ad networks and the ISPs).

In [31], the authors investigate the recent problem of ISPs becoming strategic participants in the online advertising business. They propose a game-theoretic model of this problem to study the behavior and interactions of the ISPs and ANs. Their results show that if the users private information can improve ad targeting significantly and if ad networks do not have to pay a high share of revenue to the ISPs, ad networks and ISPs will cooperate to jointly provide targeted online ads. Otherwise, ISPs will divert part of the online ad revenue for themselves. In that case, if the diverted revenue is small, ad networks will not react. However, if their revenue loss is significant, the ad networks will invest into improving the security of the Web and protecting their ad revenue. Our work also concludes that, when facing ad fraud, ANs are willing to collaborate with ISPs in order to protect their ad revenue. However, since we consider a distributed threat (in contrast to the centralized model in [31]) we propose new collaborative approach that takes into consideration the economic incentives of the ANs and ISPs.

3   System Model

We consider a system consisting of an *online advertising system*, a number of *bots* that attempt to exploit the online advertising system and an *ISP*, as depicted in Figure 1.

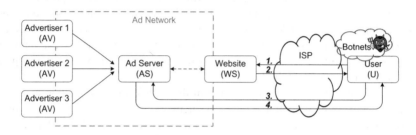

Fig. 1. System model: Online advertising system, ISP and bots exploiting the advertising system

3.1   Online Advertising Systems

The most prevalent model of serving online ads to end users is depicted in Figure 1. To have their ads appear with the appropriate web content, Advertisers (AV) subscribe with an ad network (AN) whose role is to automatically embed ads into web pages. Ad networks have contracts with Websites (WS) that want to host advertisements. When a User (U) visits a website (Figure 1, step 1) that hosts ads, while downloading the content of the web page (step 2), the user's browser will be directed to communicate with one of the Ad Servers (AS) belonging to the ad network (step 3). The AS chooses and serves (step 4) the most appropriate ads to the user, such that users' interests are matched and the potential revenue is maximized. Throughout the rest of the paper, we use the terms "user" and "user's browser" interchangeably.

In the most popular ad revenue model [5], advertisers pay a *cost-per-click* (CPC) to the ad network for each user-generated click that directs the user's browser to the advertised website. The AN gives a fraction of the ad generated revenue to the WS that hosted the ad. Popular websites that attract more visitors create more traffic towards advertised websites, thus generating more revenue for themselves and for the associated ad networks. Since we consider a single AN in our system model, we assume that all the websites that host online ads are associated to that AN.

3.2   Botnets

A botnet is a collection of software robots, or *bots*, that run autonomously and automatically. Bots are typically compromised computers running software, usually installed via drive-by downloads exploiting Web browser vulnerabilities, worms, Trojan horses or backdoors, under a common command-and-control infrastructure. Recently, a botnet of compromised wireless routers has been detected [25]. Such a botnet has the advantage of having the bots almost always connected to the Internet (compared to the typical end-user machine that is connected to the Internet only from time to time). In addition, it is more difficult to detect that a device has been compromised, due to the lack of security software for such devices (e.g., no anti-virus software) or by a user.

A bot master controls the botnet remotely, usually through a covert channel (e.g., Internet Relay Chat) and usually for nefarious purposes. According to Click Forensics, a company that produces tools to detect and filter fraudulent clicks, for the third quarter of 2009, 42.6% of fraudulent clicks came from bots (Figure 2)[1] [4]. The number is the highest in four years (since Click Forensics has been producing the reports). For the same period in 2008, botnets accounted for 27.5% of fraudulent clicks. The data show that using botnets for ad fraud is becoming more and more popular. This creates a problem for advertisers, ad networks and websites as they lose a part of the ad revenue. In the system model, we consider a number of compromised devices that run a malware that causes infected machines to participate in an advertising fraud.

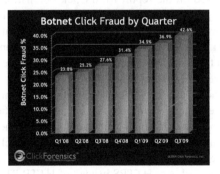

Fig. 2. Significance of botnet ad fraud: Botnet Click Fraud by Quarter

3.3   Internet Service Providers

The traditional role of an ISP is to provide Internet access to end users and to forward users' traffic in compliance with the Network Neutrality policy [11]. However, recently, ISPs have begun taking on additional roles. In the EU, ISPs have to obtain and keep the records about their users' online activities and provide them upon request to law enforcement agencies [1].

A new IETF intitiative focuses on how ISPs can manage the effects of devices used by their subscribers, detect those that have been infected with malicious bots, notify the subscribers and remediate the infection via various techniques [20]. The Internet Industry Association (IIA) has also drafted a new code of conduct that suggests ISPs should detect malware-infected machines of their subscribers and actually take the action to

---

[1] Permission for inclusion obtained from ClickForensics Inc.

address the problem [24]. Complying with these initiatives, ISPs would make it more difficult for botnets to operate, thus helping to reduce the level of online crime on the Web. However, the problem is that ISPs have to find funding for those initiatives.

One possible approach is a government-sponsored program, such as the Australian Internet Security Initiative, in which a third-party helps identify malware-infected devices, notifies the appropriate ISP which then notifies and helps the subscriber to remedy the problem. About 90% of Australian ISP subscribers are covered by this initiative. A similar program is ready to be launched in 2010 in Germany, where ISPs are cooperating with the German Federal Office for Information Security [10]. In the case governments are unwilling to fund the initiative, ISPs need to find a way to make it, at the very least, cost neutral if not cost positive. In our model, we consider an ISP that is willing to comply to the initiative, if doing so is at least cost neutral.

4   Ad Fraud: Threats and Countermeasures

Due to the immense revenues generated by online advertising, the temptation to exploit the online advertising system is high. The loss of revenue for ad networks due to ad fraud is substantial. Based on the report from Click Forensics, the overall click-fraud rate was around 14.1% in the third quarter of 2009 [4], which means that 14.1% of the clicks on the ads were bogus. Thus, click fraud alone creates a significant loss of revenue for ad networks, advertisers and publishers. In addition, ANs lose ad revenue due to new types of ad fraud, such as injecting ads into the content of webpages on-the-fly between a web server and a user [19, 26, 29].

One possible approach for ad networks to protect their revenue is to improve the security of the online advertising systems, thus making it more difficult for an adversary to successfully exploit those systems. In [31], the authors use game theory to model AN's economic incentives and show that when facing ad fraud attacks securing ad systems may maximize the revenue of a rational AN. For example, ad fraud can be reduced if webpages and ads are served over HTTPS instead of HTTP. The cost of implementing HTTPS at a web server includes the cost of obtaining a valid $X.509$ authentication certificate. Usually, website owners are not willing to bear this cost. Thus, if an ad network wants the secure protocol to be deployed, it should cover the costs itself. As explained previously, websites are not of the same value to the ad network, because of the different ad revenue they generate, but the cost of securing the ad revenue from a website is the same for all websites. Therefore, the ad network may decide to selectively secure only the websites that generate sufficient ad revenue that would compensate the costs.

Another possible approach for ad networks to protect their revenue is to cooperate with ISPs and eliminate the major cause of the revenue loss, botnets. They can do so by funding the existing initiatives for IPSs to detect and remove botnets, since ISPs are in a privileged position to fight botnets. As removing botnets would benefit ad networks, they have economic incentives to subsidize ISPs to fight botnets.

Thus, we can envision the following two scenarios of ad networks fighting ad fraud: (i) improving the security of the online advertising systems or (ii) funding ISPs to fight botnets involved in ad frauds.

5   Botnet Ad Fraud: A Case Study

Consider the system as described in Section 3, in which N_B devices (e.g., end-users's computers or routers) have been infected by a malware and participate in ad fraud. We consider exclusively the types of ad fraud: (i) that has been the most prominent lately [16,17], in which malware causes infected devices to return altered Search Engine Result Pages (SERPs) or altered content of the ads in web pages, due to DNS poisoning and (ii) in which subverted users' routers modify ad traffic on-the-fly between a web server and a user [29]. In the example of Bahama botnet, malware uses DNS poisoning by modifying HOSTS files on infected machines to redirect traffic to rogue Google servers which return altered results [9]. Thus, affected users see ads and links that are different from what they would otherwise be. When users click on the altered ads, the clicks generate revenue for the bot master instead of the ad network. Thus, the bots divert a part of the ad revenue from the ad network. For simplicity of treatment, we assume that each bot diverts an equal part of the revenue and in aggregate, all the bots together divert $\lambda \in (0, 1]$ fraction of the total ad network's revenue P. Thus, the revenue of the AN in the case of ad fraud is $P(1 - \lambda)$.

The popularity of websites, and consequently the number of user-generated clicks on ads, follow a heavy-tail distribution [6]. We infer the generated volume of clicks on ads on the 1000 most popular websites, based on the data of page views on each website in 2009, obtained from *Compete.com*. The exposure of users to online ads has been evaluated extensively in [22], showing that 58% of the top 1000 websites host advertisements and there are 8 ads per web page on average. The probability that a click occurs on an advertisement is 0.1% [2]. Consequently, to convert the number of page views into the number of clicks on ads on each website, we use the following formula: $Q(n)$=(Page views on the website n)×0.58 × 8 × 0.001. Figure 3 shows the annual number of clicks $Q(n)$ on ads , where $n \in \{1, 2, \cdots 1000\}$ is the popularity rank of a website.

Applying curve fitting to the data set, we obtain that the distribution of clicks on ads across websites corresponds to the power law $Q(n) = \alpha \cdot n^{-\beta}$, where $Q(n)$ is the annual number of clicks on ads that occurred at the website with the n-th rank. The obtained

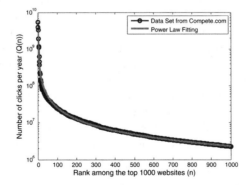

Fig. 3. Annual number of clicks for the 1000 most popular websites and the power law fitting curve, $Q(n) = \alpha n^{-\beta} = 3.18 \cdot 10^9 n^{-1.044}$

parameters of the power law are $\alpha = 3.18 \cdot 10^9$ and $\beta = 1.044$ as shown in Figure 3. In general, we assume that the number of clicks on ads follows the power law distribution $Q(n) = \alpha \cdot n^{-\beta}$, where $Q(n)$ is the annual number of clicks on ads that occurred at the website with the n-th rank and $\beta > 1$ [6]. Note that the value of parameters α and β are a characteristic of a given AN and depend on the number and the type of associated websites. In order to extend our analysis and investigate what would be the effect on the entire Web (i.e., for all websites), we extrapolate the data set we have obtained from *Compete.com* with the obtained power law.

Given the power law distribution of the clicks, the ad revenue generated by the top x websites can be estimated by[2]

$$ k \int_1^x \alpha n^{-\beta} dn = k \frac{\alpha}{(\beta - 1)} (1 - x^{1-\beta}), $$

where k is the amount of revenue that each click on ads generates for the ad network[3]. If P is the total revenue of the ad network, generated by all the websites (i.e., when $x \to \infty$), then per-click revenue can be calculated by $k = \frac{P(\beta-1)}{\alpha}$. According to the reports [5], the total ad revenue P in 2009 in the US is 22.4 billion dollars.

In the following two subsections, we analyze the two proposed strategies (i.e., improving the security of the online advertising system and cooperation between the AN and the ISP) to fight botnet ad fraud. Table 1 shows the used notation.

5.1   Securing Websites

As a countermeasure to the considered type of ad fraud (i.e., rogue servers delivering altered ads due to DNS poisoning attack on users machines or on-the-fly traffic modifications by compromised users' routers), the AN can secure the communication between users and web servers as well as between users and ad servers. For example, secure communication can be provided by the HTTPS protocol. Deploying HTTPS requires web servers to obtain an authentication certificate from a trusted third party. In the case when websites and ad servers deploy HTTPS with valid authentication certificates, even if an adversary successfully mounts a DNS poisoning attack and redirects users' communication to rogue servers, the rogues servers cannot serve valid authentication certificates that correspond to the domain names users originally wanted to visit, thus browsers will detect security issues. HTTPS also prevents on-the-fly modifications of the content. Consequently, users would receive unaltered links and ads and the clicks on unaltered ads would generate revenue for the intended AN, not the adversary.

As discussed in Section 4, website owners usually lack incentive to bear the cost of obtaining a valid certificate. Thus, to secure the communication, and consequently the ad revenue, the AN would have to pay a cost of securing the website. The cost of deploying HTTPS at ad servers can be considered negligible, given that the AN already

---

[2] Due to the impossibility of obtaining closed-form expressions in the discrete domain, we perform computations in the continuous domain. The upper bound of the error is 8% [32].

[3] Modeling auctions and different per click revenue for ad networks is out of the scope of this paper, thus we assume that all the clicks are of the same quality.

Table 1. Table of symbols

Symbol	Definition
N_B	Number of bots
λ	Fraction of diverted ad revenue by the botnet
P	Total online advertising revenue of the AN
k	Amount of generated revenue for each click
$Q(n)$	Number of clicks per year for the top 1000 websites
n	Popularity rank of the websites
α and β	Estimated parameters of power law distribution for $Q(n)$
c_S	Cost of securing a website
N_S	Optimal number of secured websites with S strategy
N_{SC}	Optimal number of secured websites with $S + C$ strategy
P_D	Fraction of bots detected by the ISP
c_D	Cost of the botnet detection system
c_R	Cost for the ISP per remediated infected device
R	Cost for the AN per remediated infected device
N_R	Optimal number of remediated infected devices
C	Cooperation strategy (employed by the ISP or the AN)
S	Secure websites strategy by the AN
$S + C$	Simultaneous Secure and Cooperation strategy by the AN
A	Abstain Strategy (employed by the ISP or the AN)

has a valid certificate and that there are typically only a few ad servers (compared to the number of web servers).

Let c_S be the cost of securing a website, i.e., the cost of obtaining a certificate and deploying HTTPS at a web server. Then the AN should pay $N_S \cdot c_S$ to secure N_S websites. N_S is the optimal number of websites that AN secures to maximize its payoff in the presence of N_B bots diverting fraction λ of the revenue. It can be calculated by the following lemma.

Lemma 1. *If the ad network fights botnet ad fraud by securing the websites, the optimal number of those secured websites is equal to* $N_S = \left(\frac{P}{c_S} \lambda (\beta - 1) \right)^{\frac{1}{\beta}}$.

Proof. The total amount of revenue for the ad network (u_{AN}) when it secures x websites, due to the attack of N_B bots diverting fraction λ of the revenue, can be estimated by

$$u_{AN} = k \int_1^x \alpha n^{-\beta} dn + (1 - \lambda) k \int_x^\infty \alpha n^{-\beta} dn - c_S x.$$

Recall that k is the revenue generated per each click and can be calculated as $\frac{P(\beta-1)}{\alpha}$. The first term in the revenue equation represents the revenue that the AN obtains from clicks generated on secured websites. The second term shows that the AN obtains only the remaining fraction $(1 - \lambda)$ of the revenue from clicks generated on unsecured websites, as the bots divert the fraction λ of the revenue.

After simplifications we obtain: $u_{AN} = P(1 - \lambda x^{1-\beta}) - c_S x$, which is a concave function of x. We can obtain the optimal N_S by finding the root of the first derivation of u_{AN} with respect to x, that is $\left(\frac{P}{c_S}\lambda(\beta - 1)\right)^{\frac{1}{\beta}}$. $\square$

5.2   ISP and Ad Network Cooperation

In addition to the just described countermeasure of securing websites, the AN can offer the ISP to cooperate in the fight against botnets. The AN has an economic incentive to fund the ISP to perform detection of the botnets and remediation of the infected devices, as discussed in Section 3. To detect bots in the network, the ISP must deploy a detection system [20,24]. We note the deployment cost of the detection system as c_D and we assume that such a system can successfully detect a fraction P_D of the bots in the network. The proposed initiatives [20,24] envision an online help desk where all the subscribers whose devices have been detected as bots can obtain instructions on how to remediate the problem and restore the functionality of their devices. Thus, the ISP has a cost per each remediated infected device, which we note as c_R.

For the ISP to cooperate with the AN, the AN has to provide a sufficient reward such that the detection and remediation is at least cost neutral for the ISP. Let R represent the reward the AN should pay to the ISP for the remediation of each infected device.[4]

If the AN and the ISP agree to cooperate, the outcome is that the ISP remediates N_R infected devices and the AN pays $N_R \cdot R$ to the ISP. The optimal N_R that maximizes both revenues, of the ISP and the AN, can be calculated by the following lemma.

Lemma 2. *The cooperative ISP and the cooperative AN can maximize their revenues by remediation of $N_R = P_D N_B$ infected devices.*

Proof. The total amount of revenue that the ISP can obtain by cooperation and remediation of x infected devices is $x(R - c_R) - c_D$ which is a linear function of x. Therefore, the ISP can maximize its revenue by remediating all of detected bots $P_D N_B$. Remediation of x infected devices reduces the aggregate power of the bots in the network, and together they can divert only a fraction $\lambda(1 - \frac{x}{N_B})$ of the revenue. The total amount of revenue that the AN can obtain by cooperation is then $P(1 - \lambda(1 - \frac{x}{N_B})) - xR = \left(\frac{P\lambda}{N_B} - R\right)x + P(1 - \lambda)$, which is a linear function with respect to x and will be maximized at $x = N_R = P_D N_B$, i.e., for all of the detected bots. $\square$

In summary, the ad network can use one of the above two actions to fight botnet ad fraud in the Internet. Each strategy has different benefits and costs for the ISP and the AN. In the next section, we use game theory to model this situation and consequently predict the behavior of the AN and the ISP in different situations.

6   Game-Theoretic Model

In this section, we introduce a static game **G** to analyze the interaction between the ISP and the AN. Our model considers potential strategies of the ISP and the AN to protect

---

[4] Our model also applies to the case when ISPs and ANs jointly bear the costs (i.e., when it is cost negative for ISPs to thwart the botnets) by adapting the values R or c_R.

Table 2. Static game: ISP chooses an action from $\{A, C\}$; AN from $\{A, C, S+C, S\}$. Strategy profiles (C, A) and $(S + C, A)$ are not applicable unless when ISP plays C.

		ISP	
		A	**C**
	A	$(0, P(1 - \lambda))$	$(-c_D, P(1 - \lambda))$
AN	**C**	N/A	$\left(N_R(R - c_R) - c_D, P(1 - \lambda(1 - \frac{N_R}{N_B})) - N_R R\right)$
	S+C	N/A	$\left(N_R(R - c_R) - c_D, P(1 - \lambda(1 - \frac{N_R}{N_B})N_{SC}^{1-\beta}) - N_{SC}c_S - N_R R\right)$
	S	$\left(0, P(1 - \lambda N_S^{1-\beta}) - N_S c_S\right)$	$\left(-c_D, P(1 - \lambda N_S^{1-\beta}) - N_S c_S\right)$

the systems against the above defined threats. Considering the benefits and the costs of different strategies we also present the equilibria for the defined game. The key points of our game-theoretic analysis is that by using the computed equilibria it is possible to choose the optimal countermeasure protocol for different situations. Note that our game is a perfect and complete information game. We assume that the players have common knowledge about their strategies and payoffs and can observe the actions of each other.

6.1 Game Model: Strategies and Payoffs

Table 2 shows the normal form of the proposed static game **G**. In this game, the players play simultaneously. The ISP can choose between the following two actions: *Abstain* (*A*) and *Cooperate* (*C*). The *Abstain* action models the behavior of the ISP that is not willing to participate in the detection and rememediation of the bots . Hence the payoff of the ISP is 0, when it plays A. The cooperative ISP (that plays C) first detects the bots and then remediates the infected devices. In return, the ISP receives a reward $N_R R$ from the AN. Recall that the cost for the ISP to remediate all detected devices is $c_R N_R$. Consequently, when the ISP and the AN cooperate, the payoff of the ISP is $N_R(R - c_R) - c_D$.

In our model, the AN can choose one of the following four possible actions: *Abstain* (*A*), *Cooperate* (*C*), *Secure and Cooperate* (*S + C*), and *Secure* (*S*). With the *Abstain* action we model the behavior of the AN that is not willing to perform any countermeasures. In this case, the payoff of the AN will decrease to $P(1 - \lambda)$. Recall that $\lambda \in [0, 1]$ is the fraction of diverted ad revenue by the bots.

If the AN cooperates with the ISP, its utility will increase to $P\left(1 - \lambda(1 - \frac{N_R}{N_B})\right)$, where N_R is the optimal number of infected devices remediated by the ISP, which can be calculated by Lemma 2. However, the AN should pay $N_R R$ to the ISP for N_R remediated devices. As a result, the total payoff of the AN when both players are cooperative is $P\left(1 - \lambda(1 - \frac{N_R}{N_B})\right) - N_R R$.

The AN can also secure the websites by choosing the action S, as discussed in Section 5.1. The AN should pay $N_S c_S$ to secure N_S websites. The benefit of the AN will then increase to $P(1 - \lambda N_S^{1-\beta})$. Consequently, the total payoff of the AN when it plays S is $P(1 - \lambda N_S^{1-\beta}) - N_S c_S$, independently of whether the ISP plays C or A.

Finally, the AN can choose to simultaneously secure some of the websites and cooperate with the ISP to remediate some of the infected devices. This action is represented by $S + C$ and the total payoff of the AN in this case is $P(1 - \lambda N_{SC}^{1-\beta}(1 - \frac{N_R}{N_B})) - N_{SC}c_S - N_R R$, where N_{SC} is the optimal number of secured websites when the AN plays $S + C$ and can be obtained by the following lemma.

Lemma 3. *If the AN fights botnet ad fraud with both countermeasures (action $S + C$), the optimal number of secured websites is equal to* $N_{SC} = \left(\frac{P}{c_S} \lambda (\beta - 1)(1 - \frac{N_R}{N_B}) \right)^{\frac{1}{\beta}}$.

Proof. The proof is similar to Lemma 1. We can obtain the optimal N_{SC}, by maximizing the total payoff of the AN when it plays $S + C$.   □

Lemma 3 shows that when the AN plays $S + C$ a smaller number (N_{SC}) of websites is secured, compared to the number (N_S) of secured websites when the AN plays S (i.e., $N_{SC} = N_S(1 - \frac{N_R}{N_B})^{\frac{1}{\beta}} < N_S$).

6.2 Game Results

In order to predict and choose the optimal action for the ISP and the AN, we investigate all Nash equilibrium strategy profiles of the defined game. In other words, we are interested in finding the strategy profiles, where neither the ISP nor the AN can increase their payoffs by unilaterally changing their strategies. We will check the existence of Nash equilibria by comparing the payoffs obtained in the game **G**.

The following theorem states conditions when the AN does not provide sufficient incentive to the ISP, such that the ISP will abstain at the Nash equilibrium.

Theorem 1. *In **G**, if $R < \frac{c_D}{N_R} + c_R$, the best response of the ISP is to play action A.*

Proof. By comparing the ISP's payoff when it plays C (i.e., whether $-c_D$ or $N_R(R - c_R) - c_D$) with that of A (i.e., 0) we obtain that the best response of the ISP is A if $N_R(R - c_R) - c_D < 0$ or $R < \frac{c_D}{N_R} + c_R$.   □

This means that if the reward for remediation of the infected devices is small, the ISP will not be willing to cooperate with the AN to fight the bots.

The following theorem states when the revenue loss due to ad fraud is not significant enough to cause the AN and the ISP to perform any countermeasure against the bots.

Theorem 2. *In **G**, if $R < \frac{c_D}{N_R} + c_R$ and $\lambda \leq \frac{N_S c_S}{P(1 - N_S^{1-\beta})}$, the action A by the ISP and the AN result in a Nash equilibrium.*

Proof. Considering Theorem 1, the ISP chooses A as its best response. The AN also plays A if its payoff when playing A (i.e., $P(1 - \lambda)$) is bigger than its payoff when playing S (i.e., $P(1 - \lambda N_S^{1-\beta}) - N_S c_S$). Comparing these two payoffs results in the second condition of this theorem, i.e., $\lambda \leq \frac{N_S c_S}{P(1 - N_S^{1-\beta})}$.   □

In other words, if the reward provided by the AN does not generate sufficient incentives for the ISP to cooperate, and the amount of revenue diverted by the bots is smaller than a given threshold, both the ISP and the AN choose A to be at Nash equilibrium.

Theorem 3 shows when the AN fights the bots alone by securing some of the websites.

Theorem 3. *In* **G**, *if* $R < \frac{c_D}{N_R} + c_R$ *and* $\lambda > \frac{N_S c_S}{P(1 - N_S^{1-\beta})}$, *action A by the ISP and action S by the AN result in a Nash equilibrium.*

Proof. The proof is similar to Theorem 2.   □

This result shows that the amount of diverted ad revenue is significant such that a countermeasure should be deployed, but the ISP does not have enough incentive to cooperate and fight bots at this equilibrium. Consequently, the AN secures some of the websites.

Let us assume that λ is very small. Considering all the possible actions and the corresponding payoffs for the AN, the *Abstain* results in maximum payoff for the AN. In fact, action A avoids unnecessary costs for the AN, such as $N_S c_S$ or $N_R R = P_D N_B R$. These results are also in line with Theorem 2 meaning that playing A by both players results in a Nash equilibrium when λ is very small.

When λ increases (i.e., more ad revenue is diverted by the bots) the AN should deviate from A and select one of the three remaining actions as its best response. The following lemma states when the AN should begin securing N_S websites.

Conjecture 1. In **G**, the AN should start securing the websites (Play S) when $\lambda > \frac{N_S c_S}{P(1 - N_S^{1-\beta})}$, which corresponds to the equilibrium presented by Theorem 3.

Proof. We should compare the payoffs of the AN when it plays S or C, with the one obtained by playing the action A. The AN should then play C if $\lambda > \frac{R N_B}{P} = \lambda_1$ and should play S if $\lambda > \frac{N_S c_S}{P(1 - N_S^{1-\beta})} = \lambda_2$. One can show that $\lambda_1 > \lambda_2$ when λ, and consequently N_S, is small enough. This means that the AN switches from A to S at equilibrium, when λ increases.   □

Note that the AN does not switch from the A to the $S + C$, when λ increases, because the AN can protect the revenue first by playing S. In other words, the AN does not need to pay $N_R R$ to the ISP, since the cost would exceed the revenue loss. Consequently, the equilibrium of **G** corresponds to the one presented by Theorem 3. Finally, the following conjecture shows when the AN plays $S + C$ in the response to the cooperative ISP.

Conjecture 2. In **G**, if the ISP is cooperative, the best response of the AN is action $S + C$ if $\lambda > \frac{N_R R - N_S c_S G}{P N_S^{1-\beta} G}$, where $G = 1 - (1 - \frac{N_R}{N_B})^{\frac{1}{\beta}}$.

Proof. The above threshold can be obtained by comparing the payoffs of the AN when it plays $S + C$ and S.   □

Conjecture 2 shows that if the bots divert even more revenue from the ad network, the AN will cooperate with the ISP and pay $N_R R$ to the ISP to remediate N_R bots. It will then secure a smaller number of websites compared to the case when it plays S.

7 Numerical Analysis

In order to understand implications of the analytical results (presented in Section 6) in reality, we simulate the game using the real data. We compute numerically the payoffs

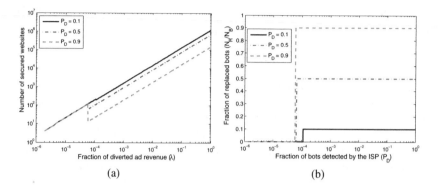

Fig. 4. Outcomes of the game applied to real data when $N_B = 10^4$: (a) Number of the most popular websites that should be secured; (b) Fraction of infected devices remediated by the ISP

of the static game (Table 2), identify the resulting equilibria and present conclusions. To investigate the effect on the entire Web (i.e., for all websites), we extrapolate the data set we have obtained from *Compete.com* with the obtained power law, as explained in Section 5.

We use the following estimated costs in our evaluations: (i) the cost of deploying HTTPS at a web server is $c_S = \$400$ [28]; (ii) the cost of remediating an infected device $c_R = \$100$ (given that it is done via online support [20], it is the estimated cost of human labor for remediating one device per hour); (iii) the cost of the intrusion detection system $c_D = \$100k$ [27].

We take into account different values of the fraction $\lambda \in (0, 1]$ of the ad revenue that the AN loses due to botnet ad fraud and the number of bots N_B. Given that the largest botnets detected so far [3] had several million bots each, we consider the total number of infected devices that participate in the ad fraud considered in our case study to be up to 100 million (regardless of whether they form a single or multiple botnets).

We represent the outcomes of the game for $N_B = 10^4$ in Figure 4. Figure 4(a) shows the number of secured websites depending on the level of threat λ. When the AN cooperates with the ISP, the fraction of remediated devices depending on the level of threat λ is shown in Figure 4(b). We consider three scenarios (the three curves in Figure 4), for three different efficiencies of the detection system employed by the ISP (i.e., when the fraction of detected bots is $P_D = 0.1$, $P_D = 0.5$ and $P_D = 0.9$).

When the threat of the botnet ad fraud is very small, $\lambda < 2 \cdot 10^{-6}$, the AN does not perceive the need to perform any countermeasure against bots. Thus, there are no websites that are secured ($N_S = 0$ in Figure 4(a))[5] and no devices are remediated ($N_R = 0$ in Figure 4(b)). This result corresponds to Theorem 2.

When the bots divert a higher fraction of ad revenue, $\lambda > 2 \cdot 10^{-6}$, the AN first secures a number of websites (Figure 4(a)). As there is no cooperation with the ISP ($N_R = 0$ in Figure 4(b)) the number of secured websites does not depend on P_D, thus it is the same in all three scenarios. The result corresponds to the finding of Theorem 3,

---

[5] Absence of curves in Figure 4(a) signifies log(0), i.e., that zero websites are secured.

i.e., the best choice for the AN is to play *Secure* and for the ISP to *Abstain*. The intuition behind this result is that the relatively small threat λ is distributed over N_B infected devices, thus each bot diverts a small amount of ad revenue. The cost of remediating the infected device would be higher than the loss of ad revenue the bot causes, thus it does not pay off for the AN to cooperate with the ISP. However, the loss is significant enough that the AN has to deploy a countermeasure, hence it secures some of the websites. The number of secured websites corresponds to the Lemma 1.

We observe that the higher λ is, the higher is the number of websites to be secured (Figure 4(a)), until λ reaches a threshold value ($\lambda_1 = 1.12 \cdot 10^{-4}$, $\lambda_2 = 6.6 \cdot 10^{-5}$ and $\lambda_3 = 6 \cdot 10^{-5}$ for $P_D = 0.1$, $P_D = 0.5$ and $P_D = 0.9$, respectively). At the threshold values the AN starts cooperating with the ISP (N_R becomes greater than zero, Figure 4(b)). Thus, the threshold value of λ represents the level of threat after which it is not enough to only secure the websites, but the AN will also cooperate with the ISP to fight bots (i.e., plays $S + C$). This result corresponds to Lemma 2.

When the AN plays $S + C$, each countermeasure protects a given part of the revenue that is otherwise diverted by the bots. The total loss of revenue for the AN due to ad fraud committed by N_B bots is $P\lambda$. The remediation of N_R infected devices reduces the loss of revenue to $P\lambda(1 - \frac{N_R}{N_B})$. As the part of the revenue loss is now eliminated by the ISP, the remaining part is smaller and consequently the AN secures a smaller number of websites. This explains the drop in the number of secured websites (Figure 4(a)), which happens at the threshold value of λ when the AN starts cooperating with the ISP. When λ increases (for values of λ greater than the thresholds), since N_R is constant for a given P_D (Figure 4(b)), in order to eliminate the increasing loss, the AN secures an increasing number of websites for the increasing λ (Figure 4(a)).

In Figure 4(b), we observe that the number of remediated devices is equal to $P_D N_B$, which confirms analytical results stated by Lemma 2. The higher the P_D is, the bigger the benefit of cooperation is, because a larger number of devices is remediated. Consequently, the AN secures a smaller number of websites for a higher P_D (Figure 4(a)).

In summary, the obtained results illustrate that: (i) For a very low level of threat λ, no countermeasures will be taken against bots; (ii) When the fraction λ of the diverted revenue increases, the AN secures a number of websites; (iii) Securing websites is not sufficient for an even higher level of threats, thus the AN will cooperate with the ISP to remediate infected devices.

Next, we analyze the effect of the number of bots N_B in the system on the equilibrium outcomes of the game.

Figure 5 represents the outcomes of the game, in the case of $N_B = 10^7$. Figure 5(a) shows the number of secured websites depending on the level of threat λ. The fraction of remediated devices depending on the level of threat λ is shown in Figure 5(b). As before, the three curves in Figures 5(a) and 5(b), correspond to the three scenarios ($P_D = 0.1$, $P_D = 0.5$ and $P_D = 0.9$).

We observe the same behavior as in the case of $N_B = 10^4$ bots in the system. The difference in the results for the case of $N_B = 10^7$ (Figure 5) compared to results for the case of $N_B = 10^4$ (Figure 4 in Section 7) is that the threshold values of λ, for which the AN begins to cooperate with the ISP, are higher. The explanation for this results is the following.

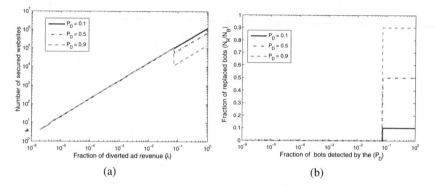

Fig. 5. Outcomes of the game applied to real data when $N_B = 10^7$: (a) Number of the most popular websites that should be secured; (b) Fraction of infected devices remediated by the ISP

When cooperating, the ISP remediates $P_D N_B$ devices, and the AN pays $P_D N_B \cdot R$ to the ISP. Therefore, the cost of cooperation for the AN is higher when N_B is higher. Whereas, the benefit for the AN, due to remediation of $N_R = P_D N_B$ devices is $P\lambda \frac{N_R}{N_B} = P\lambda P_D$, which does not depend on N_B. For a given P_D, the cooperation benefit for the AN is higher only for the higher threat λ. Hence, when the number of bots N_B is high, the AN agrees to cooperate and pay the high cost $P_D N_B R$, only when the fraction λ of the revenue bots divert is high. Because only for the high λ the cooperation benefit $P\lambda P_D$ is high enough to justify the costs of cooperation.

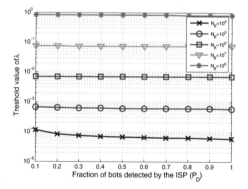

Fig. 6. Threshold values of λ for which the AN begins cooperating with the ISP, in addition to securing the websites

Figure 6 illustrates the threshold values of λ for different numbers of bots N_B in the system and for different efficiencies P_D of the detection system. For example, in the system with $P_D = 0.5$ and $N_B = 10^4$ the AN is cooperative when $\lambda > 6.6 \cdot 10^{-5}$. Whereas, if N_B is much higher, $N_B = 10^8$, the AN is cooperative only if the fraction of diverted revenue is much higher, $\lambda > 0.8$. The results confirm that for a system with a given P_D, when the number of bots is high, the AN is cooperative only when the revenue loss is very high. Based on the results in Figure 6, we also observe that the threshold value of λ does not vary much for different values of P_D. Hence, we can conclude that the value of N_B is the dominant factor in the decision of the AN whether to cooperate with the ISP or not. These results are also confirmed by Lemma 2.

8   Conclusion

In this paper, we have investigated the novel situation of ISPs and ad networks behaving as strategic participants in the efforts to fight botnets. Due to the revenue loss caused by botnet ad fraud, ad networks have economic incentives to protect their revenue by either: (i) improving the security of the online advertising systems or (ii) fighting the major cause of the revenue loss, botnets. To fight botnets, ad networks might need help from ISPs, who are in a better position to deploy detection and remediation mechanisms. We have proposed a game-theoretic model to study the behavior and interactions of the ISPs and ad networks. We have applied our model to the real data to understand the meaning of the results in practice. Our analysis shows that cooperation between the AN and the ISP could emerge under certain conditions that mostly depend on: (i) the number of infected devices (ii) the aggregate power with which bots divert revenue from the ad network and (iii) the efficiency of the botnet detection system. The cooperation is a win-win situation where: (i) users benefit from the ISP's help in maintaining the security of users' devices; (ii) the AN protects its ad revenue as the botnet ad fraud is reduced; (iii) it is at least cost neutral, if not cost positive for the ISP to fight botnets. Cooperation between the AN and the ISP would help to reduce the level of online crime and improve the Web security in general.

Acknowledgements

We would like to thank Wojciech Galuba, Julien Freudiger, and Mathias Humbert for their insights and suggestions on earlier versions of this work, and the anonymous reviewers for their helpful feedback.

References

1. Directive 2006/24/EC of the European parliament and of the council. Official Journal of the European Union (2006)
2. 2008 Year-in-Review Benchmarks. DoubleClick Research Report (2009)
3. Biggest, Baddest Botnets: Wanted Dead or Alive. PC World (2009),
   http://www.pcworld.com/article/169033/biggest_baddest_
   botnets_wanted_de_or_alive.html
4. Click Fraud Index. ClickForennsics Inc. (2009)
5. Internet Advertising Revenue Report. Interactive Advertising Bureau (2009)
6. Adamic, L.A., Huberman, B.A.: The Web's hidden order. Communication ACM (2001)

7. Australian Internet Security Initiative (AISI), A.C., Media Authority: (2010), `http://www.acma.gov.au/WEB/STANDARD/1001/pc=PC_310317`
8. Mungamuru, B., Weiss, S., Garcia-Molina, H.: Should Ad Networks Bother Fighting Click Fraud? (Yes, They Should.). Technical report, Stanford InfoLab (2008)
9. Click Forensics Discovers Click Fraud Surge from New Sophisticated Bahama Botnet: (2009),
    `http://www.clickforensics.com/newsroom/press-releases/144-bahama-botnet.html`
10. Constantin, L.: German Government to Help Rid Computers of Malware (2009), `http://news.softpedia.com`
11. Crowcroft, J.: Net Neutrality: The Technical Side of the Debate: A White Paper. SIGCOMM Computer Communication Review (2007)
12. Daswani, N., Stoppelman, M.: The Anatomy of Clickbot.A. In: Hot Topics in Understanding Botnets (HotBots) (2007)
13. Edelman, B.G.: Securing Online Advertising: Rustlers and Sheriffs in the New Wild West. SSRN eLibrary (2008)
14. Edelman, B.G.: Deterring Online Advertising Fraud Through Optimal Payment in Arrears. SSRN eLibrary (2009)
15. Gandhi, M., Jakobsson, M., Ratkiewicz, J.: Badvertisements: Stealthy Click-Fraud with Unwitting Accessories. Digital Forensic Practice 1(2) (2006)
16. Viral Web infection siphons ad dollars from Google,
    `http://www.theregister.co.uk/2009/05/14/viral_web_infection/`
17. Botnet caught red handed stealing from Google (2009), `http://www.theregister.co.uk/2009/10/09/bahama_botnet_steals_from_google`
18. Grossklags, J., Christin, N., Chuang, J.: Secure or insure?: a game-theoretic analysis of information security games. In: International Conference on World Wide Web (WWW) (2008)
19. Growing number Of ISPs Injecting Own Content Into Websites (2008),
    `http://www.techdirt.com/articles/20080417/041032874.shtml`
20. Livingood, J., Mody, N., O'Reirdan, M., and Comcast Communications: Recommendations for the Remediation of Bots in ISP Networks. Internet-Draft Version 3, IETF (2009)
21. Jakobsson, M., Ramzan, Z.: Crimeware. Addison-Wesley, Reading (2008)
22. Krishnamurthy, B., Wills, C.E.: Cat and Mouse: Content Delivery Tradeoffs in Web Access. In: International conference on World Wide Web (WWW) (2006)
23. Lelarge, M., Bolot, J.: Economic Incentives to Increase Security in the Internet: The Case for Insurance. In: INFOCOM (2009)
24. Livingood, J., Mody, N., O'Reirdan, M., and Comcast Communications: ISP: Voluntary Code of Practice for Industry Self-regulation in the Area of e-security. Internet industry code of practice, Internet Industry Association (2009)
25. Network Bluepill: Stealth Router-based Botnet (2009), `http://dronebl.org/blog`
26. Reis, C., Gribble, S.D., Kohno, T., Weaver, N.C.: Detecting In-Flight Page Changes with Web Tripwires. In: USENIX Symposium on Networked Systems Design & Implementation (NSDI) (2008)
27. Cisco Intrusion Detection Systems, `http://www.google.com/products?q=cisco+intrusion+detection+system&aq=3&oq=cisco+in`
28. VeriSign Inc.,
    `http://www.verisign.com/ssl/buy-ssl-certificates/secure-site-services/index.html`
29. Vratonjic, N., Freudiger, J., Felegyhazi, M., Hubaux, J.P.: Securing Online Advertising. Technical report 2008-017, EPFL (2008)

30. Vratonjic, N., Freudiger, J., Hubaux, J.P.: Integrity of the Web Content: The Case of Online Advertising. In: Usenix CollSec (2010)
31. Vratonjic, N., Raya, M., Hubaux, J.P., Parkes, D.C.: Security Games in Online Advertising: Can Ads Help Secure the Web? In: Workshop on the Economics of Information Security (WEIS) (2010)
32. Weisstein, E.: Euler-maclaurin integration formulas. MathWorld (2010), http://mathworld.wolfram.com/Euler-MaclaurinIntegrationFormulas.html
33. Zhao, X., Fang, F., Whinston, A.B.: An economic mechanism for better Internet security. Decision Support Systems 45(4) (2008)

A Localization Game in Wireless Sensor Networks

Nicola Gatti[1], Mattia Monga[2], and Sabrina Sicari[3]

[1] Politecnico di Milano
Dip. di Elettronica e Informazione
Piazza Leonardo da Vinci, 32
I–20133 Milan, Italy
ngatti@elet.polimi.it
[2] Università degli Studi di Milano
Dip. di Informatica e Comunicazione
Via Comelico, 39
I–20135 Milan, Italy
mattia.monga@unimi.it
[3] Università degli Studi dell'Insubria
Dip. di Informatica e Comunicazione
Via Mazzini, 5
I–21100 Varese, Italy
sabrina.sicari@uninsubria.it

Abstract. Wireless Sensor Networks (WSNs) support data collection
and distributed data processing by means of very small sensing devices
that are easy to tamper and clone: therefore classical security solutions
based on access control and strong authentication are difficult to deploy.
In this paper we look at the problem of assessing the reliability of node
localization data from a game theoretical viewpoint. In particular, we
analyze the scenario in which Verifiable Multilateration (VM) is used to
localize nodes and a malicious node (*i.e.,* the adversary) try to masquerade as non-malicious. We resort to non-cooperative game theory and we
model this scenario as a two-player game. Thus, we were able to compute
an upper bound to the error that an attacker can induce in localization
data, given the number of available verifiers. We focused on the *maximum deception,* that is the distance between the inferred position of
an unknown node and the actual one: we found that if the verifiers are
placed opportunely, the deception is at most 25% of the power range,
and can be halved by triplicating the number of the verifiers.

1 Introduction

Wireless sensor networks (WSNs) [1,2] become increasingly popular in many application domains: indoor/outdoor surveillance systems, traffic monitoring and
control systems for urban and sub-urban areas, systems supporting tele-medicine,
attendance to disable or elderly people, environment monitoring, localization and
recognition of services and users, monitoring and control of manufacturing processes in industry, etc. Most of these activities greatly rely on data about the

T. Alpcan, L. Buttyan, and J. Baras (Eds.): GameSec 2010, LNCS 6442, pp. 168–179, 2010.

positions of sensor nodes, which is not necessarily known before hand. In fact, nodes are often deployed randomly or they even move, and one of the challenges is computing localization at time of operations. Several localization approaches have been proposed (for example, [4,5,7,8,15,11,13,14]), but most of the current approaches omit to consider that WSNs could be deployed in an adversarial setting, where hostile nodes under the control of an attacker coexist with faithful ones. In fact, wireless communications are easy to tamper and nodes are prone to physical attacks and cloning: thus classical solutions, based on access control and strong authentication, are difficult to deploy.

A well defined approach to localize nodes even when some of them are compromised was proposed in [6] by Čapkun *et al.* and it is known as *Verifiable Multilateration* (VM). VM computes an unknown location by leveraging on a set of trusted landmark nodes, named *verifiers*. Although VM is able to recognize reliable localization measures (known as *robust* computations) and sure malicious behaviors, it allows for undecided positions (*unknown nodes*), *i.e.*, cases in which localization data do not provide enough information to support a certain marking as robust or malicious. A conservative approach could be to discard every undecided measure, but this could be unfeasible in some scenarios. This weakness could be exploited by a malicious node to masquerade as an unknown one, pretending to be in a position that is still compatible with all verifiers' information. To the best of our knowledge, the analysis of this scenario, in terms of how a malicious, on the one side, could act and, on the other side, could be faced, has not been explored so far in the literature. This constitutes the original contribution of our work.

To study the properties of a system based on VM deployed in an adversarial setting, we resort to non-cooperative game theory. More precisely, we model it as a two-player game, where the first player employs a number of *verifiers* to do VM computations and the second player is a *malicious node*. The verifiers act to securely localize the malicious node, while the malicious node acts to masquerade as unknown, since when it is recognized as malicious its influence on the system is ruled out by VM. As is customary in game theory, the players are considered rational (*i.e.,* maximizers). This amounts to say that the malicious node is modeled as the strongest adversary. Thanks to game theory model the potentialities of VM are analyzed in depth showing interesting information that should improve the defender's strategy. In [9] we studied the game wherein the verifiers and the malicious node act simultaneously characterizing the players' equilibrium strategies. In this paper, we model the game in extensive form assuming that the malicious node, at first, observes the verifiers' actions and, then, takes its action. This model captures more satisfactorily real world situations. We show how the verifiers should be placed to put a bound on the error the attacker might induce if the defender accepted also unknown positions. Moreover, the study of VM by a game theoretical approach improved our insight in VM properties giving us a tool to quantify the overall robustness of the localization protocol.

The paper is organized as follows: Section 2 provides a short overview about Verifiable Multilateration; Section 3 shortly describes secure localization game, providing some basic concepts; Section 4 analyzes the game in its extensive form and discusses the impact of multiple verifiers in an ad-hoc topology. Section 5 draws some conclusions and provides hints for future works.

2   Verifiable Multilateration

Multilateration is a technique used in WSNs to estimate the coordinates of the unknown nodes, given the positions of some given landmark nodes, sometimes called anchor nodes, whose positions are known. The position of the unknown node U is computed by geometric inference based on the distances between the anchor nodes and the node itself. However, the distance is not measured directly; instead, it is derived by knowing the speed of the signal in the medium used in the transmission, and by measuring the time needed to get an answer to a beacon message sent to U.

Unfortunately, if this computation is carried on without any precaution, U might fool the anchors by delaying the beacon message. However, since in most settings a malicious node can delay the answer beacon, but not speed it up, under some conditions it is possible to spot malicious behaviors. VM uses three or more anchor nodes to detect misbehaving nodes. In VM the anchor nodes work as *verifiers* of the localization data and they send to the sink node B the information needed to evaluate the consistency of the coordinates computed for U. The basic idea of VM is shown in Figure 1: each verifier V_i computes its *distance bound* [3] to U; any point $P \neq U$ inside the triangle formed by $V_1 V_2 V_3$ has necessarily at least one of the distance to the V_i enlarged. This enlargement, however, cannot be masked by U by sending a faster message to the corresponding verifier.

Fig. 1. Verifiable multilateration

Under the hypothesis that verifiers are trusted and they can securely communicate with B, the following verification process can be used to check the localization data in a setting in which signals cannot be accelerated:

1. Each verifier V_i sends a beacon message to U and records the time τ_i needed to get an answer;

2. Each verifier V_i (whose coordinates $\langle x_i, y_i \rangle$ are known) sends to B a message with its τ_i;

3. From τ_i, B derives the corresponding distance bound db_i (that can be easily computed if the speed of the signal is known) and it estimates U's coordinates by minimizing the sum of squared errors

$$\epsilon = \sum_i (db_i - \sqrt{(x - x_i)^2 + (y - y_i)^2})^2$$

where $\langle x, y \rangle$ are the (unknown) coordinates to be estimated[1];

4. B can now check if $\langle x, y \rangle$ are feasible in the given setting by two incremental tests: (a) δ-test: For all verifiers V_i, compute the distance between the estimated U and V_i: if it differs from the measured distance bound by more than the expected distance measurement error, the estimation is affected by malicious tampering; (b) *Point in the triangle test:* Distance bounds are reliable only if the estimated U is within at least one verification triangle formed by a triplet of verifiers, otherwise the estimation is considered unverified.

If both the δ and the *point-in-the-triangle* tests are positive, the distance bounds are consistent with the estimated node position, which moreover falls in at least one verification triangle. This means that none of the distance bounds were enlarged. Thus, the sink can consider the estimated position of the node as ROBUST; else, the information at hands is not sufficient to support the reliability of the data. An estimation that does not pass the δ test is considered MALICIOUS. In all the other cases, the sink marks the estimation as UNKNOWN. In an ideal situation where there are no measurement errors, there are neither malevolent nodes marked as ROBUST, nor benevolent ones marked as MALICIOUS. Even in this ideal setting, however, there are UNKNOWN nodes, that could be malevolent or not. In other words there are no sufficient information for evaluating the trustworthiness of node position. In fact, U could pretend, by an opportune manipulation of delays, to be in a position P that is credible enough to be taken into account. No such points exist inside the triangles formed by the verifiers (this is exactly the idea behind verifiable multilateration), but outside them some regions are still compatible with all the information verifiers have.

Consider N verifiers that are able to send signals in a range R. Let x_0 and y_0 the *real* coordinates of U. They are unknown to the verifiers, but nevertheless they put a constraint on plausible fake positions, since the forged distance bound to V_i must be greater than the length of $\overline{UV_i}$.

Thus, any point $P = \langle x, y \rangle$ that is a plausible falsification of U has to agree to the following constraints, for each $1 \leq i \leq N$:

---

[1] In an ideal situation where there are no measurement errors and/or malicious delays this is equivalent to finding the (unique) intersection of the circles defined by the distance bounds and centered in the V_i (see Figure 1) and $\epsilon = 0$. In general the above computation in presence of errors is not trivial: this has several consequences on the trust model; see [10].

$$\begin{cases} (y - y_i)^2 + (x - x_i)^2 < R^2 \\ (y - y_i)^2 + (x - x_i)^2 > (y_0 - y_i)^2 + (x_0 - x_i)^2 \end{cases} \tag{1}$$

The constraints in (1) can be understood better by looking at Figure 2, where three verifiers are depicted: the green area around each verifier denotes its power range, and the red area is the bound on the distance that U can put forward credibly. Thus, any plausible P must lay outside every red region and inside every green one (and, of course, outside every triangle of verifiers).

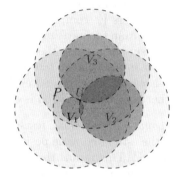

Fig. 2. Plausible falsification region: P is a plausible fake position for U since lays outside every red region and inside every green one whose radius is R (moreover it is outside the triangle of verifiers)

3   Secure Localization Game

Our aim is the study of the behavior of a possible malicious node that acts to masquerade as an unknown node and, at the same time, how the malicious node can be faced at best by the verifiers. This is a typical non-cooperative setting that can be analyzed by leveraging on game theoretical models. A *game* is described by a couple: *mechanism* and *strategies*. The mechanism defines the rules of the game in terms of number of players and actions available to the players. When the mechanism prescribes that the players act simultaneously (*e.g.*, rock-paper-scissors game), the game is said to be in *strategic form*. Instead, when the mechanism prescribes that the players act sequentially (*e.g.*, chess) the game is said to be in *extensive form*. The strategies describe the behaviors of the players during the game in terms of played actions. Strategies can be pure, when a player acts one action with a probability of one, or they can be mixed, when a player randomizes over a set of actions. The players' strategies define an outcome (if the strategies are pure) or a randomization over the outcomes (if mixed). Players have preferences over the outcomes expressed by utility functions and each player is *rational*, acting to maximize its own utility. Solving a game means to find a profile of strategies (*i.e.*, a set specifying one strategy for each player) such that the players' strategies are somehow in equilibrium. The most known equilibrium concept is *Nash* where each player cannot improve its own utility by deviating unilaterally (a detailed treatment of Nash equilibrium can be found in [12]): a fundamental result in the study of equilibria is that every game admits at least one Nash equilibrium in mixed strategies, while pure strategy equilibrium might not exist.

We now formally state our secure localization game as a two-step extensive-form game where the first player to act is the defender (*i.e.*, the verifiers) and then the attacker (*i.e.*, the malicious node) acts. This capture real-world settings where, usually, at first the verifiers are placed and subsequently nodes whose position has to be determined appear. The game is a tuple $\langle Q, A, u \rangle$. Set Q contains the players and is defined as $Q = \{\mathbf{v}, \mathbf{m}\}$ ($\mathbf{v}$ denotes the verifiers and $\mathbf{m}$ denotes the malicious node). Set A contains the players actions. More precisely, given a surface $S \subseteq \mathbb{R}^2$, the actions available to $\mathbf{v}$ are all the possible tuples of positions $\langle V_1, \ldots, V_n \rangle$ of the n verifiers with $V_1, \ldots, V_n \in S$, while the actions available to $\mathbf{m}$ are all the possible couples of positions $\langle U, P \rangle$ with $U, P \in S$ (where U and P are the same as defined in Section 2). We denote by $\sigma_{\mathbf{v}}$ the strategy of $\mathbf{v}$ and by $\sigma_{\mathbf{m}}$ the strategy of $\mathbf{m}$. Given a strategy profile $\sigma = (\sigma_{\mathbf{v}}, \sigma_{\mathbf{m}})$ in pure strategy, it is possible to check whether or not constraints (1) are satisfied. The outcomes of the game can be {MALICIOUS, ROBUST, UNKNOWN}; we denote respectively with $\sigma_M, \sigma_R, \sigma_U$ any strategy profile that has one of the stated outcome. Set u contains the players' utility functions, denoted by $u_{\mathbf{v}}(\cdot)$ and $u_{\mathbf{m}}(\cdot)$ respectively, that define their preferences over the strategy profiles. We define $u_i(\sigma_M) = u_i(\sigma_R) = 0$ for $i \in \{\mathbf{v}, \mathbf{m}\}$; since 0 will be the maximum for $\mathbf{v}$ and the minimum for $\mathbf{m}$, this captures the fact that an outcome in {MALICIOUS, ROBUST} impedes the malicious node from influencing the knowledge of the verifier. $u_i(\sigma_U)$ can be defined differently according to different criteria. A simple criterion could be to assign $u_{\mathbf{v}}(\sigma_U) = -1$ and $u_{\mathbf{m}}(\sigma_U) = 1$. However, our intuition is that the UNKNOWN outcomes are not the same for the players, because $\mathbf{m}$ could prefer those in which the distance between U and P is maximum. In particular we propose three main criteria to characterize UNKNOWN outcomes:

1. *maximum deception*, $u_{\mathbf{m}}$ is defined as the distance between U and P, while $u_{\mathbf{v}}$ is defined as the additive inverse;
2. *deception area*, $u_{\mathbf{m}}$ is defined as the size of the region $S' \subseteq S$ such that $P \in S'$ is marked as UNKNOWN, while $u_{\mathbf{v}}$ is defined as the opposite;
3. *deception shape*, $u_{\mathbf{m}}$ is defined as the number of disconnected regions $S' \subseteq S$ such that $P \in S'$ is marked as UNKNOWN, while $u_{\mathbf{v}}$ is defined as the opposite.

Players could even use different criteria, *e.g.*, $\mathbf{v}$ and $\mathbf{m}$ could adopt the maximum deception criterion and the deception shape respectively. However, when players adopt the same criterion, the game is *zero-sum*, the sum of the players' utilities being zero. This class of games is easy and has the property that the maxmin, minmax, and Nash strategies are the same. In this case calculations are simplified by the property that $u_\mathbf{v} = -u_\mathbf{m}$; in the following we shall adopt this assumption.

4   Game Analysis

For the sake of simplicity, we focus on the case in which both players adopt the maximum deception criterion (a reasoning on the same lines can be applied to the other possibilities). In this section we build upon our previous work [9] to analyze the game its extensive form and we finally draw some conclusions valid in the multiple verifier case.

4.1   Maxmin Solution with Three Verifiers

We focus on the case with three verifiers. In our analysis of the game, we consider only the case in which

$$\forall i, j \ \overline{V_i V_j} \leq R \qquad (2)$$

In fact, if we allowed $\overline{V_i V_j} > R$, then there could be several unreasonable equilibria. For instance, an optimal verifiers' strategy would prescribe that the verifiers were positioned such that only one point satisfied constraints (1). This strategy would assures the verifiers the largest utility (*i.e.*, zero), no UNKNOWN positions being possible. However, it is not reasonable because the area monitored by the verifiers has a null measure (in the sense of Lebesgue).

At first, we can show that for each action of the verifiers (under the assumption (2)), there exists an action of the malicious node such that this is marked as UNKNOWN. Therefore, there is no verifiers' strategy such that, for all the malicious node's actions, the malicious node is marked as ROBUST or MALICIOUS.

Theorem 1. *For each tuple $\langle V_1, V_2, V_3 \rangle$ such that $\overline{V_i V_j} \leq R$ for all i, j, there exists at least a couple $\langle U, P \rangle$ such that $u_\mathbf{m} > 0$.*

Proof. Given V_1, V_2, V_3 such that $\overline{V_i V_j} \leq R$ for all i, j, choose a V_i and call X the point on the line $\overline{V_k V_j}$ $(k, j \neq i)$ closest to V_i. Assign $U = X$. Consider the line connecting V_i to X, assign P to be any point X' on this line such that $\overline{V_i X} \leq \overline{V_i X'} \leq R$. Then, by construction $u_\mathbf{m} > 0$.   □

We discuss what is the configuration of the three verifiers, such that the maximum deception is minimized.

Theorem 2. *Any tuple $\langle V_1, V_2, V_3 \rangle$ such that $\overline{V_i V_j} = R$ for all i, j minimizes the maximum deception.*

Proof. Since we need to minimize the maximum distance between two points, by symmetry, the triangle whose vertexes are V_1, V_2, V_3 must have all the edges with the same length. We show that $\overline{V_i V_j} = R$. It can easily seen, by geometric construction, that U must be necessarily inside the triangle. As shown in Section 2, P must be necessarily outside the triangle and, by definition, the optimal

P will be on the boundary constituted by some circle with center in a V_i and range equal to R (otherwise P could be moved farther and P would not be optimal). As $\overline{V_iV_j}$ decreases, the size of the triangle reduces, while the boundary is unchanged, and therefore $\overline{UP}$ does not decrease. $\square$

We are now in the position to find the maxmin value (in pure strategies) of the verifiers, *i.e.*, the action that maximizes the verifiers' utility given that the malicious node will minimize it. The problem of finding the maxmin strategy can be formulated as the following non-linear optimization problem, given V_1, V_2, V_3 such that $\overline{V_iV_j} = R$ for all i, j:

$$\max_{U, P \in S} \overline{UP}$$

constraints $(1) \wedge P$ outside $V_1V_2V_3$

We normalized the problem assigning $R = 1$ and we solved it by using conjugated subgradients. We report the solution. Let W be the orthocenter of the triangle, U and P can be expressed more easily with polar coordinates with origin in W. We assume that $\theta = 0$ corresponds to a line connecting W to a V_i. We have, $U = (\rho = 0.1394R, \theta = \frac{\pi}{6})$ and $P = (\rho = 0.4286R, \theta = \frac{\pi}{6} + 0.2952)$, and, for symmetry, $U = (\rho = 0.1394R, \theta = -\frac{\pi}{6})$ and $P = (\rho = 0.4286R, \theta = -\frac{\pi}{6} - 0.2952)$. Therefore, there are six optimal couples $\langle U, P \rangle$s. In Figure 3(a) depicts one malicious node's best action and Figure 3(b) shows all the other symmetrical positions. The value of $u_{\mathbf{m}}$ (*i.e.*, the maximum deception) is $0.2516R$. In other words, when the verifiers compose an equilateral triangle, a malicious node can masquerade as unknown and the maximum deception is about 25% of the verifiers' range R.

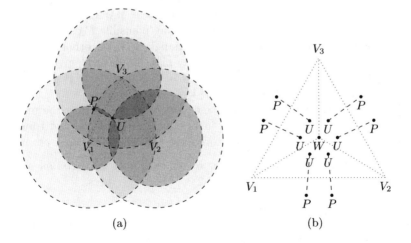

(a)   (b)

Fig. 3. Malicious node's best responses (maximum deception is $\overline{UP} = 0.2516R$)

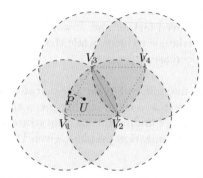 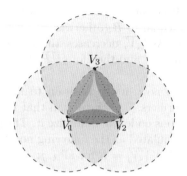

(a) Maximum deception with four veri-   (b) U positions that allows for
fiers is still $\overline{UP} = 0.2516R$ plausible Ps

Fig. 4. Impact of verifiers on U ability to fake positions

4.2 Maximum Deception with Multiple Verifiers

The result exposed in Section 4.1 are the basis to study situations with multiple verifiers. Our main result is the derivation of a bound between the maximum deception and the number of multiple verifiers.

Initially consider the simple situation in which we have four verifiers and they constitute two adjacent equilateral triangles as shown in Figure 4(a). The maximum deception does not change with respect to the case with three verifiers, since some of the best responses depicted in Figure 3(b) are still available. In fact, the fourth verifier is useful to rule out only the two positions that are on the edge V_4 faces: on this side any fake P would surely marked as MALICIOUS (or even ROBUST if $P \equiv U$) since it would be inside the triangle $V_2V_3V_4$. The proof is straightforward. Consider (without loss of generality) the triangle $V_1V_2V_3$ in Figure 4(a). In order for a node not to be marked as MALICIOUS, U must be in the areas depicted in Figure 4(b). Moreover, any plausible P cannot be neither inside the triangle $V_1V_2V_3$ nor inside the triangle $V_2V_3V_4$, otherwise the node would be marked as MALICIOUS. In fact, any plausible fake P, given a U in the blue area between V_2 and V_3 (see Figure 4(b)), cannot be in regions that are outside both the triangles $V_1V_2V_3$ and $V_2V_3V_4$.

The above observation can be leveraged to give a bound over the maximum deception with a given number of verifiers opportunely placed and tuned such that the shape of the area they monitored is a triangle.

Theorem 3. *Given a triangular area, in order to have a maximum deception not larger than $\frac{0.2516R}{2^k}$ we need at least $2 + \sum_{i=0}^{k} 3^i$ verifiers.*

Proof. Consider the basic case with three verifiers (composing an equilateral triangle) with range R and $\overline{V_iV_j} = R$. As shown in Section 4.1 the maximum deception is then $0.2516R$. Introduce now more three verifiers such that we have four equilateral triangles with edge $\frac{R}{2}$ as shown in Figure 5. The range of all the verifiers is set equal to $\frac{R}{2}$ (*i.e.*, they could just ignore any beacon message

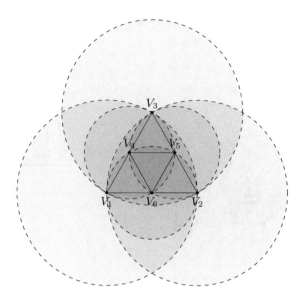

Fig. 5. Maximum deception with six verifiers is $\overline{UP} = \frac{0.2516R}{2}$

that takes longer than needed to cover the distance $\frac{R}{2}$). Since the edge of the small triangles is now $\frac{R}{2}$, the maximum deception here is $\frac{0.2516R}{2}$ and no U positions are possible in the central triangle $V_4V_5V_6$: indeed all the edges of the central triangle are adjacent to the edge of other triangles. This last result allows us not to consider the central triangle when we want to reduce the maximal deception, the malicious node never positioning itself within it. The basic idea is that if we want to halve the maximum deception we need to decompose all the triangles vulnerable by the malicious node by introducing three verifiers. By introducing three new verifiers per triangle we obtain four sub-triangles with an edge that is the half of the original triangle and therefore the maximum deception is halved. In general, in order to have a maximum deception of $\frac{0.2516R}{2^k}$, the number of required verifiers[2] is $\frac{3}{2}(1+3^k)$, as shown in Table 6(b). In Figure 6(a) we report an example with $k = 2$ and 15 verifiers. Notice that when we introduce new verifiers we need to halve the range. In general, we will have verifiers with multiple different ranges.   □

The number of verifiers increases according to the formula $n(k) = n(k-1) + 3^k$. Asymptotically $\lim_{k\to\infty} \frac{n(k+1)}{n(k)} = 3$, thus we need to multiply by three the number of verifiers to divide by two the maximum deception. Notice that, as shown in the proof of Theorem 3, the verifiers are required to have different ranges. Increasing the number of verifiers require to add new verifiers with range smaller than those already present in the network.

---

[2] That is the number of vertices in Sierpinski triangle of order k; see [16].

(a) 15 verifiers $(k = 2)$ give a maximum deception
$\overline{UP} = \frac{0.2516R}{4} = 0.0629\ R$

(b) Maximum deception

k	# ver.	max. deception
0	3	0.2516 R
1	6	0.1258 R
2	15	0.0629 R
3	42	0.02145 R
4	123	0.015725 R
5	366	$7.8625 \cdot 10^{-3} R$

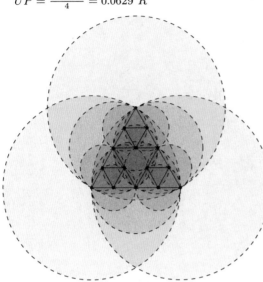

Fig. 6. Maximum deception is reduced by increasing the number of verifiers

5   Conclusion

The trust we put on wireless sensor node localization information is the fundamental base to provide trust to context aware applications and data. Verifiable Multilaterion is a secure localization algorithm, which is able to deal with nodes that falsify their data. VM defines two tests for evaluating node behavior as malicious, or robust, or in the worst case as unknown. Unknown nodes could be simply ignored since VM has not enough information for evaluating the trustworthiness of the node. But unknown nodes could also be faithful, thus by ignoring that source of information, the system loses some opportunities. However, by considering it, we give to a potential attacker the chance of introducing false data into the system. In this paper, by modelling the localization behavior of VM as non-cooperative game we were able to compute an upper bound to the error that an attacker can induce in localization data, given the number of available verifiers. We focused on the maximum deception, that is the distance between the inferred position of an unknown node and the actual one: we found that if the verifiers are placed opportunely, the deception is at most 25% of the power range, and can be halved by triplicating the number of the verifiers. Currently, our results are valid only with a single attacker, since this is key to the assumption that signals cannot be accelerated. In future, we shall consider situations where a malicious attacker can manipulate more nodes.

Acknowledgment

This research has been partially funded by the European Commission, Programme IDEAS-ERC, Project 227977-SMScom.

References

1. Akyildiz, I.F., Su, W., Sankarasubramaniam, Y., Cayirci, E.: A survey on wireless sensor network. IEEE Wireless Communications 40(8), 102–114 (2002)
2. Baronti, P., Pillai, P., Chook, V.W.C., Chessa, S., Gotta, A., Hu, Y.F.: Wireless sensor networks: A survey on the state of the art and the 802.15.4 and zigbee standards. Computer Communications 30(7), 1655–1695 (2007)
3. Brands, S., Chaum, D.: Distance-bounding protocols. In: De Santis, A. (ed.) EUROCRYPT 1994. LNCS, vol. 950. Springer, Heidelberg (1995)
4. Bulusu, N., Heidemann, J., Estrin, D.: GPS-less low-cost outdoor localization for very small devices. IEEE Personal Communications 7(5), 28–34 (2000)
5. Čapkun, S., Hamdi, M., Hubaux, J.-P.: GPS-free positioning in mobile ad-hoc networks. Cluster Computing 5(2), 157–167 (2002)
6. Čapkun, S., Hubaux, J.P.: Secure positioning in wireless networks. IEEE Journal on Selected Areas in Communications 24(2), 221–232 (2006)
7. Chen, J., Yao, K., Hudson, R.: Source localization and beamforming. IEEE Signal Processing Magazine 19(2), 30–39 (2002)
8. Doherty, L., Pister, K., Ghaoui, L.E.: Convex position estimation in wireless sensor networks. In: Proceedings of the IEEE Conference on Computer Communications (INFOCOM) (2001)
9. Gatti, N., Monga, M., Sicari, S.: Localization security in wireless sensor networks as a non-cooperative game. In: International Congress on Ultra Modern Telecommunications and Control Systems, ICUMT 2010, Moscow, Russia. IEEE, Los Alamitos (to appear, 2010) ISBN: 978-1-4244-7286-4
10. Monga, M., Sicari, S.: On the impact of localization data in wireless sensor networks with malicious nodes. In: Kamra, A. (ed.) Proceedings of the 2nd SIGSPATIAL ACM GIS International Workshop on Security and Privacy in GIS and LBS (SPRINGL 2009), ACM SIGSPATIAL, pp. 63–70. ACM, New York (2009)
11. Niculescu, D., Nath, B.: Ad-hoc positioning system. In: Proceedings of the IEEE Global Communication Conference (GLOBECOM) (2001)
12. Osborne, M., Rubinstein, A.: A course in game theory. MIT Press, Cambridge (1994)
13. Ramadurai, V., Sichitiu, M.: Localization in wireless sensor networks: A probabilistic approach. In: Proceedings of the International Conference on Wireless Networks (ICWN) (2003)
14. Savvides, A., Park, H., Srivastava, M.: The bits and flops of the n-hop multilateration primitive for node localization problems. In: Proceedings of the ACM International Workshop on Wireless Sensor Networks and Application (WSNA) (2002)
15. He, T., Huang, C., Blum, B.M., Stankovic, J.A., Abdelzaher, T.: Range-free localization schemes for large scale sensor networks. In: Proceedings of the ACM International Conference on Mobile Computing and Networking (MOBICOM) (2003)
16. Wessendorf, M.: The on-line encyclopedia of integer sequences, aT&T Labs Research (2002), http://www.research.att.com/~njas/sequences/A067771

Effective Multimodel Anomaly Detection Using Cooperative Negotiation

Alberto Volpatto, Federico Maggi, and Stefano Zanero

Dipartimento di Elettronica e Informazione
Politecnico di Milano
{volpatto,fmaggi,zanero}@elet.polimi.it

Abstract. Many computer protection tools incorporate learning techniques that build mathematical models to capture the characteristics of system's activity and then check whether live system's activity fits the learned models. This approach, referred to as *anomaly detection*, has enjoyed immense popularity because of its effectiveness at recognizing unknown attacks (under the assumption that attacks cause glitches in the protected system). Typically, instead of building a single complex model, smaller, partial models are constructed, each capturing different features of the monitored activity. Such *multimodel* paradigm raises the non-trivial issue of combining each partial model to decide whether or not the activity contains signs of attacks. Various mechanisms can be chosen, ranging from a simple weighted average to Bayesian networks, or more sophisticated strategies. In this paper we show how different aggregation functions can influence the detection accuracy. To mitigate these issues we propose a radically different approach: rather than treating the aggregation as a calculation, we formulate it as a decision problem, implemented through cooperative negotiation between autonomous agents. We validated the approach on a publicly available, realistic dataset, and show that it enhances the detection accuracy with respect to a system that uses elementary aggregation mechanisms.

Keywords: Anomaly detection, cooperative negotiation.

1 Introduction

As a growing number of business and personal activities are conducted over the Internet, cybercrime has became a growing concern [1]. In the current threat landscape, 0-day exploits and site-specific attacks are at the same time the most challenging and frequent threats [2]. 0-day exploits take advantage of vulnerabilities not publicly disclosed, whereas site-specific attacks target a specific application, such as custom product, rather than a widely deployed system. Among the defense tools, anomaly detectors leverage a description of the normal activity of the protected system, and are therefore potentially effective against attacks never seen before. The assumption (and, to some extent, the limitation) is that attacks would leave traces that can be automatically recognized as anomalous. On the other hand, misuse-based detectors, which rely on a list of signatures of

T. Alpcan, L. Buttyan, and J. Baras (Eds.): GameSec 2010, LNCS 6442, pp. 180–191, 2010.
© Springer-Verlag Berlin Heidelberg 2010

known attacks, only offer protection against attack vectors that have been publicly disclosed. Despite its remarkable precision at detecting known threats, the misuse-based paradigm offer no help against 0-day and custom-made attacks.

A long-standing approach in the design of complex systems is the use of multiple models [3, 4]. This paradigm is also effective for anomaly detection, as proven by the recent literature detection [5, 6, 7, 8, 9, 10, 11, 12] and lies in the decomposition of a complex problem (i.e., capturing the "normal behavior" of a potentially large and complex system) into simpler sub-problems, each tackled by one mathematical model. For example, there are systems that analyze the observed activity (e.g., the HTTP requests and responses) and build models such as the average length of a request, or the relative frequency of ASCII symbols into an HTTP request body. Usually, such models have a numerical representation that contributes to an overall evaluation of the degree of anomaly. The models can be encapsulated into autonomous agents [13]. This technique have been shown to be as effective as the traditional, multimodel approach [14, 15, 16, 17, 18, 19], with the benefit of a more natural design and the existence of a plethora of multiagent programming frameworks [20]. Unfortunately, while the use of multiagent systems may improve the ease of design of anomaly detectors, the problem of aggregating the models persists.

The core point of this paper is that the aggregation phase is crucial to achieve good precision. We propose to tackle this problem and its issues (described in Section 2.1) by exploiting the multiagent framework beyond its architectural properties. More precisely, we use cooperative negotiation to implement the aggregation task as a decision problem. This approach is inspired by the seminal study described in [21], where cooperative negotiation has been used to classify the payloads of TCP packets. However, the study was explicitly a toy example, tested on artificial, outdated traffic (i.e., IDEVAL [22]). We further develop the idea, adapt it to real-world settings, and examine carefully the impact of the negotiation protocol's parameters. The contributions of this paper can be summarized as follows.

- In Section 2 we discuss the issues caused by simple aggregation strategies, and motivate why this problem should be addressed to achieve reliable attack detection.
- In Section 3 we present a simple, yet very effective, technique that leverage cooperative negotiation between autonomous agents to decide whether or not a given event is an attack and detail how we incorporated this technique in an anomaly detector.
- In Section 4 we evaluate the detection capabilities of the tool obtained over a realistic, publicly-available data set, and discuss its limitations.

The results of our experiments show that the proposed technique alleviates the detection errors caused by inaccurate combinations of models. Our approach applies to any multimodel system. However, due to the popularity of web applications and web-based attacks, we validate it on a recent anomaly detector designed to protect web applications.

2   Multimodel Anomaly Detection

A learning-based anomaly detector learns the normal activity by observing a system's activity. In the representative, simple example of HTTP, such activity consists in the HTTP requests and responses exchanged between servers and clients. Requests, or *queries*, $Q = \{q_1, q_2, \ldots, q_j, \ldots\}$, are usually decomposed into resources (i.e. paths) and parameters. For instance, the request 'GET /page?uid=u44&p=14&do=delete' contains the resource '/page' and the parameters $\{\langle \text{uid}, \text{'u44'}\rangle, \langle g, 14\rangle, \langle \text{do}, \text{'delete'}\rangle\}$.

During an initial *learning* (or training) phase, a multimodel detector instantiates different mathematical models, $m^{(1)}, \ldots, m^{(Z)}$, to compute certain features, $1, \ldots, Z$, on each training sample (e.g., an HTTP request, response, or a sequence of requests-responses). The specific models and the strategy to combine their output determine the classes of attacks that can be detected. Typical models proposed in literature capture features such as the average length of the string parameters, their character distribution and probabilistic grammar, or the order in which parameters appear across the requests against the each resource. Unfortunately, due to space limitations, we must refer the interested reader to [12,9,23] for more details. During *detection*, the model instances computed during training are used as maps, $m^{(z)} : Q \mapsto [0,1], \forall z$, and their outputs are aggregated into an overall *anomaly score*. This score is checked against thresholds and alerts are fired accordingly. Thresholds are fixed a priori or, usually, computed during learning. The improvements proposed in this work — validated using the set of models implemented in Masibty [12] — are independent from the particular models, and thus can be easily applied to any learning-based detector.

2.1   Drawbacks of Model Aggregation

In the design of a multimodel anomaly detector, the value aggregation [24] phase has a significant impact on the quality of detection results. Just for simplicity in illustrating our point, we can roughly distinguish between simple and complex aggregation strategies, and in both cases such strategies can be parametric or non-parametric.

An example of simple, non-parametric aggregation strategy is the arithmetic mean. While this type of approach requires no user intervention and is very simple to understand, the lack of a differentiation between the models does not allow to control the impact of models with poor performance on the overall detection. The most natural solution to this issue is the use of parametric aggregation functions (e.g., a weighted average), with different parameters assigned to the models in order to optimize detection quality. These mechanisms obviously create the non-trivial problem of choosing the parameters (e.g., the weights), which clearly influence the final value. In addition, if the system employs a large set of models, it might be difficult to properly set all the weights. Although the optimal weights can be computed automatically from the data, as demonstrated by the results of the experiment described in Section 4.1, this is not sufficient to achieve good detection accuracy if a simplistic aggregation method is used. More complex strategies, such as Bayesian networks (or ad-hoc aggregation criteria), can perform very well under the conditions they have been designed for, but unfortunately their inherent complexity makes manual tuning and improvements difficult for the end-users. Our aggregation approach, which performances are comparable to such methods, have the advantage of being easier to manually configure.

From the previous observations, it follows that the design of a simpler, generic method to reduce detection errors caused by model aggregation is necessary to obtain a reasonable level of protection from multimodel anomaly detectors.

3   Exploiting Cooperative Negotiation

We implemented our model aggregation approach by modifying the training and detection algorithms of Masibty [12].

3.1   Modifications to the Learning Phase

To fully implement the detection phase detailed in Section 3.2, the learning phase of a traditional multimodel detector requires some minor modifications. In particular, every model, $m^{(z)}$, must implement a *trust model*, $T_j^{(z)} : Q \mapsto [0, 1]$, that assesses the "reliability" of $m^{(z)}$, up to the j-th training step (i.e., after the j-th training has been analyzed). As motivated in Section 3.2, this model is required for completing the cooperative negotiation correctly. The final decision indeed depends upon the agents with higher trust level, while the impact of poorly-trained agents (which may lead to overfitting) is minimized. For this reason, the trust functions must be designed in such a way that models that have received ample training are assigned high trusts, since they are likely to produce accurate and reliable detections. For example, one of the models checks for the presence of parameters in each HTTP request and computes their appearance ratio across all the requests. The anomaly score is calculated as $1 - \min\left(\frac{M}{R}, \min\left(\frac{P}{R}\right)\right)$, where M and P, respectively, indicate the number of missing and present parameters, while R is the total number of requests. The trust level is high if the presence is nearly constant, while it decreases if an application exhibits variations. The trust function is thus $1 - \frac{M}{R}$. Due to space limitations, we refer the reader to [12] for details on the models implemented on the prototype (see Fig. 1 in [12]) used for our experiments.

Optionally, the trust level can also be exploited to optimize training, by stopping it automatically when a sufficient amount of data is received by each model. For example, we adopted $\delta_W(j) := \max_{j \in W} T_j^{(z)} - \min_{j \in W} T_j^{(z)}$, where W is a sliding window[1]. By choosing a small $\varepsilon > 0$ (we used $\varepsilon = 0.003$ and $W = 5$), a model is considered stable after the j-th training sample if $\delta_W(j) \leq \varepsilon$. Although more sophisticated criteria can be designed, we noticed that, under the

---

[1] The sliding window size influences the training duration: smaller values tend to stop training early, while higher values result in a longer and more conservative training.

conditions described in Section 4, this is sufficient to achieve good detection results. This optimization, however, is not necessary to adopt our cooperative negotiation approach, and we will evaluate its impact separately.

3.2   Modifications of the Detection Phase

We translate the value aggregation problem that arises during detection into a decision problem implemented via *cooperative negotiation* between autonomous agents. In artificial intelligence, an *agent* is the abstraction of an entity capable of reading inputs by observing the environment, and to perform actions toward the achievement of certain goals. In our context, the *environment* is the network segment which the agent receives HTTP messages from. No other input than the HTTP messages is passed to the agents. The *goal* is to find the correct degree of anomaly, that is, the numeric value that would minimize the detection errors. *Multiagent* systems comprise a *coordination* protocol, implemented by the agents (and a mediator, if present) to achieve a *global* goal. In general, these protocols can be *competitive*, when the goals of the agents are conflicting, or *cooperative*, when the agents pursue a common goal. A form of coordination is *negotiation*, where the agents tend to "harmonize" conflicting goals toward the achievement of the global goal.

We specifically propose to use a *cooperative negotiation protocol*, described in the following, to reach an agreement on the anomaly score used to classify HTTP messages as benign or malicious.

Cooperative negotiation protocol. Our system comprises n agents, $M_1, \ldots,$ $M_i, \ldots, M_n$, and a mediator M. Each agent embeds exactly one of the partial models implemented in the original prototype [12, Section IV], and communicates only with the mediator (which embeds no models).

The protocol is initiated every time a new sample, q_j, is observed. The j-th session iterates multiple times. Each iteration is denoted with t and begins at $t = 0$. We define the partial degree of anomaly $p_i^t \in [0, 1]$ as the degree of anomaly computed by the i-th agent using its embedded model $m^{(z_i)}$ at iteration t. The protocol proceeds as follows:

1. each M_i receive the sample q_j and calculates its offer, $p_i^t = m^{(z_i)}(q_j)$.
2. p_i^t is sent to M along with the agent's trust level, $w_i = T^{(z_i)}$.
3. M receives p_i^t and w_i, $\forall i = 1, \ldots, n$, and calculates an agreement a^t by using the agreement function $a^t = A(p_1^t, w_1, p_2^t, w_2, \ldots, p_n^t, w_n)$.
4. M sends its counter offer a^t to all the agents.
5. Each M_i calculates the new offer p_i^{t+1} by using a negotiation function $p_i^{t+1} = F_i(p_i^t, a^t)$.
6. Each M_i sends p_i^{t+1} to M, and steps 3. to 6. are repeated until an agreement, a, for the j-th session, $a = a_j$, is reached (i.e., until $p_i = p_{i'}, \forall i \neq i'$).

The global evaluation of the anomaly score for the sample q_j is thus a_j.

We define the *agreement function* to be the weighted average of the offers, where weights are the trust levels: $A(p_1^t, w_1, p_2^t, w_2, \ldots, p_n^t, w_n) := \frac{\sum_{i=1}^n p_i^t w_i}{\sum_{i=1}^n w_i}$. The trust level is *constant* throughout the negotiation.

The *negotiation function* is $F_i(p_i^t, a^t) = p_i^t + \alpha_i(a^t - p_i^t)$, where $a^t - p_i^t$ expresses a measure of disagreement between agent M_i and the global system. Note that p_i^{t+1} is determined only by values at time t: for this reason, and because each agent does not change the evaluation of its embedded model over time (for a given q_j), it is unnecessary to actually perform a two-way communication between the agents and the mediator. Instead, upon receiving all the initial offers, the mediator can run the negotiation protocol without communicating with the agents. The cooperative negotiation protocol described does not impose any negotiation strategy to the agents, that is indeed related to the implementation of both functions detailed above.

The *agreement coefficient*, $\alpha_i \in (0, 1)$, expresses the willingness of agent M_i to propose its offer versus the counter-offer received from the mediator. When $\alpha_i \to 0$ each agent tends not to modify its offer, while $\alpha_i \to 1$ causes each agent to agree with the counter-offer. We compute the agreement coefficient as a function of the trust level of the agent's embedded model. If the trust level is close to 0, then its evaluation would not be reliable; thus, during the negotiation session its influence should be minimized to "ignore" its offers. On the other hand, a trust level close to 1 means maximum reliability. A sigmoid-shaped function $f_\alpha(w_i) = \frac{1}{1 + e^{h(w_i - k)}}$ is a good implementation of the above rationale, where h is the smoothness and k is the central value.

It can be shown that h has negligible impact on the final agreement (in our experiments, we used $h = 7.5$) and only influences the speed of the negotiations. We adopted $k = 0.5$ to express that all the agents have neutral impact on the agreement against the mediator, and such impact only depends on the trust levels computed from data during training. In other words, 0.5 is the natural value of k if one needs to avoid biased detections. It can be shown that also k has a negligible impact on the results. We demonstrate both observations with an experiment described in Section 4.2.

The cooperative negotiation mechanism described is proved to be *connectively stable* in [25]. This means that agents will reach a stable agreement on attack probability, regardless of their initial offer and for any n. In order to prove our point that just by modifying the negotiation protocol we could improve the quality of the results, we did not modify the original detection phase of the system any further. In particular, we have not modified the way global anomaly thresholds are calculated and used to raise alerts.

In addition to the traditional sensitivity parameter, used in every anomaly-based detector to trade off *False Positive Rate* (FPR) versus *Detection Rate* (DR), h and k are the only parameters strictly required by our approach, and their choice has a very limited impact on the system performance. W and ε are required only for optimized learning.

4   Evaluation

To validate our approach, we conducted two experiments on the traffic[2] captured during the International Capture the Flag 2008, organized by the University of California, Santa Barbara. The traffic, mostly HTTP, contains 0-day vulnerabilities, and the majority of the players were skilled hackers, able to prepare custom and diverse exploits. Unfortunately, this causes the lack of any ground truth other than the list of (known) attacks detectable with the Snort misuse-based system. To alleviate this issue we used the portion of the dataset that contains clean background traffic, and injected custom, real-world attacks during detection. In this way, the ground truth is perfectly known. The background traffic contains 44,102 HTTP messages, i.e., 22,051 request-response couples. We cross-validated our system by using 14,961 request-response couples for training and 7,090 for detection. More precisely, we injected instances of the three most common types of attacks against web applications[3], such as cross-site scripting (XSS) (e.g., CVE-2009-0781), SQL injections (e.g., CVE-2009-1224), and command injections (e.g., CVE-2009-0258). The XSS attacks are variations on those listed in [26], the SQL injections were created similarly from [27], and the command execution exploits are variations of common command injections against the Linux and Windows platforms. In addition to cross-validation, to avoid biased experiments on the same attack instances, the injected strings are randomly drawn from a set of alternatives. In particular, we used 14 different SQL injection vectors, 4 command injections, and 94 XSSs. Note that, this is similar to the use of variants of the same dataset, as it avoids using the same exact set of attacks over and over. To this end, using a uniform probability distribution, we randomized (1) the type of attack, (2) the HTTP parameter to use for the injection, and (3) the vector (of a given type) to inject. The resulting traffic contains 1,000 randomized attacks in every experiment.

4.1   Benefit of Cooperative Negotiation

This experiment aims at showing that our approach can effectively mitigate detection errors better than a technique based solely on value aggregation (i.e., weighted average). To this end, we ran the original prototype in its original configuration, as described in [12], then with the modifications described in Section 3, and finally implementing also the optimization for the learning phase briefly described in Section 3.1 (which is optional).

The comparison is show in Fig. 1, with a ROC curve showing DR and FPR for different working points. As it can be seen, the cooperative negotiation dramatically improves the classification accuracy with respect to the simplistic weighted average. For instance, at $FPR = 0.1$, the cooperative negotiation yields an increment of about 54% on the DR. It must be noted that, in the paper that

---

[2] Available for download at http://ictf.cs.ucsb.edu/data/ictf2008/

[3] http://owasptop10.googlecode.com/files/OWASPTop10-2010.pdf

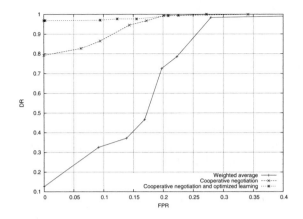

Fig. 1. ROC curve of the original prototype (solid line) and with the modifications we propose (dashed line), and with optimized training (dotted line)

describes the original prototype, the experiments leveraged a completely artificial, small dataset, comprising only a limited number of attacks: this explains why the original system performs significantly worse in this experiment.

The accuracy can be further improved by adopting the optimization of the learning phase (see Section 3.1), although this is not required to apply the cooperative negotiation.

4.2   Influence of the Parameters

This experiment shows that the parameters introduced by our approach, i.e., the smoothness, h, and the central value, k, of the alpha function, f_α, only marginally impact the detection quality. In addition, we provide guidance to choose these parameters to minimize the negotiation overhead. To this end, we used all of the modifications described in Section 3 (with optimized learning) at a fixed ROC working point, and varied $h \in \{2.5, 5, 7.5, 10\}$, $k \in \{0.1, 0.3, 0.5, 0.7, 0.9\}$. As shown in Fig. 2, the influence of these parameters on (a) DR and (b) FPR is barely noticeable.

Rather than providing a theoretical complexity boundary to the computational overhead induced by the negotiation process, we estimate it under real conditions. As shown in Fig. 2(c), h and k have significant impact on the computational overhead (e.g., number of iterations I), necessary to complete the negotiation. However, as seen in Fig. 2(a-b), these parameters have almost *no* impact on the detection accuracy. Thus, setting $k \geq 0.5$ and choosing a safe value of h allows to limit the number of iterations. For example, in the experiment discussed in Section 4.1 we used $h = 7.5$.

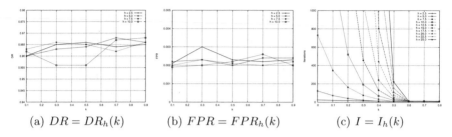

(a) $DR = DR_h(k)$ (b) $FPR = FPR_h(k)$ (c) $I = I_h(k)$

Fig. 2. Negligible influence of k and h on the detection quality of our system (a-b) and noticeable influence on the length of the negotiation, I

4.3   Discussion and Limitations

It is important to clarify the main limitations of our approach. Firstly, an incorrect choice of h and k may cause the negotiation protocol not to terminate in a reasonable amount of time. However, in Section 4.2, we show that these parameters do not influence the detection quality and, more importantly, provide a guidance for setting h and k to safe values. Secondly, as described in Section 3.2, the trust level is constant throughout all the negotiations. In our opinion, this does not impact the detection accuracy significantly, although this needs to be assessed thoroughly.

From our experiments we conclude, however, that the technique described in Section 3 effectively improves the detection capabilities of a multimodel anomaly detector.

5   Related Work

Anomaly-based approaches have been proposed to protect computer systems from attacks by exploiting learning algorithms in different veins. Ensembles of simple models (e.g., character distribution) are effective at capturing the normal characteristics of computer programs [10,11], network traffic [7], or HTTP messages [9,12], respectively. Unfortunately, due to space limitations, we must refer the reader to a comprehensive survey on anomaly detection [28].

Multiagent systems are very useful for implementing complex systems [14] and have been applied also to intrusion detection as a handy replacement of classic multimodel architectures. For instance, in [13] an agent is assigned to each system component or task (e.g., network sniffing, stream reassembly). However, in this type of approaches, only the architecture of a multiagent system is exploited. Instead, our approach exploits multiagent systems as a *paradigm*: not only each agent embeds a detection procedure, but proper algorithms drawn from artificial intelligence are leveraged to perform the decision. One of the first attempts of translating the intrusion detection problem to an interaction between agents appeared in [15], where the use different classes of agents, which do not communicate to each other, is used to detect different types of anomalies, and to aggregate the decisions through a special agent. Unfortunately, some malicious activity (e.g., distributed attacks or evasion attacks) are difficult to detect if the agents do not communicate, since each agent has a limited view of the attacks. This idea is improved in [16, 18] by introducing communication between agents and by embedding a Bayesian network into each agent. Different decision techniques have been embedded in the multiagent detector described in [17], where special agents called "decisors" use fuzzy inference to bid for the most appropriate actions to counteract the anomalies reported by other agents. CAMNEP [19] has two important analogies with our work, since each agent (1) learns a different model (similar to those cited in Section 2) and, (2) is assigned a trust level that reflects the completeness of its training. Unfortunately, this approach do not fully exploit artificial intelligence algorithms as results from each agents are simply averaged. The exploratory work described in [21] uses the intrusion detection task as a case study to apply cooperative negotiation algorithms to detect attacks. Although the results are promising, the approach reduces the computation of the trust to a constant function. More importantly, the agreement coefficient devised by the authors causes the agents to agree on results that once again tend to approximate a weighted average. Recent proposals focused on updating the learned specifications dynamically at run-time, hence requiring no or little human intervention also in the case of concept drifts [29, 30]. Among this research line, the technique described in [31] is applied to the aforementioned CAMNEP multiagent anomaly detector. However, although these model-updating techniques can certainly improve the accuracy of an existing detector such as the one described in this paper, they do not constitute a new *detection* mechanism *per sé*, while in this work we focused on multiagent algorithms for designing effective detection strategies.

6 Conclusions

In this work, we have analyzed an issue that occurs in virtually any multimodel, anomaly-based intrusion detection system: the detection errors caused when the outputs of each partial model are aggregated together to form a global evaluation, used to decide whether or not a certain event is an attack. We proposed to embed detection models into separate agents, in a multiagent system, and then to exploit cooperative negotiation to implement a more robust value-aggregation strategy. To test our approach, we modified a web anomaly detection system which used a simple weighted average, to use cooperative negotiation.

The results obtained by testing the original versus the modified system are promising. More precisely, our approach can improve the detection rate dramatically at parity of false positive rate. In addition, the detection quality is not influenced by new parameters we introduced, thus our approach does not require any further tuning effort and could be effortlessly applied to any learning-based anomaly detector that employs simpler aggregation approaches.

As future work, besides addressing the limitations discussed in Section 4.3, we plan to evaluate the performance overhead introduced by our approach more thoroughly, with particular attention to the time necessary to complete the negotiation phase and the comparison to other simpler, yet less formal, aggregation methods.

Acknowledgments

The authors are thankful to N. Basilico and F. Amigoni. This work has been partially supported by the European Commission through IST-216026-WOMBAT funded by the 7th FP. The opinions expressed in this paper are those of the authors and do not necessarily reflect the views of the European Commission.

References

1. Carr, J.: Inside Cyber Warfare: Mapping the Cyber Underworld. O'Reilly Media, Inc., Sebastopol (2009)
2. The SANS Institute: Zero-day vulnerability trends (September 2009), http://www.sans.org/top-cyber-security-risks/zero-day.php
3. Fishwick, P.A.: An integrated approach to system modeling using a synthesis of artificial intelligence, software engineering and simulation methodologies. ACM Trans. Model. Comput. Simul. 2(4), 307–330 (1992)
4. Fishwick, P.A., Zeigler, B.P.: A multimodel methodology for qualitative model engineering. ACM Trans. Model. Comput. Simul. 2(1), 52–81 (1992)
5. Denning, D.E.: An Intrusion-Detection Model. IEEE Transactions on Software Engineering 13(2), 222–232 (1987)
6. Lee, W., Stolfo, S.J.: A framework for constructing features and models for intrusion detection systems. ACM Transactions on Information and System Security 3(4), 227–261 (2000)
7. Kruegel, C., Toth, T., Kirda, E.: Service-Specific Anomaly Detection for Network Intrusion Detection. In: Proceedings of the Symposium on Applied Computing (SAC 2002), Spain (March 2002)
8. Kruegel, C., Mutz, D., Valeur, F., Vigna, G.: On the detection of anomalous system call arguments. In: Proceedings of the 2003 European Symp. on Research in Computer Security, Gjøvik, Norway (October 2003)
9. Kruegel, C., Robertson, W., Vigna, G.: A Multi-model Approach to the Detection of Web-based Attacks. Journal of Computer Networks 48(5), 717–738 (2005)
10. Mutz, D., Valeur, F., Kruegel, C., Vigna, G.: Anomalous System Call Detection. ACM Transactions on Information and System Security 9(1), 61–93 (2006)
11. Maggi, F., Matteucci, M., Zanero, S.: Detecting intrusions through system call sequence and argument analysis. IEEE Transactions on Dependable and Secure Computing 99(PrePrints) (2008)
12. Criscione, C., Maggi, F., Salvaneschi, G., Zanero, S.: Integrated detection of attacks against browsers, web applications and databases. In: European Conference on Computer Network Defence - EC2ND 2009 (2009)
13. Helmer, G., Wong, J.S.K., Honavar, V.G., Miller, L., Wang, Y.: Lightweight agents for intrusion detection. J. Syst. Softw. 67(2), 109–122 (2003)
14. Jennings, N.R.: An agent-based approach for building complex software systems. Commun. ACM 44(4), 35–41 (2001)
15. Spafford, E., Zamboni, D.: Intrusion detection using autonomous agents. Computer Networks 34(4), 547–570 (2000)
16. Ghosh, A., Sen, S.: Agent-based distributed intrusion alert system. In: Sen, A., Das, N., Das, S.K., Sinha, B.P. (eds.) IWDC 2004. LNCS, vol. 3326, pp. 240–251. Springer, Heidelberg (2004)

17. Dasgupta, D., Gonzalez, F., Yallapu, K., Gomez, J., Yarramsettii, R.: CIDS: An agent-based intrusion detection system. Computers & Security 24(5), 387–398 (2005)
18. Gowadia, V., Farkas, C., Valtorta, M.: PAID: A probabilistic agent-based intrusion detection system. Computers & Security 24(7), 529–545 (2005)
19. Rehak, M., Pechoucek, M., Celeda, P., Novotny, J., Minarik, P.: Camnep: agent-based network intrusion detection system. In: AAMAS 2008: Proceedings of the 7th International Joint Conference on Autonomous Agents and Multiagent Systems, Richland, SC, International Foundation for Autonomous Agents and Multiagent Systems, pp. 133–136 (2008)
20. Allan, R.J.: Survey of agent based modelling and simulation tools. Technical report, STFC Daresbury Laboratory, Daresbury, Warrington WA4 4AD (May 2010)
21. Amigoni, F., Basilico, F., Basilico, N., Zanero, S.: Integrating partial models of network normality via cooperative negotiation: An approach to development of multiagent intrusion detection systems. In: WI-IAT 2008, Washington, DC, USA, pp. 531–537. IEEE Computer Society, Los Alamitos (2008)
22. Lippmann, R., Haines, J.W., Fried, D.J., Korba, J., Das, K.: The 1999 DARPA off-line intrusion detection evaluation. Comput. Networks 34(4), 579–595 (2000)
23. Song, Y., Stolfo, S., Keromytis, A.: Spectrogram: A Mixture-of-Markov-Chains Model for Anomaly Detection in Web Traffic. In: Proc. of the 16th Annual Network and Distributed System Security Symposium, NDSS (2009)
24. Kittler, J., Hatef, M., Duin, R.P., Matas, J.: On combining classifiers. IEEE Transactions on Pattern Analysis and Machine Intelligence 20, 226–239 (1998)
25. Amigoni, F., Gatti, N.: A formal framework for connective stability of highly decentralized cooperative negotiations. Autonomous Agents and Multi-Agent Systems 15(3), 253–279 (2007)
26. Robert Hansen (RSnake): XSS (Cross Site Scripting) Cheat Sheet (June 2009), http://ha.ckers.org/xss.html
27. Robert Hansen (RSnake): SQL Injection cheat sheet (June 2009), http://ha.ckers.org/sqlinjection/
28. Chandola, V., Banerjee, A., Kumar, V.: Anomaly detection: A survey. ACM Computing Surveys (CSUR) 41(3), 15 (2009)
29. Cretu-Ciocarlie, G.F., Stavrou, A., Locasto, M.E., Stolfo, S.J.: Adaptive anomaly detection via self-calibration and dynamic updating. In: RAID 2009: Proceedings of the 12th International Symposium on Recent Advances in Intrusion Detection, pp. 41–60. Springer, Heidelberg (2009)
30. Maggi, F., Robertson, W., Kruegel, C., Vigna, G.: Protecting a moving target: Addressing web application concept drift. In: RAID 2009: Proceedings of the 12th International Symposium on Recent Advances in Intrusion Detection, pp. 21–40. Springer, Heidelberg (2009)
31. Rehák, M., Staab, E., Fusenig, V., Pěchouček, M., Grill, M., Stiborek, J., Bartoš, K., Engel, T.: Runtime monitoring and dynamic reconfiguration for intrusion detection systems. In: RAID 2009: Proceedings of the 12th International Symposium on Recent Advances in Intrusion Detection, pp. 61–80. Springer, Heidelberg (2009)

The Password Game: Negative Externalities from Weak Password Practices

Sören Preibusch and Joseph Bonneau

University of Cambridge, Computer Laboratory
Cambridge CB3 0FD, UK
{sdp36,jcb82}@cl.cam.ac.uk

Abstract. The combination of username and password is widely used as a human authentication mechanism on the Web. Despite this universal adoption and despite their long tradition, password schemes exhibit a high number of security flaws which jeopardise the confidentiality and integrity of personal information. As Web users tend to reuse the same password for several sites, security negligence at any one site introduces a negative externality into the entire password ecosystem. We analyse this market inefficiency as the equilibrium between password deployment strategies at security-concerned Web sites and indifferent Web sites.

The game-theoretic prediction is challenged by an empirical analysis. By a manual inspection of 150 public Web sites that offer free yet password-protected sign-up, complemented by an automated sampling of 2184 Web sites, we demonstrate that observed password practices follow the theory: Web sites that have little incentive to invest in security are indeed found to have weaker password schemes, thereby facilitating the compromise of other sites. We use the theoretical model to explore which technical and regulatory approaches could eliminate the empirically detected inefficiency in the market for password protection.

1   Password Practices on the Web

Computer systems have traditionally authenticated users by prompting for a username and textual password. This practice has proliferated on the Web as users now create new login credentials on a monthly basis. More than 90% of the 100 most visited destinations on the Web and more than 80% of the top 1000 Web sites collect passwords from their users. Well-known weaknesses of password schemes were inherited by the Web, including weak user-chosen passwords and password-reuse across domains. These problems were exacerbated by self-enrolment on the Web and the proliferation of password accounts per user (estimated to be over 25 [9]).

Numerous password enhancement schemes have been proposed, including improved cryptographic protocols and alternative means for creating and entering passwords such as graphical passwords. The merit of such alternative authentication schemes remains contested as users continue to pick easily guessable

T. Alpcan, L. Buttyan, and J. Baras (Eds.): GameSec 2010, LNCS 6442, pp. 192–207, 2010.
© Springer-Verlag Berlin Heidelberg 2010

secrets and fall for simple social engineering attacks. Automatic management tools beyond simple browser password caches have seen limited deployment [14].

In parallel, with the advent of global social networks as new identity providers, single sign-on schemes (SSO) have gained new interest after previous attempts such as Microsoft Passport have failed. Service providers have started to accept Facebook Connect for general-purpose Web authentication [2], whilst the non-proprietary OpenID [13] has gained an enthusiastic following but limited deployment.

In an earlier investigation, we set out to quantitatively assess the current state of password implementations on the Web [5]. In the first large-scale survey of current password practices, we surveyed 150 sites equally sampled from mainstream and less popular Web sites. We concluded that a large number of technical issues remain on the Web, with almost all sites studied demonstrating some security weaknesses and many sites implementing unique and inconsistent password security policies.

In this work, we present a fresh game-theoretic analysis into the economics behind these technical failures. Building on our previous identification of two key negative externalities in the password market, we model firms' decisions to invest in password security as a simple game between security-interested Web sites and security-indifferent Web sites. This partition is supported by the existing password landscape. We demonstrate that the empirical evidence about inefficient password practices matches the theoretical prediction from the game-theoretic model. We affirm the validity of our empirical evidence by analysing a second sample of public Web sites which is larger by an order of magnitude. We conclude that market failures provide a pertinent economic explanation as to why password security weaknesses are so commonly observed.

2   Incentives for Password Deployment

From a service provider's perspective, passwords enable authentication by ensuring that a specific identity can only be used by parties knowing the correct password. Passwords may also sustain real-world authorisation decisions if a username/password combination is tied to a pre-established offline identity, as in the case of online banking or corporate email facilities. Merchant Web sites, which often enable users to store payment details for future shopping, have direct incentives to secure password authentication, as they can be held liable for any purchases made using stolen password credentials.

In other sites, interaction with other users of the same site (e.g., social networking) or beyond (e.g., webmail) introduces the risk of harmful actions if authentication fails, such as fraud, spam, or blackmail which may damage the reputation of the operator of the Web site as well as harm users. Passwords are intended to make users accountable for their actions and eliminate plausible deniability. In some European jurisdictions at least, operators of interactive sites can be held responsible for their users' actions if they fail to take appropriate counter-measures. Banning users, locking accounts, and keeping evidence from electronic discovery for court trials would become impossible if access were not restricted. Some operators therefore have an intrinsic motivation to secure their Web sites to protect their own assets, notwithstanding the negative business impact which careless practices can bring.[1]

These incentives to secure password authentication do not hold for all Web sites using passwords, such as free news Web sites, which use passwords to provide access to a stored identity for customisation. From a security point of view, it is not clear why a service that is offered for free requires authentication. As long as users can create an unlimited number of free accounts, password authentication seems dispensable. It places no restrictions on the user other than the hassle of account creation—which is a rather weak throttling mechanism. The motivation for password deployment may be revenue rather than security. In combination with usernames, passwords act as persistent identifiers; they increase the accuracy of behavioural profiling. Password collection also initiates account creation during which further personal information is collected, including contact information such email address and socio-demographic details. This data may be used directly for targeted advertising or indirectly when aggregated to construct an audience profile, based on which advertising businesses book slots on the Web site. Password-protected accounts facilitate the extraction and monetarisation of customer data even when they have dubious security value.

Regardless of the motivation, there are costs associated with deploying password authentication which rise with the sophistication of password security mechanisms. These include at least the need for skilled programmers to implement and maintain the system and operating costs of password storage servers. Passwords also add communication overhead, which can be exacerbated by the need to transmitted encrypted data.

3   Negative Externalities from Password Deployment

Password deployment by one company generates negative externalities in two ways: first, consumers are asked to remember yet another password; second due to password reuse each password server is an additional failure point for the entire system.

3.1   Too Many Passwords: Tragedy of the Commons

Given the low adoption of password management tools [14], mental storage capacity has the characteristics of a common good to password-collecting companies. A common good is a good from which no consumer can be excluded, as

---

[1] For example, a well-publicised password database compromise at the social gaming Web site RockYou brought both a storm of negative publicity and pressure from Facebook, a key business partner [15].

consumption is not or cannot be regulated and there is no cost for consuming the good. Unlike public goods, however, the quality of a common good decreases as more people consume it. There is an incentive to overuse the common good to the detriment of society as a whole, as one is not held accountable for the deterioration of the good's value. The resulting inefficiency is described as the "tragedy of the commons" in reference to pastures which were often over-grazed by the local community of farmers. Any given farmer could fully appropriate the benefits from sending another sheep to the commons, but the negative effects as the pasture depleted affected the entire community equally.

Humans have limited physiological capacity to remember random data. Passwords consume this capacity and as the total number of passwords remembered grows, the risk of forgetting typically increases. Indeed, the majority of surveyed users indicate that they do not use stronger passwords because they simply have too many to remember [11]. Yet companies can continue to ask consumers to register new passwords at no expense. Mental password storage capacity, therefore, exhibits characteristics of a common good: access to it is neither limited nor priced; with increased usage, its quality in terms of recall decreases. To cope with this password overload and maintain an acceptably low rate of forgotten passwords, consumers must lower their overall password quality. With per-password strength requirements in place, password reuse is a common coping strategy to lower the burden of remembering strong passwords.

3.2 Password Reuse: Negative Externalities of Insecurity

Password reuse enables each site to introduce a further negative externality on the password market. As companies have differing incentives to invest in password security, the resulting security of their password implementations inevitably differs. As the vast majority of consumers reuse the same password across multiple sites [10, 9], passwords for high-security accounts such as banking and email are often held by less secure Web sites as well. The accumulation of high-security login credentials at Web sites with weaker incentives for security presents an attractive attack strategy: compromise a low-security site and attempt to use the login credentials at higher security sites. The feasibility of this attack has been strengthened by the trend towards all Web sites using email addresses as user identifiers, as observed at 87% of sites in our recent survey [5].

Indeed, within the past year strong evidence has emerged that this attack is not only feasible but is occurring in the wild. The January 2010 compromise of social gaming Web site RockYou! leaked over 32 million email/password pairs of which over 10% were claimed to be directly usable at PayPal [15]. The reuse of passwords from compromises at low-security sites is acknowledged as a common means of attack in documentation at both Windows Live [3] and Yahoo! [4]. The strongest evidence comes from Twitter, which in January 2010 forced millions of users to reset their passwords after detecting a large scale attack using a stolen list of credentials from a torrenting Web site [12].

Web sites with weak password security can thus create a real negative externality for more secure Web sites. As operators of low-security site do not bear the social costs of weak password practices, there is a tendency for under-investment in good password practices.

4   Password Implementers: Game-Theoretical Model

To understand the net incentive for companies to deploy password schemes and to invest into their security, we understand the Web as an arena where a password requirements game is played. We model password design choices as actions in a game with two players: one Web site operator with an intrinsic incentive to invest in security, the other without such an incentive, as detailed in Section 2. In the following, these two players will be labelled the security-sensitive Web site ($\mathcal{W}_s$) and the security-indifferent Web site ($\mathcal{W}_i$), respectively. The firms' payoffs in this game reflect how their own decisions to enforce passwords as well as those of the other firm affect their performance, risk exposure, and implementation costs. The resulting game-theoretic model will also serve to analyse the efficiency of the overall password practices on the Web.

4.1   Action Spaces and Timing

The action each Web site can take is to specify the set S of permissible passwords. When first using the site's service, users will need to create a password from within the set of permissible passwords. For simplicity, we assume there are weak passwords (wp) and strong passwords (sp). Although in reality password strength is a spectrum and little consensus exists as to how exactly two medium-strength passwords should be ranked in terms of security, the extremes can be identified quite easily.[2] Web sites $\mathcal{W}_s$ and $\mathcal{W}_i$ thus choose their action from the powerset of $\{wp,sp\}$: only strong passwords ($\{sp\}$), only weak passwords ($\{wp\}$), any password ($\{wp,sp\}$) or no password protection at all ($\varnothing$).

For simplicity, the Web sites choose their password schemes simultaneously. Still, the following analysis and conclusions would not differ much if the password requirements game is understood as a market entry game, where a newly created Web site and an incumbent Web site have to respond to the other player's past or pre-empted decisions. The timing of the game effectively determines how users can reuse their passwords, since only the password created with incumbent at an earlier point in time could be reused with the entrant at a later point in time. This directedness of the externality effect is removed in a simultaneous game. We believe sequential moves are contrary to the observation that, first, users' ability to change their passwords removes the relevance of timing, and second, the effects of password leakage do not depend on whether it was used originally or reused at the leaking site.

---

[2] Conflicting estimates of expected password strength given a set of requirements are found in NIST's Electronic Authentication Guideline [7] and the BSI IT-Grundschutz Methodology Catalogues [6], although both agree that phrases such as `pass` or `password` are weak.

4.2   Payoffs

The pay-offs in the password requirements game are determined by the summed positive and negative effects of password implementations, including interaction effects. Quantification of each of these seven effects is difficult. The rationale is as follows, expressed in terms of costs and benefits; a summary is provided in Table 1.

protection benefit. The security-concerned player receives a strictly greater payoff from requiring some password than from requiring no password at all. For a security-concerned Web site, the protection effect mitigates liability and convinces users they interact securely with the site.

data collection benefit. For a security-indifferent Web site, requiring any form of user accounts triggers the inflow of personal information that can be re-purposed, as detailed in Section 2.

fortification benefit. Following a similar argument, the payoff for $\mathcal{W}_s$ decreases if weak passwords are accepted ($S_s = \{wp\}$ or $S_s = \{wp, sp\}$), since these provide less protection against fraudulent access. The benefit from strengthening password protection applies to $\mathcal{W}_s$ only.

development costs. Any password scheme will require resources to be developed and/or deployed. Password implementation costs are nil if no passwords are collected, they are low if passwords are collected but no restrictions are enforced.

strength assessment costs. More restrictive password schemes require technical skill; higher implementation costs make payoffs decrease. Implementation costs do not differ in the required strength (only weak versus only strong), since only allowing strong passwords is equivalent to filtering out weak passwords.

sharpness benefit. If $\mathcal{W}_s$ is the only player who enforces passwords (isolated deployment), it can make people care more about passwords and there is no obtunding effect. The sharpness effect eliminates the tragedy of the commons.

negative externality costs. The payoff for $\mathcal{W}_s$ is severely affected if passwords created with $\mathcal{W}_i$ are also permissible at $\mathcal{W}_s$ and vice versa, as detailed in Section 3.2. We note that the positive sharpness effect will not be observed together with the negative externality.

The negative externality effect on payoffs captures the interaction of password practices: an indifferent company $\mathcal{W}_i$ as described above will take inadequate measures to protect user account details, because the site has no incentive to do so. The interest an indifferent site has in the data is mainly their suitability for advertising or resale. Leaking that data through security holes corresponds to giving the data away for free, which would still not be a direct problem for the indifferent Web site. Subject to compatible password requirements, users may reuse passwords created with $\mathcal{W}_i$ for $\mathcal{W}_s$ ($S_i \subseteq S_s$) or vice versa ($S_s \subseteq S_i$). The negative externality of bi-directional reuse thus dissipates when sets of permissible passwords overlap ($S_i \cap S_s \neq \varnothing$).

Table 1. Summary of costs (−) and benefits (+) summing up to the payoffs in the password game. Summands negative in value are underlined. The last two columns enumerate the player's own and the opponent's strategies for which the respective effect applies.

payoff component		applies to	if own strategy	if opponent's strategy
$\underline{d}$	development −	$\mathcal{W}_s, \mathcal{W}_i$	$\{wp\}, \{sp\}, \{wp,sp\}$	any
$\underline{s}$	strength assessment −	$\mathcal{W}_s, \mathcal{W}_i$	$\{wp\}, \{sp\}$	any
c	data collection +	$\mathcal{W}_i$	$\{wp\}, \{sp\}, \{wp,sp\}$	any
p	protection +	$\mathcal{W}_s$	$\{wp\}, \{sp\}, \{wp,sp\}$	any
f	fortification +	$\mathcal{W}_s$	$\{sp\}$	any
i	sharpness +	$\mathcal{W}_s$	$\{wp\}, \{sp\}, \{wp,sp\}$	$\{\}$
$\underline{r}$	neg. externality −	$\mathcal{W}_s$	$\{wp\}, \{sp\}, \{wp,sp\}$	any overlapping

4.3   Security-Indifferent Web Sites $\mathcal{W}_i$

For the security-indifferent Web site, player $\mathcal{W}_i$, strategies $\{wp\}$ and $\{sp\}$ are dominated by $\{wp,sp\}$, because $d + c - |s| \leqslant d + c$. A security-indifferent Web site either implements no password scheme at all ($S_i = \{\}$), or a scheme that accepts weak and strong passwords alike ($S_i = \{wp, sp\}$). The latter will be beneficial if the opportunity costs of missing out on collecting personal information from Web users prevail over deployment costs of a password scheme.

Ignoring future maintenance, password deployment can be thought of as an investment for $\mathcal{W}_i$: initial costs of d, result in accruing inflow of personal information. This data needs to be monetised to amortise the investment. c would therefore be the expected net present value from turning collected data into profits.

Deploying no password scheme will become the profit-maximising strategy if, first, $\mathcal{W}_i$ is unable to monetise the personal information it collects through own or third-party use, or second, if personal information can be acquired without requiring passwords.

4.4   Security-Concerned Web Sites $\mathcal{W}_s$

Given that $\mathcal{W}_i$ will play $\{\}$ or $\{wp,sp\}$, the security-concerned Web site $\mathcal{W}_s$ will always prefer $\{sp\}$ over $\{wp\}$ as it provides extra fortification f. This preference would only be affected by the relative impact of r, the negative externality from password reuse, which does not make a difference except when $\mathcal{W}_i$ would play $\{sp\}$—a dominated strategy. Quite intuitively, allowing weak passwords only is a dominated strategy for Web sites sensitive to securing their systems.

As its opponent plays neither $\{sp\}$ nor $\{wp\}$, the difference in payoff for $\mathcal{W}_s$ between $\{sp\}$ and $\{wp,sp\}$ lies in $s + f$, that is the extra profit from strong passwords versus the costs to identify and to enforce them. An argument could even be made that this sum actually is a negligible quantity: if the effort in identifying strong passwords is expertise rather than the technical implementation, the difficulty lies in determining what non-trivial characteristics a strong password should have. Then, if f materialises in the form of reduced liability for the Web site when strong passwords are used, this is a result from a regulation prescribing "strong passwords" which would also hint at how to make technical systems compatible with the strength requirements.

4.5   Equilibria

In the light of the foregoing argument, two kinds of market equilibria are expected. First, the security-indifferent Web site accepts any password, and the security-concerned Web site requires strong passwords and potentially allows weak passwords as well, depending on the relative importance of the fortification effect compared to costs of assessing and enforcing password strength. This group of equilibria is inefficient due to negative externalities from password reuse. The second group of equilibria is reached if the security-indifferent Web site renounces password collection. Again, the concerned Web site may enforce strong passwords or accept weak and strong passwords. Collectively, this group of equilibria is socially better than the first (Pareto-superior): because $S_i = \{\}$ instead of $S_i = \{wp, sp\}$ requires $c + d \leqslant 0$ by definition, and in addition it holds that $i \geqslant 0 \geqslant r$, the social costs are lower if no password scheme is put in place at indifferent Web sites.

We now assume the likely case that benefits of protection are high for the security-concerned Web site (p is very high and f outweighs s). We also assume the indifferent Web site is similarly keen on collecting passwords ($c+d$ is positive). The password game then exhibits a single Nash equilibrium. This equilibrium is in dominant strategies, $(S_s, S_i) = (\{sp\}, \{wp, sp\})$. There is a unique combination of strategies that maximises social welfare, the sum of payoffs: $(\{sp\}, \{wp\})$; the payoff for the security-sensitive site is maximised in $(\{sp\}, \{\})$.

The Nash equilibrium is found, when the security-sensitive site enforces strong passwords, and also the indifferent site requires passwords but any password—weak or strong—will do there. In the Nash equilibrium, W_i will receive a high payoff, since personal information (which translates to a revenue stream for the indifferent site) is collected, but little development effort is necessary. As the indifferent site will accept any password, weak or strong ($\{wp, sp\}$), practically no validity checking needs to be performed on the user-chosen passwords, eliminating the development costs s. The security-sensitive firm suffers from the negative externalities r this behaviour dissipates: due to password reuse, passwords from the security-sensitive site spill over to the indifferent site, where they are then subject to inadequate protection.

The security-sensitive site would get a maximum payoff if no indifferent site implemented any passwords ($S_i = \{\}$). Both the fortification effect and absence

Table 2. Algebraic expression of payoffs for the row player ($\mathcal{W}_s$, above the line) and the column-player ($\mathcal{W}_i$, below the line). Letters denote the cost/benefit effects from Table 1.

$\frac{\mathcal{W}_s}{\mathcal{W}_i}$	{}	{wp}	{sp}	{wp,sp}
{}	$\frac{0}{0}$	$\frac{0}{c+\underline{d}+\underline{s}}$	$\frac{0}{c+\underline{d}+\underline{s}}$	$\frac{0}{c+\underline{d}}$
{wp}	$\frac{p+\underline{d}+\underline{s}+i}{0}$	$\frac{p+\underline{d}+\underline{s}+r}{c+\underline{d}+\underline{s}}$	$\frac{p+\underline{d}+\underline{s}}{c+\underline{d}+\underline{s}}$	$\frac{p+\underline{d}+\underline{s}+r}{c+\underline{d}}$
{sp}	$\frac{p+\underline{d}+\underline{s}+f+i}{0}$	$\frac{p+\underline{d}+\underline{s}+f}{c+\underline{d}+\underline{s}}$	$\frac{p+\underline{d}+\underline{s}+f+r}{c+\underline{d}+\underline{s}}$	$\frac{p+\underline{d}+\underline{s}+f+r}{c+\underline{d}}$
{wp,sp}	$\frac{p+\underline{d}+i}{0}$	$\frac{p+\underline{d}+r}{c+\underline{d}+\underline{s}}$	$\frac{p+\underline{d}+r}{c+\underline{d}+\underline{s}}$	$\frac{p+\underline{d}+r}{c+\underline{d}}$

of negative externalities make this market outcome attractive for $\mathcal{W}_s$. Its only costs are the implementation effort for making only strong passwords acceptable. However, an indifferent site will not renounce password schemes since they act as a means to collect email addresses, for instance, and thereby realise positive payoffs c.

Comparing the Nash equilibrium with the social optimum is instructive. Both are Pareto-optimal; not both companies can improve their payoffs at once, compared to these strategy combinations. Maximum differentiation in the password requirements space maximise social welfare: the indifferent firm accepts only weak passwords and the security-sensitive firm accepts only strong passwords.[3] The reason this equilibrium is not realised is the disincentive for the indifferent site to invest in restricting permissible passwords. The combined payoff in the social optimum is higher by $|s - r|$ than in the Nash equilibrium.

Interestingly, if the security-sensitive site would subsidise the indifferent site to help it in developing desirable password schemes which only accept weak passwords, the social optimum could be achieved. $\mathcal{W}_s$ would need to transfer an amount between $|r|$ and $|s|$ to $\mathcal{W}_i$. Alternatively, $\mathcal{W}_s$ could lower the development costs for $\mathcal{W}_i$, to make this site prefer $S_i = \{wp\}$ over $S_i = \{wp, sp\}$. Whilst this approach is equivalent to a subsidisation, it provides a technical rather than a transfer solution to the market failure. Indeed, the security-concerned Web site has already invested s in a password-strength assessment technology. These alternatives for regulation are discussed in greater detail in Section 6.

---

[3] The reversed maximum differentiation yields a lower overall payoff since the fortification effect f does not apply to $\mathcal{W}_i$.

5   Password Implementers: Empirical Evidence

Analysis of the password practices at 150 public Web sites in the areas of electronic commerce, identity services such as emailing and social networking, and content services such as news, reveals market-wide technical and organisational shortcomings [5]. While problems are widespread, there is evidence that weak password practices are more common in the news industry and lower-tier Web sites. This observed dichotomy is confirmed in a larger sample of 2184 automatically analysed public Web sites.

5.1   Notes on the Datasets

Password practices in the wild are quantitatively assessed using two datasets, called the "Password Thicket" dataset and the "BugMeNot / Alexa" dataset respectively.

The *Password Thicket* dataset comprises 150 manually surveyed Web sites, equally sampled from mainstream and less popular Web sites [5]. The sample is organised by industry into three equally sized groups: identity sites (which use passwords to protect a user's identity for interacting with other users, notably webmail and social networking), electronic commerce sites (designed for purchasing goods with no interaction with other users), as well as news and content sites (using passwords to customise site layout or limiting access to account holders only). Our methodology to assess the quality of password implementations was based on manual inspection of the security measures during enrolment (account creation), login, password reset, and password recovery. The details of our experimental setup and our main technical findings are reported elsewhere [5]. This dataset is characterised by the depth of investigation for each Web site in the sample.

The "BugMeNot / Alexa" (*BMN*) dataset comprises 2184 Web sites listed on www.bugmenot.com. BugMeNot is a major credential-swapping site "created as a mechanism to quickly bypass the login of Web sites that require compulsory registration and/or the collection of personal/demographic information", a practice considered a "pointless" exercise [1]. Users of BugMeNot can upload new username/password combinations for a given Web site. Fellow visitors of BugMeNot can retrieve these credentials for free. Web site operators can request to be removed from BugMeNot, and it is possible to check which sites have taken this step. BugMeNot explicitly bans pay-per-view sites, community sites for interaction amongst members, sites with a fraud risk due to banking/commerce details stored with the user accounts. When browsing the BugMeNot repository, sites which were blocked can be told apart from sites for which no passwords have been submitted yet.

We crawled BugMeNot in a two-step process. First, the 3172 most popular Web sites were retrieved from Alexa Top Sites, an Amazon Web service. Popularity is measured by the proprietary Alexa rank to which traffic rank is a paramount ingredient. Popularity in the USA is used rather than the international ranking to avoid having localised versions of the same site showing up twice (e.g., google.de on position 15 and google.com.hk on position 16). For each Web site in this top list, the Alexa Web Information Service was queried for the date the site went online, for the median load time, for whether it contains adult content, and for the first three categories this site pertains to.

A Web site's category correspond to the classification in the Open Directory Project. These categories are used for coarse, keyword-based binary identification of news/weather/magazine Web sites. Whilst classification was automatic, one of the authors and another non-expert but skilled rater also classified a random subset of 388 of the sites (12%). These raters agreed on the classification at an intraclass correlation coefficient 0.64 (model 2, single measure). Although agreement with the automated classification was at an ICC of 0.49 and 0.42 respectively, the programmatic rating was judged acceptable, since inspection of diverging ratings indicated that manual classification resulted in a broader set of news sites (i.e. the automated rating was typically stricter). In summary, 274 sites are classified as news sites, 2882 as non-news sites; for 16 sites, the category information was missing and no classification was performed.

In a second step, each of the top Web sites was matched mechanically against BugMeNot. The listing status was recorded as one of 'ok' (there are accounts listed for this sites), 'blocked' (the Web site is barred from BugMeNot), or 'missing' (the Web site is not listed on BugMeNot). Missing Web sites may simply have never been registered by the community and never been blocked by the site operators, or may not collect passwords at all. For the sites in our survey, the majority of missing Web sites appear to not collect passwords. In total, 988 of the top sites were not listed in BugMeNot; the remaining 2184 sites making up the *BMN* dataset are divided into 531 'blocked' Web sites and 1653 'ok' Web sites. In summary, the *BMN* dataset is characterised by the breadth of investigation and focuses on the most popular destinations on the Web.

5.2   Practices at Content Sites

Content sites are the prototypical example of security-indifferent Web sites that use passwords as a trigger for harvesting profile information (Section 2). They are significantly more likely to collect personal information at the time when a new password is created. All but 2% collect email addresses on a mandatory basis and they are more likely to verify these with very high significance.

Password carelessness is more prevalent at news sites with high significance. This is manifested in very significantly lower adoption of TLS to protect passwords and sending the cleartext passwords in a welcome email or "forgotten password" responses. Also, news sites place fewer limits on guessing passwords, so they could be abused as password oracles in brute-force attacks against more security-sensitive sites. Users of news sites are rarely given password advice or hints on how to make a password more secure, for instance by including digits or via a graphical password strength indicator.

5.3   Practices at Sites with Merchant Facilities

Merchant facilities are not exclusive to e-commerce sites but may be used at other Web sites as well. Sites which store users' payment details perform significantly better on overall password security and in several key measures, including TLS deployment and notification to users about password reset events. Although the lack of encrypted data transmission is widespread, the dichotomy between content sites and e-commerce sites is very strongly significant and TLS deployment is strongly correlated with merchant facilities. It holds that sites which process and store fraud-prone details such as payment information take more care in handling passwords securely and offer more advanced security features overall.

Critically, we observe that sites with merchant facilities are also significantly more likely to impose minimum password lengths and to blacklist common passwords. If sites with merchant facilities are interpreted to represent the class of security-interested Web sites, this directly supports the predictions of the password game.

5.4   Prevention of Credential Sharing

Web site operators' eagerness to stop the sharing of credentials amongst their users can be interpreted as an indication of whether or not password compromise is deemed a serious risk. In merging BugMeNot data with Alexa traffic rank data, we observe that very-high traffic Web sites are much more likely to block listing of their credential of BugMeNot (Fig. 1; 28% blocked in upper half versus 20% in lower half, highly significant).

After the top 50 Web sites, there is a small and gradual decrease in the rate of blocking. However, blocking is still quite common among the lower ranked sites, at above 25% in the band of the 500th to 1000th most popular sites, and 17% around rank 3000. Password collection remains at a high level of above 80% through the entire range up to the 1000th rank and is still at 69% for up to the 3000th rank. Thus, accross a range of Web sites BugMeNot blocking is a useful indicator of Web sites's real security motivations for collecting passwords.

We see a pronounced trend towards news Web sites blocking BugMeNot sharing significantly less often. We observed 18% of non-news sites to block BugMeNot sharing, while only 8% of news Web sites did so. If we restrict ourselves to only Web sites which exist in BugMeNot's database the divide is even stronger, with 26% of non-news Web sites blocking BugMeNot sharing and only 9% of news Web sites doing so. Both results are highly statistically significant ($p < 0.0001$ using a two-tailed G-test). Thus, the BugMeNot data provides strong evidence that news Web sites do have lower security concerns than other password-collecting Web sites.

In summary, a Web site's password practices improve with popularity, industry-specific leadership, and growth. Password deployments are prevalent across popularity strata, but the most popular sites are also most likely to exhibit password care in taking measures to prevent password sharing amongst users.

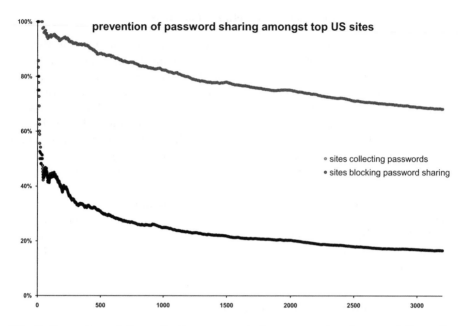

Fig. 1. Proportion of sites collecting passwords (upper curve) and amongst these, the proportion of sites blocking password sharing (lower curve). Ratios given for top k US sites with k up to 3200. Bumps are artefacts of the increasing window size for the arithmetic mean.

5.5 Limitations

Despite these consistent results, we acknowledge there are limitations inherent to our empirical and theoretical methodologies.

First, the game-theoretic model may not have captured all intricacies of the players' payoffs. A notable simplification is the symmetry in the cost structures. It seems convincing that security-concerned Web sites exhibit a structural difference compared to other Web sites which would lower their costs for developing standard or sophisticated password schemes. Yet, the current model would then be an underestimation of the resulting inefficiency on the market. A more fundamental question seems to be whether password deployment decisions can be modelled as a game which has the underlying assumption of rational players.

Second, insofar as we base our analysis of data retrieved from Alexa and BugMeNot, we are limited by our resources in fully ascertaining their accuracy. All popularity ranks are based on Web usage behaviour from a sample of users who are not representative of the entire online population. We nonetheless deem the rankings reliable. Further, the BugMeNot database suffers from quality problems inherent to a crowd-sourced endeavour. Although we did not systematically probe the accuracy of the BugMeNot repository, we have tried to reduce recording errors by a full manual inspection of suspicious cases.

6   Observed Market Inefficiency and Pathways for Regulation

The ubiquity of password schemes on the Web contrasts with the variety of security flaws found in implementations across industries and Web site popularity levels. Failure to provide adequate password security cannot be attributed to technical problems alone; mis-aligned incentives on the market explain why some groups of Web sites score significantly better on password security.

In the early 1960s, the former US security advisor McGeorge Bundy noted: "if we guard our toothbrushes and diamonds with equal zeal, we will lose fewer toothbrushes and more diamonds". In 2010, the Windows Live help pages echo: "Access to your financials and email account is a critical thing; therefore your passwords for these accounts should be un-guessable, even by a computer. If it is to post a response on a gamer forum, perhaps it doesn't need to be so complex." [3]

We have presented empirical evidence that password-equipped Web sites fall in two broad categories: on the one hand, there are those sites which have a self-interest to invest in security, on the other hand, some sites only use passwords as a trigger to collect personal information from their users for secondary purposes. Password carelessness is significantly more prevalent amongst the latter and encompasses phenomena such as sending out passwords in cleartext, allowing unlimited password guessing, lack of secure data transmission, and failure to fight sharing of credentials.

The observed lack in technical safeguards can be explained by a lack of business interest to invest in password security. At content sites, passwords do not serve a genuine authentication purpose but rather vest information profile creation with credibility. The resulting dichotomy in the market corresponds to the prediction of a game-theoretical model where a security-indifferent and a security-sensitive Web site operators make password deployment choices. Their differing motivations lead to differing optimal choices as to what combination of weak and strong passwords should be accepted, bearing in mind the required implementation efforts as well as increased levels of protection.

The password game exhibits a unique Nash equilibrium where security-indifferent Web sites will accept any password regardless of its security but security-sensitive Web sites allow strong passwords only. This outcome can also be observed empirically on the Web. It is not socially optimal, since the overlap in permissible passwords creates room for password reuse. Overuse of consumers' mental ability to remember passwords is similar to the tragedy of the commons and makes consumers use the same passwords across sites, which differ in security practices. Careless handling of strong passwords at a security-indifferent site dissipate a negative externality on security-sensitive Web sites, where leaked or fraudulently acquired credentials can be reused by an attacker. As a result, the market allocation of password strength is inefficient.

Due to the negative externalities from password reuse, Web sites for which security is critical find themselves exposed to threats created by not accounted for by security-indifferent sites. This calls for regulation. Effectively controlling password deployments and their strength by a global, supra-national Internet authority remains unrealistic. Certification could become mandatory for new deployments, which are then watermarked so that any alterations break the certificate. Obviously, exactly those Web sites currently exhibiting weak password practices would have little incentive to participate.

Password deployment needs to be priced to internalise the externalities on other password implementers and consumers. Pricing need not be direct, although a pay-per-stored-password is the most direct solution and could be realised through charged-for database storage as a service (similar to the emailing facility offered by Amazon Web services for instance). Requiring yearly, printed data statements sent to users would involve costs that reduce the incentive to collect passwords without genuine authentication purposes [8]. These approaches could be complemented and supported by targeted legislation. A purely legal approach may increase the costs of leaking passwords by making the weaker-security Web site liable for account breaches at higher-security Web sites. Regulation could further reduce the ability for Web sites to monetise personal information collected via password schemes and thus make their deployment less attractive. Security-sensitive sites could also be given the opportunity to sue security-indifferent Web sites for unfair competition, creating a dynamic similar to the one that made German Web sites observe the requirement of an imprint with contact details. Abiding by a password scheme would not preclude content Web sites to tailor their services on a per-account basis; federated identities such as OpenID provide a viable and mutually beneficial alternative [13]. Insofar as password collection is merely a trigger to ask users for personal details, programmatic access to people repositories such as Facebook Connect makes this anchor redundant and may reduce the proliferation of password collection.

On the technical side, password kits could lower the costs for password-implementers to secure their systems. For security-indifferent Web sites, it should be cheaper to use an out-of-the-box password solution with known, weak security than implementing its own, arbitrarily secure system. Security-sensitive Web sites would have an incentive to sponsor the development and provision of such support tools, as demonstrated in Section 4. They would also have the beneficial side-effect of unifying the password experience for consumers and could potentially be branded. Open questions as to how such a password kit should look precisely and what password standards should be incorporated are important research challenges. Our limited consensus of what parameters constitute an appropriate password scheme calls for further, large-scale experimental research.

Acknowledgements

Katarzyna Krol provided valuable help in data collection and enhancement.

References

1. BugMeNot (February 2010)
2. Facebook Connect (2010), http://www.facebook.com/advertising/?connect
3. Windows Live Solution Center: Creating a strong password for your e-mail account (September 2010), http://windowslivehelp.com/solution.aspx?solutionid=3ca67154-2ee7-4da4-%8b95-f8aef17a71bc
4. Yahoo! Password Help (September 2010), http://help.yahoo.com/l/us/yahoo/abuse/password/faq.html
5. Bonneau, J., Preibusch, S.: The password thicket: technical and market failures in human authentication on the web. In: The Ninth Workshop on the Economics of Information Security, WEIS 2010 (2010)
6. Bundesamt für Sicherheit in der Informationstechnik (BSI) (Federal Office for Information Security). IT-Grundschutz Catalogues (2005)
7. Burr, W.E., Dodson, D.F., Timothy Polk, W.: Electronic Authentication Guideline. NIST Special Publication 800-63 (April 2006)
8. Chaos Computer Club (CCC). Datenbrief (January 2010), http://www.ccc.de/datenbrief
9. Florêncio, D., Herley, C.: A large-scale study of web password habits. In: WWW 2007: Proceedings of the 16th International Conference on World Wide Web, pp. 657–666. ACM, New York (2007)
10. Gaw, S., Felten, E.W.: Password Management Strategies for Online Accounts. In: SOUPS 2006: Proceedings of the Second Symposium on Usable Privacy and Security, pp. 44–55. ACM, New York (2006)
11. Notoatmodjo, G., Thomborson, C.: Passwords and Perceptions. In: Brankovic, L., Susilo, W. (eds.) Seventh Australasian Information Security Conference (AISC 2009), Wellington, New Zealand. CRPIT, vol. 98, pp. 71–78. ACS (2009)
12. Prince, B.: Twitter Details Phishing Attacks Behind Password Reset. eWeek (January 2010)
13. Recordon, D., Reed, D.: OpenID 2.0: a platform for user-centric identity management. In: DIM 2006: Proceedings of the Second ACM Workshop on Digital Identity Management, pp. 11–16. ACM, New York (2006)
14. Riley, S.: Password Security: What Users Know and What They Actually Do. Usability News 8(1) (2006)
15. Vance, A.: If Your Password Is 123456, Just Make It HackMe. The New York Times (January 2010)

Towards a Game Theoretic Authorisation Model

Farzad Salim[1], Jason Reid[1], Uwe Dulleck[2], and Ed Dawson[1]

[1] Information Security Institute
[2] School of Economics and Finance,
Queensland University of Technology, Brisbane, Australia
{f.salim,jf.reid,uwe.dulleck,e.dawson}@qut.edu.au

Abstract. Authorised users (insiders) are behind the majority of security incidents with high financial impacts. Because authorisation is the process of controlling users' access to resources, improving authorisation techniques may mitigate the insider threat. Current approaches to authorisation suffer from the assumption that users will (can) not depart from the expected behaviour implicit in the authorisation policy. In reality however, users can and do depart from the canonical behaviour. This paper argues that the conflict of interest between insiders and authorisation mechanisms is analogous to the subset of problems formally studied in the field of game theory. It proposes a game theoretic authorisation model that can ensure users' potential misuse of a resource is explicitly considered while making an authorisation decision. The resulting authorisation model is dynamic in the sense that its access decisions vary according to the changes in explicit factors that influence the cost of misuse for both the authorisation mechanism and the insider.

1 Introduction

The three well known cornerstones of information security are confidentiality, integrity and availability. Each of these properties is defined by reference to an exogenous notion of *authorised*. For instance, confidentiality (integrity) is preserved if and only if a resource is read (modified) by an authorised user. Therefore, the complexity of preserving information security is directly dependant on *authorisation*, which is the process of mediating every requested access to resources maintained by the system and determining whether the request should be *authorised* or *denied*.

Authorisation proves to be a complex task in practice. It is based on a prediction of the users who may require access to resources to perform a job, while the correctness of this prediction appears to be inherently dependant on the future behaviour of the user. Despite this, all existing authorisation approaches inherently attempt to predict both the system's future needs (i.e., to determine who needs access) and the future user behaviour (i.e., in terms of the satisfaction of the need). To make the problem tractable, so far these two concepts have been conflated into a single construct. For example, in Multilevel Security (MLS) users are assigned clearances, or in Role Based Access Control (RBAC)

T. Alpcan, L. Buttyan, and J. Baras (Eds.): GameSec 2010, LNCS 6442, pp. 208–219, 2010.

to roles. Inherent in both of these assignments is the concept of 'need' and to some extent 'user's behaviour' (i.e., if users were assumed to misuse their access, they wouldn't be assigned to the role or been given the clearance). Authorisation decisions within these approaches are based on a *security policy*, that constitutes a set of rules binding access rights to users on the basis of need and assumed unlikelihood of misuse. The main shortcoming of the current policy-based approaches is their use of static criteria to determine a dynamic phenomena: future needs and future users' behaviour[1].

The adverse implication of this is significant as is discussed below under two streams of criticism, one arguing for more flexible authorisation models, another for an optimal access rights assignment. First, there is evidence suggesting that static policy may not be effective in today's dynamic environment [12,14]. As a result, a user's legitimate access request to perform a job that is beneficial for the organisation will be rejected. To address some of the rigidity of such authorisation models, risk-based approaches have been proposed [12,4,14], where an unauthorised user may be given access to resources when the risk of doing so is estimated to be below a predefined threshold. Second, there are industry surveys suggesting that a significant portion of security incidents are due to authorised users (insiders) [6]. There are several proposals to detect and prevent insider misuse by inferring user's intention through behavioural indicators captured using intrusion detection or computer forensic techniques [15,13,3].

Our research is motivated by the gap between these two perspectives: one identifies the need for more flexible authorisation models to facilitate resource sharing in dynamic environments. The other suggests, even with the current pessimistic rights assignment, misuse remains commonplace. At the heart of both lies the uncertainty about future behaviour of users. Such uncertainty is traditionally buried under an informal tradeoff analysis a priori to constructing an authorisation policy. Our goal is to make this tradeoff a dynamic decision based on explicit factors. To this end, we believe authorisation is in nature close to the principal-agent problem in the field of economics [7]. The theory has been extended to discuss the issues of delegation, especially the incentives of employers (principals) and employees (agents) to invest effort into finding the most profitable ways of using employer's resources if incentives are not, or are only partially aligned [1], a problem very similar to that of authorisation, where not to authorise a legitimate request implies that the employer has to spend additional effort to carry out the job. The implication of this perspective is profound for authorisation. It suggests that users are to be considered as self-interested; they attempt to increase their objective function without caring about the objectives of the authorisation system. Therefore, it is no longer sensible to assume users' behaviour based purely on constructs such as role, clearance or trustworthiness. For instance, a high clearance user may be more likely to misuse an access right when he is confident that it can go undetected. Whilst, a low

---

[1] In authorisation literature, user's compliance with policy is external to the authorisation model - assumption has been the existence of policy enforcement mechanisms.

clearance user may be less likely to misuse the same access right when she is certain about being detected and the punishment that follows.

To formally reason about potential user behaviour while making authorisation decisions we utilise techniques from game theory [5] which provide a mathematical foundation for reasoning about conflicts of interest between rational self-interested individuals. The principal contribution of this paper is the proposal of a formal game theoretic authorisation model. In this paper we deliberately introduce strong assumptions to emphasize the effectiveness of this novel approach. We introduce four types of users that an authorisation system may be interacting with, namely, benevolent, malicious, selfish and inadvertent. The type of a user defines their objective function. Further, we show that given a selfish user, under some strong assumptions, the authorisation decision is reduced to solving an inequality, representing the user's tradeoff about the misuse of a resource.

The rest of the paper is organized as follows. Section 2 discusses the related work, focusing on those employing game theoretic techniques in information security. Section 3 introduces the authorisation problem and narrows the scope of our work. Section 4 presents a game theoretic authorisation model and briefly discusses the implications of Nash equilibrium for such model. Section 6 enumerates the simplifying assumptions made in this paper and outlines possible directions for future work. Finally, Section 7 provides the concluding remarks.

2   Related Work

The marriage between economics and information security has attracted considerable attention recently. Game theory provides a mathematical framework for studying the behaviour of rational agents in a multi-player decision problem where players with different objectives can compete and interact with each other on the same system to increase their objective function. The use of game theory in modelling the interaction between users within a system has appeared in several areas of information security research, though not explicitly in addressing the authorisation problem.

Liu et al., in [10] suggest that the concept of incentives can be employed to express attackers' intentions, while the concept of utilities may be used to integrate incentives and costs in such a way that the system as well as attackers' objectives can be practically modelled. They introduce a conceptual model for determining attacker intent, objectives and strategies rather than using a specific type of game for modelling attacks; further they introduce conditions under which a specific type of game model will be feasible and desirable. Alpcan and Basar in [2] have also investigated a security game as a two player, non-cooperative, non-zero-sum game. Their work is related to ours as the game is assumed to be a complete information game and the player's optimal strategy depends only on the payoff function of the opponent. Lye et al., [16] has shown how the network security problem can be modelled as a general-sum stochastic game between attacker and the administrator. They also showed how to compute Nash equilibria, however, their approach is specific to network security applications and they assume the benefit of attackers arises from harming the network, hence only dealing with malicious users. In [11] the authors introduce some of the problems in performing tradeoff analysis in network security. They formulate both static and dynamic Bayesian games to demonstrate the suitability of game theory for the development of various control algorithms in intrusion detection. Further, they discuss the existence of Nash equilibria for these games. However, like [16], they only deal with potentially malicious users, who may only have a positive payoff through attacking the system.

In [9], Liu et al. introduce stochastic game theoretic model for the analysis of the behaviour of malicious insiders. They suggest such a game to be a zero-sum game, where the loss of the employer is the gain of the insider. However, the zero-sum assumption is restrictive as most security games are non-zero sum [11]. Further, their model only deals with malicious insiders. It ignores the circumstances where an insider may also benefit from not attacking, which is the case for selfish insiders as will be discussed in this paper.

In another work, Liu et al., [8] propose a risk-based approach to deal with inadvertent insiders, those users who do not deliberately intend to harm the system. They propose assigning a risk budget to tasks and rewarding those employees who perform their tasks while consuming less than the allocated budget. The reward value is equal to the remaining risk tokens for the task. On the other hand, those employees who consume all their risk budget before completing their job are punished. In this way, the risk is communicated to the inadvertent insiders and the cost of risky actions is shifted from the organisation to them. However, the proposed approach is not abstract and falls short of a formal model. Further, they assume the punishment of the users is a certainty, when in reality punishment is a function of the ability to both detect an attack and administer punishment, neither of which is certain. They also assume the benefit to the user from misusing a resource is less than the punishment cost which implies the punishment is assumed to always be an effective deterrent.

The focus of our work is specifically on authorisation, where the users are not necessarily adversaries. This makes our problem distinct from the above works, because users' benefit is not always driven from attacking, as the organisation may reward actions that advance its objectives. Further, sometimes the expected cost of denying access exceeds the expected cost of authorising the access. This is contrary to the underlying belief behind existing authorisation approaches where the cost of denying access is not accounted for within the model. To the best of our knowledge, all the existing approaches to authorisation make implicit assumptions about how users will behave rather than explicitly reasoning about the users' use/misuse of resources.

3   Authorisation Problem

Let I, A, R, P respectively denote a set of all *Individuals, Actions, Resources* and *Purposes* in a system. We say $\mathbb{U} = I \times A \times R \times P$ is a set of all the possible *uses* - all the actions that can be performed by individuals on resources for any purpose. Given this, the authorisation problem revolves around the design of an authorisation function that determines a subset of uses, $\mathbb{A}^+ = \{(i, a, r, p)\} \subseteq \mathbb{U}$, referred to as the *authorised space*.

The aim of the authorisation function is to reduce the probability of an *attack*, that is defined as a user's action on a resource for a purpose other than for which the user was authorised. Formally, a usage (i, a, r, p') is an attack by user i if $\forall a, r, p \in \mathbb{U}, \exists(i, a, r, p) \in \mathbb{A}^+ \wedge \nexists(i, a, r, p') \in \mathbb{A}^+$. By definition our authorisation problem is focusing on the scenarios where resources provided to users may be used for purposes other than those intended by the authorisation system. For instance, Alice using her access permission to copy sensitive records for the purpose of financial benefit (by selling them) is considered as an attack.

An attack inherently suggests an unwelcome usage by the user regardless of the potential damage they may incur to the system. From this, we define a *user threat* (threat for short) as a probability $\rho \in [0, 1]$ of attack by a user. This expresses the unpredictability of the users' actual purpose of using the resource.

3.1 Insider Types

One of the major complexities involved in dealing with users' attacks is the fact that such attacks may be *intentional* as well as *accidental*. The former may occur for reasons such as revenge, financial gain, policy workarounds, and the latter may be an honest mistake or due to a user's lack of knowledge about the risk of their actions for the organisation [13]. Even though knowledge about intentions provides an important criteria for detecting and preventing attacks, teasing out intentions is a challenging task as there are no uniquely identifying indicators associated with attack actions [10]. Despite this, there are already several tools and approaches for detecting attacks as well as predicting them based on behavioural patterns and sequences of actions executed by a user. Although such tools are still in their infancy, the empirical results show several signs of improvement. For instance, Bishop et al., in [3] introduced an architecture for a tool that attempts to identify certain behavioural changes that may be alarming. Others [15] have suggested the use of Intrusion Detection Systems (IDS) to identify the deviations from "normal" usage patterns by users.

Here we assume such a tool exists as a function $\Gamma_{\{\Theta\}} \in [0, 1]$ that provides a probability of user's *type*, given a *type space* $\Theta = \{benevolent, selfish, malicious, inadvertent\}$. For example, $\Gamma_\Theta = \{0, 0.5, 0.5, 0\}$ suggests that that a given user might be either selfish or malicious with the probability of $1/2$. A user's type embodies their *private information* that is relevant to both the user and the authorisation mechanism - each user type specifies what the user considers a utility, hence their intention:

Malicious: those who consider the loss (increased cost) of the organisation as their gain. They would like to incur as much cost to the organisation as possible. Most of the existing works deal with detecting and preventing malicious insiders [9].

Selfish: those whose aim is to maximise their own (financial) payoff. Their aim is not to incur cost to the organisation, even though this may happen as a result of their selfish choices. Hence, this type will respond to appropriate incentives (e.g., financial).

Benevolent: those who consider the loss of the organisation their loss (their utility function is the organisation's utility function), hence they do not attack.

Inadvertent: those with incomplete or incorrect information about the outcome of their actions. They are not misusing the resources (attack) to harm the organisation or doing so to increase their financial gain, but they may be careless or negligent or uninformed [8].

While our ultimate goal is to design a general authorisation model that can make an optimum authorisation decision (i.e., to reduce threat) under uncertainty about user's type, here for simplicity we will assume that for a given user the sum of her/his type probabilities is 1 and that users are only selfish ($\Gamma_\Theta = \{0, 1, 0, 0\}$). This focuses our attention on how to design an authorisation mechanism that explicitly takes into account (selfish) user's potential misuses of their access rights before making authorisation decisions.

4   Game Theoretic Authorisation Mechanism

In this section we formulate an authorisation mechanism as a game between a *selfish employee* (i) and the *benevolent employer* (j) who is the sole authority in making authorisation decisions. The game starts with a request from the employee for access to a resource. Along with the request, the employee indicates the outcome of such action for the employer, denoted as *proposal* (p)[2]. Given this request, the employer shall decide whether to *authorise* or *deny* the access to the resource[3]. On the other hand, the strategy space of the employee consists of either *attack* or *not attack*.

Such a binary description of employees' alternatives simplifies our model, however, it is no longer possible to differentiate attacks based on their consequences. For instance, given a sensitive record and a disgruntled employee with two alternative attack actions, i.e., *destroy* or *sell* the record to competitors, there may be a great difference between the two attacks from the employer's perspective, particularly if a backup of the record exists (i.e., selling it may incur a great financial loss while destroying it may merely interrupt a service).

The authorisation game centres around a resource valuable to both players. An employee may use the resource to either make a personal profit (i.e., attack) or perform a job that actions the proposal (p) for the employer. The employer,

---

[2] We deliberately reuse p that represented a purpose in Section 3 to draw the connection between the notion of proposal and purpose.

[3] The employer is actually the representation of our authorisation function, that given an access request $(i, a, r, p) \in \mathbb{U}$ decides whether $(i, a, r, p) \in \mathbb{A}^+$ (*authorise*) or $(i, a, r, p) \notin \mathbb{A}^+$ (*deny*) the request.

hence, is concerned about the expected cost of the attack, which causes the resource to transition from a *secure* state to a *compromised* state. Such a transition is associated with a specific monetary *cost* for the employer, denoted by C_j^r. Depending on the resource the cost of being compromised changes. In reality there may be several compromised states including loss of confidentiality, integrity, availability, privacy or reputation. Further, as we have mentioned, there may be several attacks based on the employee's action space and each may incur a specific cost depending on the causal relationship between an attack action and resource transition to a costly state. However, since *attack* generalises any single/sequence of undesirable employee actions, we assume an employee's attack incurs a cost (C_j^r) to the employer.

The employer is also susceptible to *opportunity* cost: the benefit forgone by denying a request. The quantification of opportunity is determined by the proposal (p) made by the employee to access the resource. Through such a formulation, distinct from the existing authorisation approaches, denying an access request as well as authorising it may incur a cost for the authorisation system.

Now we turn to the employee's cost factors. An employee may incur cost through *fines*, denoted by C_i^f (i.e., given they attack). However, usually a fine is not certain - it is only applicable if the employer can *detect* the attack, which is a function of the accuracy of detection techniques and the ability to *enforce* the fine. For now the ability to detect and enforce the fine is combined and referred to as the probability of being fined, denoted by $\psi \in [0, 1]$, which is assumed to be common knowledge. For example, when an employee is out-sourced from another country there may be less chance of enforcement of the fine in comparison to a circumstance when the employee is local[4]. In addition, in order to attack, the employee is assumed to incur a *preparation* cost, denoted by C_i^t. This cost abstracts the effort the employee must expend in order to acquire access to the resource to use it for personal benefit. For instance, if the resource is commercially valuable, finding a buyer requires time and effort. In other cases, the employee may need to prove to the employer that the proposal amount is attainable by him and this could require training courses and faking trustworthiness.

Sometimes the employee is given a personal benefit for the opportunity they realise. This is represented as a rate of return, $\epsilon \in [0, 1]$ on the proposed opportunity, p. We regard the predictions of the employee in terms of the actual achievement of p to be always correct if the access is granted. On the other hand, the actual personal profit for the employee from an attack is a portion $\alpha \in [0, 1]$, of the cost of resource (C_j^r), if the access is given. Note that this may not always be the case as sometimes a very costly resource for the employer has a very low value for a selfish employee or vice versa.

Given the above game setting, the game tree of employer and employee in an authorisation game is shown in Figure 4. The authorisation problem is, given complete information of both players about the payoffs, when should the employer authorise the access?

---

[4] For now we are not interested in the size of this fine in proportion to the loss (cost) of the employer.

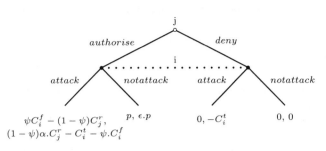

Fig. 1. Players Payoffs

4.1 Nash Equilibrium and Its Implications

A solution to a non-cooperative game predicts certain strategy profiles as outcomes of the game. Defining or interpreting a solution revolves around how players reason and behave or are believed to reason and behave. This inevitably leads to the need for players to attempt to understand and predict how the other player will behave. In a game of complete information, where the strategies and payoffs are common knowledge, this is then reduced to players choosing their best responses to the potential strategy of others. The well known concept of Nash equilibrium [5] provides an exit from a cycle of speculations as to what strategies the players should use, and provides an appropriate solution for the game. In the context of the proposed authorisation game, Nash equilibrium can be defined as a set of actions from the employee and the employer such that none of them has any incentive to deviate from their chosen action.

Assuming that the employer always takes the pure strategy *deny*, then the employee's best response is *not attack*. However, this is not an equilibrium as the pure strategy of *not attack* by the employee motivates the rational employer to change his strategy to *authorise* whenever $p > 0$. By switching to *authorise*, the employee is then inclined to *attack* when

$$\epsilon.p < (1 - \psi)\alpha.C_j^r - \psi.C_i^f - C_i^t. \tag{1}$$

Conversely, this can be reduced to the following: if the employer *authorises* then the employee does *not attack* if and only if

$$\epsilon.p \geq (1 - \psi)\alpha.C_j^r - \psi.C_i^f - C_i^t. \tag{2}$$

The above finding is interesting and rather counter intuitive in the context of authorisation. It suggests that in making an authorisation decision the authorisation mechanism may only need to focus on the employee's payoff instead of its own. This is contrary to the existing approaches to authorisation, where there exists a policy, assumed to incorporate access rules which result from a trade-off analysis between some implicit contextual factors, for all current and future requests. Here instead, the decision factors are explicit and abstract enough to adapt to the required application. For example, the value of ψ can depend on the existing monitoring techniques, audit, accuracy of forensic techniques, physical security employed, etc.

On the other hand, the application of such an authorisation model can introduce a prescriptive system rather than simply providing authorise/deny responses. Through such an interpretation, given an authorisation request, a game theoretic authorisation model may also attempt to meet the above inequality, if not already met, through taking either or a combination of *deterrence* or *appeasement* policies. The former is to increase the cost of attacking for the user, so that the above inequality is met. This may be achieved through either increasing any or combination of ψ, C_i^t, C_i^f, or reducing α. On the other hand the appeasement policy attempts to increase the benefit for not attacking, by aligning users' utility function with the organisation through increasing ϵ. A thorough investigation of how such policies can be implemented is left for future work.

5   Case Based Analysis of Authorisation Mechanism

In this section we will introduce two authorisation cases and compare a decision made under a traditional authorisation model (e.g., RBAC) to the potential decision from a game theoretic authorisation mechanism.

5.1   Case 1: Less Valuable Resource

Consider a hypothetical organisation with a role-based access control framework in place and an employee who requests to use a resource (e.g., printer), to which she does not currently have access. An RBAC model simply denies the request without considering the payoffs to the employee from misusing the printer or the payoffs to the organisation when the access is authorise/denied. Let us describe how a game theoretic authorisation model analyses the potential responses, given the following inputs[5]:

$\alpha = 1$	private benefit ratio of a resource value
$C_j^r = 1$	the cost of printing a document for organisation
$\epsilon = 0$	interest given to employee
$C_i^f = 1$	punishment cost if resource misused
$\psi = 1$	detection rate (e.g., through print logs)
$C_i^t = 0$	cost of preparation for using printer for personal purposes
$p = 1$	value of opportunity proposed (e.g., time saved)

Given the above setting the payoff for both the employee and the employer would be as shown in Figure 2. It can be seen that for the employee the rational choice regardless of the employer's action is *not attack*, and for the employer, authorising the request weakly dominates denying it, hence the pair (authorise, not attack) is the equilibrium state. This case exemplifies the authorisation dynamics for resources with a small intrinsic value, where misuses could also be easily detected.

---

[5] Note that the above representation of costs and punishment are ordinal numbers rather than cardinal. Hence, they show the relationship between the factors rather than their actual quantity.

		Employee	
		attack	*notattack*
Employer	*authorise*	$1, -1$	$1, 0$
	deny	$0, 0$	$0, 0$

Fig. 2. Less valuable resource and high chance of punishment

5.2   Case 2: Highly Valuable Resource

Consider a scenario where an employee of a financial firm makes stock forecasts based on some highly valuable information resources. Due to the importance of these resources, each employee only has access to a segment of the information. However, let us assume, an employee observes a good opportunity to invest in a stock, but needs some more information, which he does not have access to. Again, traditional access control models simply deny such access on the basis of their predefined policy. Let us first analyse the circumstance under the following inputs.

$\alpha =$	1	ratio of return from selling the resource
$C_j^r =$	10	the cost of selling the document to competitors
$\epsilon =$	0	employee's interest for achieving opportunity
$C_i^f =$	10	the cost of punishment
$\psi =$	.25	attack detection rate
$C_i^t =$	1	low attack preparation is needed
$p =$	3	value of opportunity proposed

Again, given the above setting the payoff for both the employee and the employer would be as shown in Figure 3.

		Employee	
		attack	*notattack*
Employer	*authorise*	$-5, 4$	$3, 0$
	deny	$0, -1$	$0, 0$

Fig. 3. Valuable resource and low chance of punishment

Given the above payoffs there is no equilibrium in pure strategies. This is because, if the employer chooses to authorise, the employee will rationally choose to attack in which case the employer switches to deny. However, when the employer chooses the pure strategy deny, then the response of the employee is not attack. Although there is no equilibrium in pure strategies, in this authorisation problem an equilibrium in mixed strategies exists: both employer and employee randomise between their pure strategies. The employer will correctly predict the employee's probabilities of possible actions, and vice versa. These probabilities make both players indifferent in choosing between their pure strategies option, thus randomizing is rational for both of them. For example, in Figure 3, if we denote by β the probability of the employer to authorise and by ρ the probability of the employee to attack, we get a mixed equilibrium for the game when $\beta = 1/5, \rho = 3/8$[6]. Given this, to prevent an attack, the authorisation mechanism authorises a request only if it believes $\rho < 3/8$.

---

[6] For details on mixed strategy equilibrium refer to the Chapter 1 of Fludenberg and Tirole's book [5].

In this case, the key factors behind the employee's decision to attack are the ability to monetize the valuable resource, as well as the small probability of punishment ($\psi = 0.25$), for instance because the employee is leaving the organisation. However, given the same scenario, this decision can swiftly change by the change in the probability of enforcing the punishment (C_i^f). For example, now assume that $\psi = .75$, to say the chance that the employer can enforce the punishment is high. Everything else being equal, we get the payoffs in Figure 4.

		Employee	
		attack	notattack
Employer	authorise	5, −6	3, 0
	deny	0, −1	0, 0

Fig. 4. Valuable resource and high chance of punishment

Hence, the rational action for the authorisation system in this circumstance is to authorise the access, even though the resource is sensitive and its misuse is costly. This is because the payoff of the employee reveals that attacking is not the rational choice, as the pair (3,0) is the equilibrium.

6   Future Work

In this paper we have made several simplifying assumptions to flag the potential manner of employment and benefits of game theoretic techniques in designing new authorisation mechanisms. So, our immediate efforts to improve the proposed abstract model focus on relaxing some of our assumptions: First, employer's complete information about user's type. In reality, the predication of a type involves uncertainty in the form of a probability distribution over types. Second, users alternative actions were modelled as binary, e.g., attack or not-attack. However, in reality a user may have several different attack alternatives which vary in likelihood as well as consequence. Finally, a more realistic authorisation mechanism may need to be modelled as a dynamic game rather than a one-shot static game. In dynamic games, players observe other players' behaviours and modify their strategies accordingly.

7   Conclusion

This paper discusses the authorisation problem and proposes a new paradigm of thinking for designing dynamic authorisation models. It suggests that the problem of authorisation is at it's core analogous to the principal and agent problem studied in the field of game theory. Based on this premise, it proposes the preliminary components and a basic but novel authorisation model that makes access decisions based on explicit reasoning about users available actions and the likelihood and consequences of choosing such actions, both for the authoriser and the user. Finally, it provides some extreme authorisation cases to illustrate the advantages of such a new authorisation model in comparison to the existing well-known authorisation models such as RBAC.

References

1. Aghion, P., Tirole, J.: Philippe Aghion and Jean Tirole. Formal and real authority in organizations. Journal of Political Economy 105(1), 1 (1997)
2. Alpcan, T., Basar, T.: A game theoretic approach to decision and analysis in network intrusion detection. In: Proceeding of the 42nd IEEE Conference on Decision and Control (CDC) (December 2003)
3. Bishop, M., Frincke, C.G.D., Greitzer, F.L.: AZALIA: an A to Z Assessment of the Likelihood of Insider Attack (2010)
4. Cheng, P.-C., Rohatgi, P., Keser, C., Karger, P.A., Wagner, G.M., Reninger, A.S.: Fuzzy multi-level security: An experiment on quantified risk-adaptive access control. In: IEEE Symposium on Security and Privacy, pp. 222–230 (2007)
5. Funderberg, D., Tirole, J.: Game Theory. MIT Press, Cambridge (1992)
6. Gordon, L.A., Loep, M.P., Lucyshyn, W., Richardson, R.: CSI/FBI computer crime and security survey. Technical report, CMP Media, Manhasset, NY (2004)
7. Holmstrom, B.: Moral hazard and observability. The Bell Journal of Economics 10(1), 74–91 (1979)
8. Liu, D., Wang, X., Camp, J.L.: Mitigating inadvertent insider threats with incentives, pp. 1–16 (2009)
9. Liu, D., XiaoFeng, W., Camp, J.L.: Game theoretic modeling and analysis of insider threats. International Journal of Critical Infrastructure Protection 1, 75–80 (2008)
10. Liu, P., Zang, W.: Incentive-based modeling and inference of attacker intent, objectives, and strategies. In: CCS 2003: Proceedings of the 10th ACM Conference on Computer and Communications Security, pp. 179–189. ACM, New York (2003)
11. Liu, Y., Comaniciu, C., Man, H.: A bayesian game approach for intrusion detection in wireless ad hoc networks. In: GameNets 2006: Proceeding from the 2006 Workshop on Game Theory for Communications and Networks, p. 4. ACM, New York (2006)
12. MITRE Corporation Jason Program Office. Horizontal integration: Broader access models for realizing information dominance. Technical Report JSR-04-132, MITRE Corporation (2004)
13. Pfleeger, S.L., Predd, J.B., Hunker, J., Bulford, C.: Insiders behaving badly: Addressing bad actors and their actions. IEEE Transactions on Information Forensics and Security 5(1), 169–179 (2010)
14. Salim, F., Reid, J., Dawson, E.: Towards authorisation models for secure information sharing: A survey and research agenda. The ISC International Journal of Information Security, ISeCure (2010)
15. Eugene Schultz, E.: A framework for understanding and predicting insider attacks. Computers & Security 21(6), 526–531 (2002)
16. Lye, K.w., Wing, J.M.: Game strategies in network security. Int. J. Inf. Sec. 4(1-2), 71–86 (2005)

Disperse or Unite? A Mathematical Model of Coordinated Attack*

Steve Alpern[1], Robbert Fokkink[2], Joram op den Kelder[2], and Tom Lidbetter[1]

[1] London School of Economics,
Department of Mathematics,
Houghton Street, London,
WC2A 2AE, United Kingdom
s.alpern@lse.ac.uk
[2] Delft University, Faculty of Electrical Engineering,
Mathematics and Information Technology,
P.O.Box 5031, 2600 GA Delft, Netherlands
r.j.fokkink@ewi.tudelft.nl

Abstract. We introduce a new type of search game that involves a group of immobile hiders rather than a single hider. We provide the mathematical framework for the game and we show that the game is useful to understand under what conditions attackers disperse or unite.

Keywords: zero-sum game, near value, probability of dispersion.

We introduce a new type of search game that involves multiple hiders. It models situations in which a collection of valuable objects are targeted by a malevolent attacker. These valuables could be a number of installations in a factory or a number of servers in a network. They are guarded by a security system, but due to a malfunction the security system is out of operation during a certain time window. Once the security system is up and running again, it is known that the security has been breached by a very well informed attacker, who knew about the duration of the time window beforehand. The attacker can decide to attack more than one target. For instance, the attacker could be infiltrating a computer network and he can place copies of a worm on several servers. He could also direct the attack against a single object. The main problem that we want to address in this paper is the following dilemma: *under what conditions does the attacker choose multiple targets instead of a single target?* If the attack is directed at multiple targets, then we say that the attacker 'disperses' and otherwise he unites. This is a problem that has been studied before, using methods from game theory and reliability theory [4]. We study it by introducing a new type of search game.

One other example that we had in mind when we developed the search game, was a terrorist attack. The September 11 attack and the transatlantic bomb

* This work was supported by NWO Visitor Grant 2010/04308 and NATO Collaborative Research Grant 983583.

T. Alpcan, L. Buttyan, and J. Baras (Eds.): GameSec 2010, LNCS 6442, pp. 220–233, 2010.

plot were examples of dispersive operations. The terrorist organization activates several cells at the same time. The authorities get to know about this and have to uncover all the cells in time. This is comparable to the situation that Scotland Yard was in, when it received intelligence from Pakistan on August 9, 2006, that a transatlantic attack by terrorist cells was imminent [12]. Several games have been proposed to explain the modern style of terrorist operations and the search game in this paper also contributes to that literature [8]. Our game is very much in the style of search games that have been proposed in [11].

The paper is organized as follows. In section 1 we describe the game and provide a mathematical framework. We prove that the game has a near value, similar to the near value of a classical search game that is usually derived from the minimax theorem in [2]. In section 2 we study a particular case of this game in detail. Section 3 connects the game to problems from combinatorial geometry and gives a lower bound on the game for certain parameter values. Section 4 summarizes results.

1 A Search Game with Multiple Agents

Work on search theory begin during the second World War and for a long time it has remained in the same form: a single evading target is lost and the problem is to find it with fixed resources [3],[6]. Recently, new forms of search games have been developed, in which finding the items requires a group decision [13] or in which the searcher acquires information about the target over time [9]. In our paper we propose a new type of search game, that involves the time-flow of information and the coordinated behavior of a group. Since this is a new game, we reserve some space to put the game in a proper mathematical framework.

1.1 Idea of the Game

We consider a search game in which a malevolent hider H, places k objects in n locations. H may place more than one object in one location, but the objects are immobile. The authorities, who play the role of searcher S, seek to uncover H's objects. The objects are well hidden. If S looks in the right location, then he does not immediately uncover the object, or objects, but has to spend time. Even if S knows exactly how many objects are hidden at each location, which would allow a perfect search, then still it would take him one unit of time to uncover all the objects. That is why S searches each location for a total time of at most one unit. If S does that, we call this an exhaustive search. S may decide to switch the search to another location, once he has found an object, but he may also decide to search that location further, since there may be more concealed objects hidden there. If S switches to another location, but returns to that location later, then the search continues as if S had not left. The total amount of time that S needs to search a location is one unit.

At the start of the game, S does not have any information on where the objects are hidden. He only knows the total number of objects k and the total number of locations n. As time goes by, however, S acquires information by uncovering objects. If S uncovers an object that required time t to find at a certain location, then the remainder of the search would take time $1 - t$ if the search were perfect.

1.2   Mathematical Description

We take some care to formalize the set-up since this is a new type of search game. We have a two-player win-loose game $G = (X, Y, \Pi)$ where X and Y are the strategy spaces of S and H respectively and

$$\Pi \colon X \times Y \to \{0, 1\}$$

is the payoff function. If Π is equal to 1 then we say that S wins and that H loses. S wins if he uncovers all hidden objects. To describe the hider strategy space Y we say that $J = \{1, \ldots, n\} \times [0, 1]$ is a labeled interval, because an element $(i, x) \in J$ consists of a value $x \in [0, 1]$ and a label i that is in between 1 and n. We say that x is the *place* and that i is the *location* of the object. In total, H hides k objects within the union of the labeled intervals and it is convenient to picture this union as a star with n spokes of unit length. For instance, if $n = 2$ this star is the interval $[-1, 1]$ and a pure Hider strategy is a subset of k elements of this interval. We discuss this case in section 1.3 below. One of the things we find in the analysis of the game is that H prefers to put the objects at discrete locations. This is illustrated in Figure 3 for $k = 2$ and $n = 3$.

H may decide to hide two objects at the same location (i, x) and (i, y) for $x < y$. The advantage of this is that the object at x can be used as a decoy. For every location i we denote the maximum value of an object hidden by H by h_i. The hider strategy space Y consist of sets of k-points:

$$\{(i_1, x_1), \ldots, (i_k, x_k) \colon h_1 + \cdots + h_n = 1, \ h_i = \max\{x_j \colon j = i\}\}$$

Let $\tau \geq 1$ be the total time that is available for the search. A pure searcher strategy is given by continuous function $f \colon [0, \tau] \to [0, 1]^n$ for which each component $f_i(t)$ records the amount of time that S has searched location i up to time t. Therefore f_i is non-decreasing and $\sum_{i=1}^{n} f_i(t) = t$. One could picture a searcher strategy as a subset of the star on n spokes that increases in time and starts at the centre. This subset is a star with end points at $f_i(t)$ in the i-th spoke.

By time t, S has found all objects (i, x) such that $x \leq f_i(t)$ and from this S acquires information on the possible position of remaining objects that he may use to adapt the search. In other words, f depends on the Hider's pure strategy ω and to be precise we could write f_ω instead of f, or as a map $\omega \mapsto f_\omega$. Let C be the space of continuous functions $f \colon [0, \tau] \to [0, 1]^n$, then the searcher strategy space is a subset of the product space C^Y. Suppose that ω' is another hider strategy, then S has to adopt the same strategy against ω' for as long as the uncovered objects are the same. To make this precise, if the strategies f_ω and $f_{\omega'}$ are different and coincide up to time

$$t_0 = \max\{t \colon \forall s \leq t \ f_\omega(s) = f_{\omega'}(s)\}$$

then there exists an object (i, x) that is only contained in one of the two hider strategies, say ω, such that $x \leq f_\omega(t_0)$. In other words, by time t_0 the two strategies f_ω and $f_{\omega'}$ have uncovered different objects and that is why the searches may diverge. The searcher strategy space $X \subset C^Y$ consists of all sequences (f_ω) that satisfy this property. It is a closed subspace under the product topology. The payoff $\Pi(f_\omega, \omega) = 1$ if and only if $f_{\omega,i}(\tau) \geq h_i$ where the h_i are as defined above.

A mixed searcher strategy σ is a probability measure on the Borel σ-algebras of X. A mixed hider strategy γ is a probability measure on the Borel σ-algebras of Y. The payoff $\Pi(\sigma, \gamma)$ of these mixed strategies is the expected value $\mathbb{E}[\Pi]$ under the product probability measure on $\sigma \otimes \gamma$. In other words, the payoff is the probability that S wins (or that H loses), if the searcher picks a random strategy according to σ and the hider picks a random strategy according to γ. A searcher strategy σ^* is optimal if it maximizes the minimal probability of a searcher win against any hider strategy. Similarly, a hider strategy γ^* is optimal if it minimizes the maximal probability of a hider loss against any searcher strategy. The game has a value $\mathcal{V}$ if the maximin searcher win equals the minimax hider loss.

The mathematical framework that is needed to define the game is intricate. The searcher strategy space is large and standard minimax theorems do not apply. A priori it is not even clear that the game has a value and that optimal strategies exist.

1.3   An Example

To clarify the picture, we give an example of the simplest non-trivial case. If $n = 1$ then we have a trivial game. There is only one location which S can search completely since we assume that $\tau \geq 1$. S always wins. If $k = 1$ then we have a standard search game with one immobile hider on a star with n spokes [10]. The optimal strategies of the players are as follows. S performs an exhaustive search a random subset of $\lfloor \tau \rfloor$ of the spokes. H puts an object at $(i, 1)$ for a random i. The value of this game is $\lfloor \tau \rfloor / n$.

The simplest non-trivial case is $n = k = 2$, which we have considered before in [1] where it forms the starting point for a study of behavioral ecology. We picture the two labeled intervals as a single interval $[-1, 1]$. H puts objects at points $p, q \in [-1, 1]$. If these points are at opposite sides of the origin, then $|p - q| = 1$. If they are on the same side, then $\max\{|p|, |q|\} = 1$. Either way, the interval that is spanned by p, q and the origin has length 1. A pure searcher strategy is a subinterval $[-x(t), y(t)]$ that grows in time such that $x(t) + y(t) = t$. S wins if and only if all hidden objects are contained in the interval $[-x(\tau), y(\tau)]$. For instance, if H adopts the strategy in Figure 1, then S can only win against one of the three pure strategies if $\tau < \frac{3}{2}$.

If $\tau \geq 2$ then S always wins. If $\tau < 2$ then H may put both objects at one of the two ends ± 1 equiprobably. Since S can only search one of the two locations fully, H loses with a probability of no more than $\frac{1}{2}$. On the other hand, if $\tau \geq \frac{3}{2}$ then S has a strategy that guarantees a win with a probability of at least $\frac{1}{2}$,

Fig. 1. A mixed Hider strategy that consists of three pure strategies: placing the two objects in the interval $[-1, 1]$ equiprobably at $\{-1, -\frac{1}{2}\}, \{-\frac{1}{2}, \frac{1}{2}\}, \{\frac{1}{2}, 1\}$

which we leave to the reader to check. So the value of the game is $\frac{1}{2}$ if $\tau \geq \frac{3}{2}$ and H hides both objects in the same location: the hider does not disperse if $\tau \geq \frac{3}{2}$.

The game gets more interesting if $\tau < \frac{3}{2}$. Then H puts the objects either at $\{-1, -\frac{1}{2}\}$ or at $\{-\frac{1}{2}, \frac{1}{2}\}$ or at $\{\frac{1}{2}, 1\}$ equiprobably. We leave it to the reader to check that any searcher strategy can win against only one of these three hider strategies. So H can limit the probability of loss to at most $\frac{1}{3}$. On his turn, S starts a search in one of the two locations at random. He continues that search until he finds an object and if he does not, then the search of the location is exhaustive. If S finds an object, then with probability $\frac{1}{3}$ he stops the search and starts an exhaustive search of the other location. With probability $\frac{2}{3}$ he continues the search. We claim that this guarantees a win for S with probability of at least $\frac{1}{3}$. To see that, first consider the case that H hides both objects in the same location. S chooses that location with probability $\frac{1}{2}$ and he continues an exhaustive search with probability $\frac{2}{3}$ upon finding the first object. S wins with probability of at least $\frac{1}{3}$ in this case. If H hides the objects in different locations, at opposite sides of the origin, then S wins if he stops the search upon finding the first object. Again, S wins with a probability of at least $\frac{1}{3}$. This is the value of the game if $\tau \leq \frac{3}{2}$.

We say that H *disperses* if he puts objects in all locations. If $\tau < \frac{3}{2}$ then H disperses with probability $\frac{1}{3}$. If $\tau \geq \frac{3}{2}$ then H does not disperse.

1.4   Discretization of the Game – Near Value and Near Optimal Strategies

We discretize the game by limiting the hider's options to a finite number of points. H may put the objects only at points (i, x) such that the place x is a rational number $\frac{j}{N}$ for a fixed natural number N. In other words, H hides the objects at *grid points*. Note that in the example above, H puts objects at grid points for $N = 2$.

We say that $t \in [0, \tau]$ is a time-grid point if it is a rational number $t = \frac{k}{N}$. We say that a searcher strategy f is a *grid search* if all its components $f_i(t)$ are grid points if t in the time-grid.

Lemma 1. *Suppose that H is limited to hiding objects at grid points. Then any searcher strategy f is dominated by a grid search.*

Proof. If $f_i(t)$ is a grid-point then we say that the strategy f uncovers this grid point at time t. The strategy uncovers the grid points in a certain order. Suppose that f uncovers a total of M grid points and let $t_1 \leq t_2 \leq \cdots \leq t_M$ the times at which these points are uncovered. Note that the inequalities are not strict since grid points that are at different locations may be uncovered at the same time. Let $i_1, i_2, \ldots, i_M$ be the locations of these grid points and let $x_1, x_2, \ldots, x_M$ be their places.

We define a new search g by induction. The search starts at location i_1 until it uncovers the grid point $(i_1, f_{i_1}(t_1))$. The search then continues at location i_2 until it uncovers $(i_2, f_{i_2}(t_2))$, etc. If g has uncovered all the grid points that f uncovered, there may be some time left (though not more than $\frac{n}{N}$). The search then continues lexicographically at location 1 or 2, etc, by exhaustive search.

We claim that g uncovers the grid at least as fast as f. To see this, picture the search space as a star with n spokes. The minimal amount of time t_j that is needed to uncover the grid points $(i_1, x_1), \ldots, (i_j, x_j)$ corresponds to the total length of the minimal connected set M that includes all these points. The end points of M correspond to the coordinates of $g(t_j)$. Now $f(t_j)$ corresponds to the end points of another connected set. This set cannot cover other grid points, since M contains the points that f uncovers first. Therefore g uncovers the grid at least as fast as f.

Now f depends on a hider strategy and we should write f_ω and similarly g_ω. We know that by the time

$$t_0 = \max\{t \colon \forall s \le t \; f_\omega(s) = f_{\omega'}(s)\}$$

there exists an object (i, x) that is uncovered by f_ω and not by $f_{\omega'}$ or vice versa. We should verify that the same applies to g_ω and $g_{\omega'}$. Until time t_0 the strategies f_ω and $f_{\omega'}$ uncover the same set of grid points S. Hence, g_ω and $g_{\omega'}$ coincide at least until the time s_0 at which both of them have uncovered S. But $(i, x) \in S$ so by time s_0 the strategies g_ω and $g_{\omega'}$ have uncovered different objects. In case $t_0 = \tau$ then $g_\omega = g_{\omega'}$ since both strategies uncover the same grid points and then continue by the same lexicographic search. $\qquad\Box$

Theorem 1. *Suppose H is restricted to N grid points. Then the strategy spaces of both players reduce to finite spaces so the restricted game has a value $\mathcal{V}_N$. The limit $\mathcal{V} = \lim_{j \to \infty} \mathcal{V}_{2^j}$ exists. For each $\epsilon > 0$, H has a mixed strategy that limits the probability of a loss to no more than $\mathcal{V} + \epsilon$.*

Proof. If H hides at grid points, then he has finitely many strategies. In response, it is optimal for S to perform a grid search. Now observe that a grid search is completely determined by the order in which the grid points are uncovered, so there are finitely many grid searches as well. Therefore $\mathcal{V}_N$ is well-defined by the minimax theorem. As the $2N$-grid refines the N grid, this increases the hider strategy space and we see that $\mathcal{V}_{2N} \le \mathcal{V}_N$. The sequence $\mathcal{V}_{2^k}$ is decreasing and hence it converges. For each $\epsilon > 0$ there exists a k such that $\mathcal{V}_{2^k} < \mathcal{V} + \epsilon$. Let γ_k^* be an optimal hider strategy for the 2^k-grid. Then the payoff of any searcher strategy against γ_k^* is at most $\mathcal{V}_{2^k}$, so H can keep its probability of a loss below $\mathcal{V} + \epsilon$. $\qquad\Box$

The limit $\mathcal{V}$ depends on the total search time τ and therefore it would be more precise to write $\mathcal{V}(\tau)$. We leave it to the reader to verify that the function $\mathcal{V}(\tau)$ is non-decreasing with τ, so it has only countably many points of discontinuity.

In our definition of the game, a pure hider pure strategy has to satisfy $h_1 + \cdots + h_n = 1$ where the h_i denote the maximum place on a location. It is equivalent to allow that $h_1 + \cdots + h_n \leq 1$. In the same vein, we may allow that S to call of the search at a time $\leq \tau$. These assumptions turns out to be convenient in the proof of the following theorem.

Theorem 2. *Suppose that τ is a continuity point of $\mathcal{V}$. For each $\epsilon > 0$, S has a mixed strategy that guarantees a probability of a win of at least $\mathcal{V} - \epsilon$.*

Proof. Choose $\tau' < \tau$ such that $V(\tau') > \mathcal{V}(\tau) - \epsilon$. For sufficiently large N we find $\mathcal{V}_N(\tau') > \mathcal{V}(\tau) - \epsilon$. We may assume that $\frac{n}{N} + \tau' < \tau$. Let ω be any pure hider strategy. S treats it as if the objects are hidden at N-grid points: let ω' be the strategy that moves an object $(i, x) \in \omega$ to the rounded down grid point $(i, \frac{j}{N})$ for $\frac{j}{N} \leq x$ with maximal j. Note that ω' is an admissible strategy, since we allow that the h_i sum up to 1 or less. Let $f_{\omega'}$ be a pure strategy in the optimal mixed searcher strategy in the discretized game. In particular, $f_{\omega'}$ is a grid search. S adapts this strategy so it can be used against ω, as follows. S starts by an initial search at each location for time $\frac{1}{N}$. This takes an extra time $\frac{n}{N}$. After that S adopts the strategy $f_{\omega'}$ and he carries out the search as if the time of the initial search has not elapsed. If S finds an object at (i, x) then he records it as being found at $x - \frac{1}{N}$, since he ignores the time that evolved in the initial search. S can do this since $\frac{n}{N} + \tau' < \tau$. Observe that S wins against ω if $f_{\omega'}$ wins against ω'. Now an optimal mixed searcher strategy in the N-grid game guarantees a probability of a win of at least $\mathcal{V}_N(\tau') > \mathcal{V}(\tau) - \epsilon$.

Now that we know that the game is well-defined and has a (near) value, it is natural to try and compute $\mathcal{V}$, or at least come up with some accurate bounds. This appears to be difficult. The next section has a full analysis of the case that $k = 2$ and $n = 3$, which turns out to be quite difficult. In the final section of the paper we give some bounds on $\mathcal{V}$ and show that our search game with multiple hiders is related to non-trivial problems from combinatorial geometry.

2   The Game for 2 Hidden Objects and 3 Locations

We study the game for $k = 2$ and $n = 3$ and $1 \leq \tau < 3$. If H hides the objects in two different locations, then we say that the strategy is *dispersive*. We compute the value of the game $\mathcal{V}(\tau)$, which is given by

$$\mathcal{V}(\tau) = \begin{cases} \frac{1}{6}, & \tau \in [1, \frac{3}{2}) \\[2mm] \frac{1}{4}, & \tau \in [\frac{3}{2}, \frac{5}{3}) \\[2mm] \frac{1}{3}, & \tau \in [\frac{5}{3}, 2) \\[2mm] \frac{2}{3}, & \tau \in [2, 3) \end{cases}$$

and we show that this is the actual value of the game, it is not a near value. The probability of dispersion decreases as τ increases. The probability of dispersion is $\frac{1}{2}$ on the first two intervals. It is zero on the last two intervals. Optimal strategies for H are grid hiding strategies. The optimal strategies for S turn out to be more involved: if $\tau \geq 2$ then S needs to search locations simultaneously.

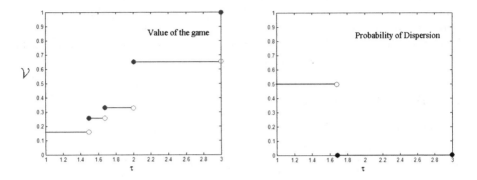

Fig. 2. Value of the game versus τ if $k = 2$ and $n = 3$. H hides objects in the same location if $\tau \geq \frac{5}{3}$.

We use the following notation. A pure hider strategy that puts the objects in locations 1 and 2 at places x and y is written as the triple $(x, y, 0)$. The coordinates of this triple correspond to the locations. A non-dispersive hider strategy in which objects are hidden at the same location, say one at x and one at 1 at location 3, is written as $(0, 0, \{x, 1\})$.

2.1   Total Time τ in $\left[1, \frac{3}{2}\right)$

H equiprobably chooses one of the following strategies:

$$\left(\left\{\frac{1}{2}, 1\right\}, 0, 0\right), \left(\frac{1}{2}, \frac{1}{2}, 0\right)$$

up to a permutation of the coordinates. These are a total of 6 pure strategies, 3 of which are dispersive, and H chooses either one equiprobably. Note that H hides at grid points and a grid point contains an object with probability $\frac{1}{2}$. We claim that any pure searcher strategy has a probability of a win of no more than $\frac{1}{6}$. To see this, first note that S optimally performs a grid search by Lemma 1. Since $\tau < \frac{3}{2}$ the searcher can only uncover two grid points. If the first grid point does not contain a hidden object, then S loses since the objects are placed at different grid points. If S finds a hidden object, which he does with probability $\frac{1}{2}$, then the remaining object can be in 3 grid points. S finds the second object with probability $\frac{1}{3}$.

S adopts the following mixed strategy. He picks a location at random and searches this until he finds an object. If S finds an object, then with probability $\frac{1}{2}$ he switches the search to another location and with probability $\frac{1}{2}$ he persists and continues the search in the same location. Suppose that H chooses a dispersive pure strategy $(x, y, 0)$. Then S wins if he starts the search in a location that contains an object (probability $\frac{2}{3}$) and decides to switch (probability $\frac{1}{2}$) to the right location (probability $\frac{1}{2}$). If H adopts a non-dispersive strategy, then S wins if he starts in the right location (probability $\frac{1}{3}$) and persists (probability $\frac{1}{2}$). In both cases, S wins with probability $\frac{1}{6}$.

2.2   Total Time τ in $\left[\frac{3}{2}, \frac{5}{3}\right)$

H equiprobably chooses one of the following strategies:

$$\left(\left\{\frac{1}{3}, 1\right\}, 0, 0\right), \left(\left\{\frac{2}{3}, 1\right\}, 0, 0\right), \left(\frac{1}{3}, \frac{2}{3}, 0\right)$$

up to a permutation of the coordinates. These are a total of 12 pure strategies, 6 of which are dispersive, and H chooses either one equiprobably. Again H hides at grid points, but the grid has increased from six to nine points, so the analysis gets a little more involved.

An object is at an end point (place 1) with probability $\frac{1}{6}$ and at an intermediate grid point with probability $\frac{1}{4}$. A grid search uncovers 4 grid points, since $\tau < 5/3$. Suppose that S uncovers an object at the first grid point (probability $\frac{1}{4}$). Then the next object can be in a grid point in either one of the three locations. S can only search one of these three grid points, so he finds the second object with probability $\frac{1}{3}$. If S does not uncover an object at the first grid point, then he can either persist or switch. Suppose that S persists. He only wins if he finds an object at the next grid point (probability $\frac{1}{3}$) otherwise he loses. If he indeed finds an object, then the remaining object can be at three grid points, of which S can uncover only two. So S wins with probability $\frac{2}{9}$ if he persists. If S switches and finds an object (probability $\frac{1}{3}$), then the remaining object can be at three locations, of which S can only uncover one. If S switches and does not find an object (probability $\frac{2}{3}$), then he can uncover two more grid points for two more objects. There are six remaining possibilities for the remaining objects and S can only uncover one such combination. So S wins with probability $\frac{1}{9} + \frac{2}{18}$. Regardless whether S decides to switch or persist, his probability of success is $\frac{2}{9}$ if he is not successful at the first grid point (probability $\frac{3}{4}$). We conclude that the probability of a searcher win is limited to $\frac{1}{12} + \frac{6}{36} = \frac{1}{4}$.

S picks a location at random and searches this until he finds an object. If S finds an object, then with probability $\frac{1}{4}$ he switches the search and with probability $\frac{3}{4}$ he persists. Against a non-dispersive pure strategy $(\{x, 1\}, 0, 0)$, S wins if he starts the search in the right location (probability $\frac{1}{3}$) and persists (probability $\frac{3}{4}$). If H adopts a dispersive strategy $(x, 1 - x, 0)$ with $x \geq \frac{1}{2}$, then

S wins if he starts in the location with the object placed at x (probability $\frac{1}{3}$) and switches (probability $\frac{1}{4}$). For in this case, S can searches both remaining locations up to place $1-x$ since $x+(1-x)+(1-x) \leq \frac{3}{2} \leq \tau$. If S persists, then he still wins if he continues the search in the location of $1-x$ after the location of x has been exhausted. If S starts in the location that contains $1-x$ then he wins if he switches to the location that contains x. So S wins with probability of at least $\frac{1}{12} + \frac{1}{8} + \frac{1}{24} = \frac{1}{4}$.

2.3   Total Time τ in $\left[\frac{5}{3}, 2\right)$

H hides both objects at one and the same end point $(\{1,1\},0,0)$. Obviously, this limits the probability of a searcher's win to $\frac{1}{3}$. S picks a location at random and performs an exhaustive search. If he finds one object at place x and if $x \geq \frac{2}{3}$ then he searches both remaining locations until place $1-x \leq \frac{1}{3}$. If $x < \frac{2}{3}$ then S picks one of the two remaining locations and searches it. Obviously, S wins with probability $\frac{1}{3}$ if H hides both objects at the same location. If H disperses $(x, 1-x, 0)$ for $x \geq \frac{2}{3}$, then S wins if he starts in the location that contains x. If H disperses $(x, 1-x, 0)$ for $\frac{1}{2} \leq x < \frac{2}{3}$ then S wins if he starts in a location that contains an object (probability $\frac{2}{3}$) and then searches the location that contains the remaining object (probability $\frac{1}{2}$). No matter what, S wins with probability $\frac{1}{3}$.

2.4   Total Time τ in $[2, 3)$

Again, H hides both objects at the same end point $(\{1,1\},0,0)$. This limits the probability of a searcher's win to $\frac{2}{3}$. The hider strategy remains the same, but the searcher strategy varies: S *picks two locations at random and searches them simultaneously.* As soon as S finds an object, he continues the search in that location and halts the search in the other location. Against a non-dispersive hider strategy if S finds the first object, then he also finds the second object. Since S starts out in two locations, he wins with probability $\frac{2}{3}$ in this case. Against a dispersive strategy, S finds one object for sure, but then continuous the search in a location that contains no further objects. If S finds an object at place $\geq \frac{1}{2}$ then he knows that the second object must be at the unsearched location and S uncovers it. If S finds an object at place $< \frac{1}{2}$ then S continues on the halted search in the other location. Against a dispersive strategy $(x, 1-x, 0)$ with $x \geq \frac{1}{2}$, S only loses if he starts out in the empty location and the location that contains $1-x$. So S has a guaranteed win of probability $\frac{2}{3}$.

This concludes our analysis of the case that $k = 2$ and $n = 3$. H adopts the strategy of hiding the objects in grid points, as illustrated below. S adopts more intricate strategies. In particular, he may search different locations simultaneously. Our analysis is very specific to the case $k = 2$ and $n = 3$ and no clear pattern emerges. It probably is difficult to devise an algorithm to compute the game.

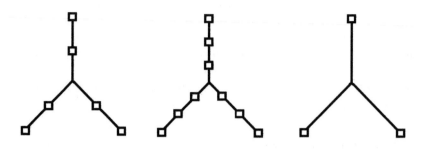

Fig. 3. H hides the objects at a finite number of locations, from left to right: positions of hidden objects in optimal mixed hider strategies for $\tau < \frac{3}{2}$, and $\frac{3}{2} \leq \tau < \frac{5}{3}$ and $\tau \geq \frac{5}{3}$

3   Simple Strategies

We simplify the game in order to find bounds on the value. Suppose S does not use any information and decides on which locations are searched, and by what amount, at the start of the game. A pure searcher strategy then corresponds to a non-negative vector $(y_1, \ldots, y_n)$ such that the sum of the coordinates is equal to τ. The coordinate y_i denotes the proportion of search in location i. If $y_i = 1$ then the location is searched fully. We say that this is a *simple strategy*. In this section we consider the game in which S uses simple strategies. Obviously, this restriction does not help S and the value of this game (probability of a win) decreases.

If S does not use any information, then H will not use decoys. So H may restrict to pure hider strategies that are given by a non-negative vector $(x_1, \ldots, x_n)$ of coordinate sum equal to 1, such that at most k coordinates are positive. In particular, the hider strategy space is equal to Δ_n^k, the k-skeleton of the unit n-simplex. The searcher strategy space corresponds to subsets V_y of simplex, given by

$$V_y = \{x \in \Delta_n^k : \sum_{i=1}^{n} y_i = \tau, x_i \leq y_i \text{ for all coordinates } i\}$$

If S picks a vector y and H picks a vector x, then S wins if and only if $x \in V_y$. The game with simple strategies is of the following general type: S may pick a certain subset of a space X and H may pick a point. S wins if and only if the subset contains the point. This game has been studied by McEliece and Posner [7] in the special case that X is a compact metric space and S may pick any closed ball of a fixed radius r. They show that the value of the game is equal to the absolute r-entropy, which can be computed from the covering number of Cartesian powers of X. Computing the covering number in this general setting is far from trivial [5, p. 1038]. In our case, the sets V_y are not balls of fixed radius in any metric on the k-skeleton. There seems to be no existing result that readily applies to the game with simple strategies.

Theorem 3. *If $k = 2$ and $\tau \leq 2$ and if S uses simple strategies, then H picks a point uniformly and S wins with probability $\frac{\tau - 1}{\binom{n}{2}}$.*

Proof. The 2-skeleton is the complete graph on n vertices (locations). It has $\binom{n}{2}$ edges of unit length. A set V_y intersects a subset of these edges in closed intervals. More specifically, V_y intersects the edge between i and j if and only if $y_i + y_j \geq 1$. The intersection is a subinterval of length $y_i + y_j - 1$. So the total length of V_y is equal to

$$\sum_{i<j} \max\{y_i + y_j - 1, 0\}.$$

Let's say that an edge of the complete graph is *heavy* if $y_i + y_j > 1$. If two edges are heavy, then they intersect since $\tau \leq 2$. Suppose that all heavy edges have a common vertex i_0. If $y_{i_0} = 1 - \epsilon < 1$ then we may add ϵ to y_{i_0} and subtract ϵ from a vertex j that is at the other end of a heavy edge. This can only increase the total length of V_y. So we may assume that $y_{i_0} = 1$. Then

$$\sum_{i<j} \max\{y_i + y_j - 1, 0\} = \sum_{j \neq i_0} \max\{y_{i_0} + y_j - 1, 0\} = \sum_{j \neq i_0} y_j,$$

which is equal to $\tau - 1$.

Now suppose that not all heavy edges have a common vertex. Suppose that $y_i + y_j > 1$ so every heavy edge contains either i or j. By assumption, there exists a heavy edge that contains i and a heavy edge that contains j. Say $y_i + y_p > 1$ and that $y_j + y_q > 1$. Then every heavy edge has a vertex in $\{i, j\}$ and in $\{i, p\}$ and in $\{j, q\}$. It is not hard to see that it must be equal to either ij or ip or jq. Then

$$\sum_{i<j} \max\{y_i + y_j - 1, 0\} = (y_i + y_j - 1) + (y_i + y_p - 1) + (y_j + y_q - 1) \leq 2\tau - 3$$

which is bounded by $\tau - 1$. So the total length of V_y is bounded by $\tau - 1$. If H chooses a point uniformly randomly from the 2-skeleton, then S finds it with probability $\leq \frac{\tau - 1}{\binom{n}{2}}$. This proves that the probability of a win for S is bounded by that fraction.

To prove that the probability is in fact equal to the fraction, it suffices to prove that V_y equals any subarc of Δ_n^2 of length $\tau - 1$. If the arc is properly contained in an edge, then it is of the form

$$\{x \in \Delta_n^2 : x_i \leq a, x_j \leq 1 - a + \tau, \text{ all other } x_k = 0\}$$

This arc is equal to V_y if $y_i = a$ and $y_j = \tau - a$ and all other $y_k = 0$. If the arc contains a vertex i in its interior and is contained in two edges ij and im, then take $y_i = 1$ and choose the proper y_j and y_m so that V_y equals that arc.   □

The value of this theorem puts a lower bound on $\mathcal{V}$ if $\tau \leq 2$ and $k = 2$. In particular, if $n = 3$ then the lower bound is $\frac{\tau-1}{6}$ which is quite sharp at the discontinuity points of the value $\mathcal{V}(\tau)$ in Figure 2.

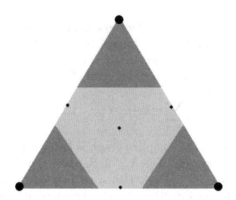

Fig. 4. Illustration of searcher and hider strategies if $k = n = 3$

If $k = n = 3$ and if $\tau = \frac{9}{5}$ then there exists a cover by four sets V_y of the triangle, as depicted in Figure 4. So S has a guaranteed win of probability $\frac{1}{4}$. There exists a subset of 7 points, consisting of the three vertices and four middle points, such that any V_y contains at most all middle points, or one vertex and one middle point. If H picks each vertex with probability $\frac{3}{13}$ and each middle point with probability $\frac{1}{13}$, then the probability that S wins is $\frac{4}{13}$ at most. It follows that the value of the game with simple strategies is between $\frac{1}{4}$ and $\frac{4}{13}$ in this particular case. A full analysis of the game with simple strategies is a challenging problem in combinatorial geometry.

4   Conclusion

We have introduced a new type of search game, with a single searcher and a team of hiders. We proved that the game is well-defined and has a (near) value but we could only solve the game for small parameter values. Our proof that the game has a value actually yields an algorithm to compute the game by discretization. We essentially showed that

$$\mathcal{V}_N\left(\tau - \frac{n}{N}\right) \leq \mathcal{V}(\tau) \leq \mathcal{V}_N(\tau)$$

and in principle we can compute $\mathcal{V}(\tau)$ up to arbitrary accuracy by taking a large enough N. Unfortunately, such an approach is not feasible as the number of grid searches is finite but very large even for small N. It is a challenging problem to find a good algorithm. A simplified version of the game, in which S does not use feedback from the search, is related to covering problems from combinatorial geometry that are known to be hard.

Our game has the qualitative feature that the team of hiders tends to disperse if the searcher's resources are low. We proved this for small parameter values only, but we conjecture that this is true for all parameter values. This is a purely mathematical problem, because of the setup of our game. A more realistic version would have incomplete information on the resources of the players. This could be an interesting line of further research.

References

1. Alpern, S., Clayton, N., Fokkink, R., Lidbetter, T.: A game theoretic model for nut caching by scatter hoarders (2010) (preprint)
2. Alpern, S., Gal, S.: A Mixed Strategy Minimax Theorem without Compactness. SIAM J. Control Optim. 26, 1357–1361 (1988)
3. Alpern, S., Gal, S.: The theory of search games and rendezvous. Kluwer, Dordrecht (2003)
4. Bier, V., Olivers, S., Samuelson, L.: Choosing what to protect. Defensive allocation against an unknown attacker. J. Publ. Economic Th. 9(4), 563–587 (2007)
5. Cormen, T.H., Leiserson, C.E., Rivest, R., Stein, C.: Introduction to Algorithms, 2nd edn. McGraw-Hill, New York (2002)
6. Garnaev, A.Y.: Search games and other applications of game theory. Springer, Heidelberg (2000)
7. McEliece, R., Posner, E.: Hiding and covering in a compact metric space. Ann. Stat. 1(4), 729–739 (1973)
8. Memon, N., Farley, J.D., Hicks, D.L., Rosenorn, T.: Mathematical methods in counterterrorism. Springer, Heidelberg (2009)
9. Owen, G., McCormick, G.H.: Finding a moving fugitive. A game theoretic representation of search. Comp. Oper. Res. 35, 1944–1962 (2008)
10. Reijnierse, J.H., Potters, J.A.M.: Search games with an immobile hider. Int. J. Game Theory 21, 385–394 (1993)
11. Ruckle, W.H.: Geometric games and their applications. Pitman (1983)
12. Statement of Peter Clarke, head of Scotland Yard's anti-terrorist branch (August 10, 2006), http://news.bbc.co.uk/2/hi/4778575.stm
13. Zoroa, N., Zoroa, P., Fernández-Sáez, M.J.: Weighted search games. European J. Oper. Res. 195, 394–411 (2009)

Uncertainty in Interdependent Security Games*

Benjamin Johnson[1], Jens Grossklags[2], Nicolas Christin[1], and John Chuang[3]

[1] CyLab, Carnegie Mellon University
[2] Center for Information Technology Policy, Princeton University
[3] School of Information, UC Berkeley
{johnsonb,nicolasc}@andrew.cmu.edu,
jensg@princeton.edu,
chuang@ischool.berkeley.edu

Abstract. Even the most well-motivated models of information security have application limitations due to the inherent uncertainties involving risk. This paper exemplifies a formal mechanism for resolving this kind of uncertainty in interdependent security (IDS) scenarios. We focus on a single IDS model involving a computer network, and adapt the model to capture a notion that players have only a very rough idea of security threats and underlying structural ramifications. We formally resolve uncertainty by means of a probability distribution on risk parameters that is common knowledge to all players. To illustrate how this approach might yield fruitful applications, we postulate a well-motivated distribution, compute Bayesian Nash equilibria and tipping conditions for the derived model, and compare these with the analogous conditions for the original IDS model.

1 Introduction

Starting with the Morris Worm in 1988, security attacks on computer systems have gradually shifted from "point-to-point" attacks, where a single attacker targets a single defender, e.g., to deny service, to propagation attacks, where the attacker attempts to compromise a few machines and, similar to an epidemic, uses these compromised machines ("bots" or "zombies") to infect additional hosts. The advantage of propagation attacks is that the miscreants behind them can commandeer reasonably quickly a very large pool of machines, which can, in turn, be monetized. Among many other activities, bots have been used to send spam email, host phishing websites [22], or acquire banking credentials [5].

Traditional security models that pit a defender (or a set of defenders) against an external attacker may not capture all the intricacies of propagation attacks, as the attacker population may vary over time. In contrast, models of interdependent security (e.g., [18]), where hosts in the network may (involuntarily or not) act on behalf of the attacker, appear more suitable to characterize propagation attacks.

Interdependent security models have been used in the context of airline security [14], and disease propagation [15]. In these contexts, it may be possible to characterize infection rates or measure attack probability based on historical data. In the context of

---

* This research was supported in part by CyLab at Carnegie Mellon under grant DAAD19-02-1-0389 from the Army Research Office, and by the National Science Foundation through award CCF-0424422 (TRUST - Team for Research in Ubiquitous Secure Technology).

T. Alpcan, L. Buttyan, and J. Baras (Eds.): GameSec 2010, LNCS 6442, pp. 234–244, 2010.

information security, on the other hand, we posit that uncertainty on the possibility of an attack, and ambiguity on the configuration of other networked hosts imposes significant challenges for the selection of effective security strategies.

For instance, networks in many organizations may be quite large, and are prone to have poorly known configuration parameters, even by their own administrators [19]. A firewall that governs the entrance to the network may have thousands of rules, some of them obsolete, some of them redundant, and thus it may be difficult to explicitly characterize the probability a given outside attack could actually succeed in penetrating the corporate network. Network configurations may be relatively complex, and two machines located close to each other geographically may be far apart in the network topology. In the end, network administrators may only have very rough estimates of the various probabilities of external attacks or of attack propagation between interior nodes [3].

The contribution of this paper is to introduce and exemplify a method for resolving risk uncertainty by means of a well-motivated probability distribution on risk parameters. We introduce the method within the context of a single interdependent security game that draws its motivation from an organizational LAN in which agents have a significant residual impact on the security of their own and their peers' resources (e.g., such as in university and many corporate networks). Our examples show that such distributions can be easily motivated, and that the resulting derived conditions for equilibria and tipping effects are reasonable, in the sense that they compare similarly to equilibrium conditions derived in the original IDS model using the distribution's expected values of the model's risk parameters.

The rest of this paper is organized as follows. We review related work in Section 2. In Section 3, we describe our model, which is directly inspired by the work of Kunreuther and Heal [18], and explain how we address risk uncertainty within this model. We provide formal and numerical analysis of interdependent security games with homogeneous and heterogeneous populations, and with or without uncertainty, in Section 4. We conclude in Section 5.

2   Related Work

2.1   Interdependent Security

In their 2003 study, Kunreuther and Heal formalize the concept of interdependent security with their primary example stemming from the airline industry [14,18]. In this case, the individual airlines are concerned about a major single attack that may originate at some point in the network, but could be propagated to another airline in the system. Airlines can defend themselves against direct attacks, however, they are powerless against dangerous loads received from other aviation entities. In follow-up work, they also consider a game in which players can protect themselves effectively against direct and indirect attacks through some protection measure (e.g., vaccination), however the benefit of the security investment diminishes with its popularity in the population [15]. This research has motivated follow-up contributions in algorithmic computation of equilibria with real-world data [17], and human-subjects experimentation in the laboratory [16].

In this paper, we refer to a complementary computer security model commented upon by Kunreuther and Heal [18]. In this scenario, a single compromised network resource can adversely impact other connected entities multiple times. We study this game more formally by deriving game-theoretic equilibrium solutions for different information conditions and network-wide behaviors (e.g., tipping point phenomena).

Concurrently to the research on interdependent security, Varian started a formal discussion on the role of security as a public good [25]. In our work, we expanded on his work by developing a security games framework including additional games and investment strategies (i.e., self-insurance) [9]. We also considered the impact of player heterogeneity [10], and the influence of strategically acting attackers on the security outcome [7].

An alternative optimization approach is pursued by Miura-Ko *et al.* who derive Nash equilibrium conditions for simultaneous move games in which the heterogeneous interactions of players can be represented with a set of piece-wise linear conditions [21]. They further enrich their basic model to develop three studies on password security, identity theft, and routing path verification. The authors verify the robustness of their approach to perturbations in the data, however, do not formally consider the role of uncertainty.

2.2   Uncertainty and Security

In the context of the value of security information, research has been mostly concerned with incentives for sharing and disclosure. Several models investigate under which conditions organizations are willing to contribute to an information pool about security breaches and investments when competitive effects may result from this cooperation [8]. Empirical papers explore the impact of mandated disclosures [4] or publication of software vulnerabilities [24] on the financial market value of corporations.

Another contribution to the security field is the preservation of location privacy in mobile networks [6]. A different approach is followed by Alpcan and Başar who present an application of game theory and stochastic-dynamic optimization to attack scenarios in the sensor network context [2].

In our prior work, we studied the impact of uncertainty in three different games [11,13]. We also developed a set of metrics to study the value of better information [12].

A more extended review of theoretical and empirical work is presented by Acquisti and Grossklags in which they discuss the moderating role of risk, uncertainty and ambiguity in the areas of privacy and security [1].

3   Model

3.1   Interdependent Network Security

We focus our attention on interdependent security games that directly model network security. For the basic setup, suppose that each of n players is responsible for operating her personal computer, and that players' computers are connected to each other through a given internal network, e.g., a corporate LAN. Each computer is also connected to an external network, e.g., the Internet. The external connection poses certain risks (e.g., infection with viruses), and if a user's resources are compromised then she will suffer a total loss, normalized to 1. In addition, some of these viruses have the ability to propagate through the internal network to compromise all the other players' computers. If a player's computer is compromised in this way, she also faces a total loss of 1.

Each player has a choice of investing in security mechanisms with a cost c to eliminate the risk of being infected by an external virus. For consistency with normalizing losses, we assume $c \leq 1$. We also assume that there is no effective way to protect from the risk of an internal contamination as the result of another player passing along a virus through the internal network. This modeling choice reflects a relatively common situation in corporate networks where security policies are set to have computers almost blindly trust contents coming from inside the corporate network (which facilitates automated patching, and software updates for instance), while contents coming from outside of the network are thoroughly inspected.

In the full heterogeneous version of the game, p_{ii} is the probability that player i becomes infected with a virus, and p_{ij} is the probability that player i causes player j to become contaminated due to virus transmission. Since a computer must be infected before contaminating another computer, we may assume that $p_{ij} \leq p_{ii}$ for every i and j. For simplicity of analysis we assume any two events involving virus contamination are independent, and similarly for any two events involving virus infection. We also consider a more specialized homogeneous version of the game in which the value of p_{kl} depends only on whether $k = l$. So that p is the probability that a given computer becomes infected with a virus, and q is the probability that a given computer contaminates another given computer in the system.

The utility of each player in this game depends not only on her choice to protect, but also on the choices of other players. If there are k players in the network who are not protecting, then player i's choice can be framed as follows. If she protects, then she pays a cost c, eliminating the risk of a direct virus infection, but she still faces the risk of internal contamination from k different players. If she fails to protect, then she does not pay c, but she faces both the risk of an internal contamination from one of the k players, as well as the risk of an external infection. The utility function for player i is derived directly from these considerations.

For the homogeneous version of the game, the expected utility of player i is given by the equation:

$$U_i = \begin{cases} -c + (1-q)^k & \text{if player } i \text{ protects} \\ (1-p)(1-q)^k & \text{if player } i \text{ does not protect} \end{cases} \quad (1)$$

where k is the number of players other than i who choose not to protect.

In the heterogeneous version of the game, the expected utility for player i is given by:

$$U_i = \begin{cases} -c + \prod_{j \neq i: e_j = 0}(1 - p_{ji}) & \text{if player } i \text{ protects} \\ (1 - p_{ii}) \prod_{j \neq i: e_j = 0}(1 - p_{ji}) & \text{if player } i \text{ does not protect} \end{cases} \quad (2)$$

where e_j in a binary indicator variable telling us whether player j chooses to protect.

3.2 Uncertainty

In the usual treatment of interdependent security (IDS) games such as the one above, the risk parameters (i.e. the p_{ij}) are known. We are interested in the case in which the risks of virus infection and contamination are unknown. Such uncertainty is especially well-motivated in the IDS computer network game since computer users in general do not know or understand well the potential risks posed by various types of viruses.

For our model with uncertainty, we assume that players do not know the risks, but they believe and agree upon some probability distribution over risk parameters. In other words, there is a probability distribution D that describes players beliefs about the relevant risks. True to rational Bayesian form, everyone believes that the relevant risk parameters are drawn from the same distribution D.

In the homogeneous case, D is a distribution on $[0, 1] \times [0, 1]$, representing the players' mutually-held beliefs about the parameters p and q. In the heterogeneous case, D is a distribution on $[0, 1]^{n \times n}$, representing players' mutually-held beliefs about the parameters p_{ij}.

4 Analysis

4.1 Overview

Our analysis focuses on determining equilibrium conditions. We start with the homogeneous version and then proceed to the heterogeneous version. In each case, we begin by looking at the game with full information and computing conditions under which various Nash equilibria exist and how they can be tipped or disrupted. We extend these results to the realm of uncertainty by positing a general distribution D and rewriting the equilibria conditions using expected values of aggregate risk parameters conditioned on D. We follow by providing and motivating a parametrized example distribution D_{ϵ} and using this distribution to compute various equilibrium conditions explicitly. In the homogeneous version, we analyze these conditions numerically and graphically, and compare the results to the original IDS game in which risk parameters are known.

4.2 Homogeneous Case: A Monoculture of Potential Failure Modes

Nash Equilibrium. We begin with the homogeneous case. Let's first assume that p and q are known. This game has two possible strong Nash equilibria, one in which all players protect, and one in which no player protects. Considering a simple cost-benefit analysis, the "everyone-protects" equilibrium is achievable if and only if the cost of protection is less than the cost of an external infection (i.e. $c < p$). Similarly, the "everyone-defects" equilibrium is achievable if and only if the cost of protection is greater than the likelihood that a player is infected, but not compromised, assuming that all of the other players are failing to protect (i.e. $c > p(1 - q)^{n-1}$). In the middle area $p(1 - q)^{n-1} < c < p$, both equilibria are possible, the protection equilibrium is Pareto optimal, and both equilibria are subject to the possibility of tipping phenomena in which forcing a certain number of players to switch strategies will effect the opposite equilibrium.

Tipping Phenomenon. To understand this game's tipping phenomenon when there are n players, it suffices to understand the game's defection equilibrium conditions when there are k players and $k < n$.

If players are in an "everyone defects" equilibrium, then to tip the equilibrium to one in which everyone protects, it is necessary (and sufficient) to force protection upon enough players so that universal defection among the remaining players is no longer an equilibrium strategy. The number of forced protections required to accomplish this is the least integer k such that $c < p(1 - q)^{n-1-k}$. In words, k is the least integer such that universal defection fails to be an equilibrium strategy in a game with only $n - k$ players.

Similarly, if players are in an "everyone protects" equilibrium, then the number of defections required to tip the equilibrium toward universal defection is the least integer k such that $c > p(1 - q)^k$. In this case k is the least integer such that, in a game with $k + 1$ players, universal defection is an equilibrium strategy.

In any case, the boundary conditions that describe the tipping phenomenon are the same conditions that describe defection equilibria in games with fewer players.

Uncertainty. When dealing with a joint probability distribution over the parameters p, q, the above reasoning applies with the exception that players compute an expected value for p and $p(1 - q)^{n-1}$ using the distribution D. Thus "everyone protects" is an equilibrium if and only if $c < E_D[p]$ and "everyone defects" is an equilibrium if and only if $c > E_D[p(1 - q)^{n-1}]$. The tipping phenomenon have an analogous translation involving these expected values.

Example distribution D_ε. To exemplify the scenario, we propose a class of distributions D_ε, parametrized by a number $\varepsilon \in [0, 1]$. To motivate this distribution, we suppose that there is a fixed number $\varepsilon \in [0, 1]$ such that players believe the risk of external infection, p, is no more than ε. D_ε then assigns a probability to the pair $p, q \in (0, 1) \times (0, 1)$ according to the following two-step procedure. First draw p from the uniform distribution on $(0, \varepsilon)$. Then draw q from the uniform distribution on $(0, p)$. Since the only thing players really know for certain about the risks are that $0 \leq q \leq p \leq 1$, the parametrized distribution D_ε represents an effort to reflect the notion that "infection is somewhat unlikely, ('somewhat' being explicitly quantified by the parameter ε), and contamination as a result of infection is even less likely, and aside from that we do not have a very good idea what the risk is."

Bayesian Nash equilibrium for D_ε. To determine the Bayesian Nash equilibrium conditions for the parametrized game with uncertainty, we must compute the expected values $E_{D_\varepsilon}[p]$ and $E_{D_\varepsilon}[p(1 - q)^{n-1}]$ explicitly. The expected value of p under D_ε is $\frac{\varepsilon}{2}$, because p is drawn from the uniform distribution on $(0, \varepsilon)$. The expected value of $p(1 - q)^{n-1}$ under D_ε can be computed by evaluating the expression:

$$\frac{1}{\varepsilon} \int_0^\varepsilon \left(\frac{1}{p} \int_0^p p(1-q)^{n-1} dq \right) dp \tag{3}$$

where the inner integral is to be evaluated assuming that p is constant relative to q. The expression evaluates to

$$\frac{1}{n} \left(1 - \frac{1 - (1-\varepsilon)^{n+1}}{\varepsilon(n+1)} \right). \tag{4}$$

When $\varepsilon = 1$ this expression simplifies to $\frac{1}{n+1}$. In practical terms, the parameter selection $\varepsilon = 1$ describes a situation in which players have so little knowledge of the risk factors, that they may as well believe the parameters are uniformly distributed across all possible options. Under such conditions and with many players, the protection costs must be very small to counteract defection incentives. On the other hand, from a social planner's point of view the situation may be manageable, as the total cost (cost per player × number of players) necessary to properly incentivize network protection is bounded by a constant independent of the network size.

Graphical analysis. Figure 1 plots the boundary conditions for Bayesian Nash equilibrium as a function of ε, for a range of N. For comparison, Figure 2 plots the boundary conditions for Nash equilibria in the full information case when $p = \frac{\varepsilon}{2}$ and $q = \frac{\varepsilon}{4}$. Figure 3 exemplifies the equilibrium tipping phenomenon in a 7-player game.

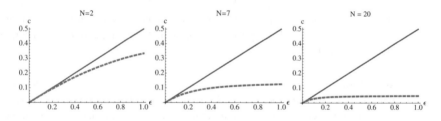

Fig. 1. Bayesian Nash equilibrium boundaries for the homogeneous game with N players. If (c, ε) is below the solid line then "everyone protects" is a Bayesian Nash Equilibrium. If (c, ε) is above the dashed line, then "everyone defects" is a Bayesian Nash equilibrium. In the middle area, there are competing equilibria, and tipping points.

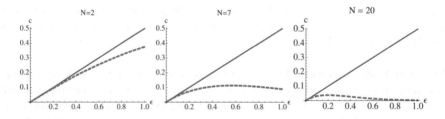

Fig. 2. Nash equilibrium boundaries for the homogeneous case with p and q common knowledge among all players. For comparison with Figure 1, we assume that $p = \frac{\varepsilon}{2}$ and $q = \frac{\varepsilon}{4}$.

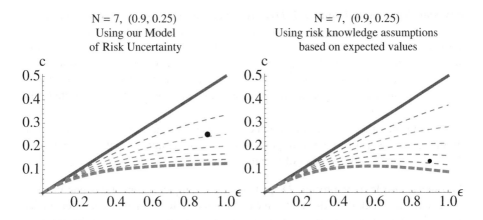

Fig. 3. Tipping point boundaries for the homogeneous game with 7 players. In this example, the risk threshold value is $\varepsilon = 0.9$ and the protection cost is $c = 0.25$. In the model incorporating uncertainty, it takes 2 defections to tip a protection equilibrium into a defection equilibrium, while in the model in which risk parameters are known, it takes 3 defections to tip the equilibrium from protection to defection. In the other direction from universal defection to universal protection, tipping the equilibrium requires 5 forced protections in the case of the uncertainty model and 4 forced protections in the case of knowledge assumptions.

4.3   Heterogeneous Case: Unknown and Diverse Configuration Problems

Nash Equilibrium. For the heterogeneous case, we begin by assuming the p_{ij} are known. Here, once again, the strategy "everyone protects" is a Nash equilibrium if and only if $c < p_{ii}$. The strategy "everyone defects" is a Nash equilibrium if and only if $c > p_{ii} \prod_{j \neq i}(1 - p_{ji})$. In the middle area $p_{ii} \prod_{j \neq i}(1 - p_{ji}) < c < p_{ii}$, both equilibria are possible, the protection equilibrium is Pareto optimal, and the situation is subject to the tipping phenomenon.

Tipping Phenomenon. The tipping phenomenon in the heterogeneous case is completely analogous to the homogeneous case. Tipping conditions for an n-player game are determined by considering the defection equilibrium conditions for games with fewer players.

Uncertainty. In the presence of uncertainty with beliefs about risk parameters governed by a general distribution D, the same analysis as above holds with $E_D[p_{ii}]$ and $E_D[p_{ii} \prod_{i \neq j} p_{ji}]$ replacing p_{ii} and $p_{ii} \prod_{j \neq i}(1 - p_{ji})$, respectively.

Example distribution D_ε. To exemplify the heterogeneous scenario, we propose a class of distributions D_ε, analogous to the homogeneous case, again parametrized by a number $\varepsilon \in [0, 1]$. As before, players believe the risk of external infection, p_{ii}, is no more than ε. D_ε then assigns a probability to the matrix $p_{ij} \in (0, 1)^{n \times n}$ according to the following procedure. First draw each p_{ii} independently from the uniform distribution on $(0, \varepsilon)$. Then draw each p_{ij} independently from the uniform distribution on $(0, p_{ii})$, so that $0 \leq p_{ij} \leq p_{ii} \leq 1$ for every $i \neq j$.

Bayesian Nash equilibrium for D_ε. To determine the Bayesian Nash equilibrium conditions for the parametrized game with uncertainty in the heterogeneous case, we must compute the expected values $E_{D_\varepsilon}[p_{ii}]$ and $E_{D_\varepsilon}[p_{ii}(1-p_{ji})^{n-1}]$ explicitly. Unlike the homogeneous case, these expected values are trivial to compute because all the variables in relevant expressions are independent (or conditionally independent, e.g. for $i \neq j$, $\frac{p_{ij}}{p_{ii}}$ is independent of p_{ii}). The expected value of p_{ii} is $\frac{\varepsilon}{2}$, and the expected value of $p_{ii}(1 - p_{ji})^{n-1}$ is $\frac{\varepsilon}{2}(1 - \frac{\varepsilon}{4})^{n-1}$.

We omit the graphical analysis for the heterogeneous case both due to space constraints and because there is no simple way to compare results with the original model due to differences in the number of free parameters.

Another example distribution, $D_{\epsilon,i}$. One final example to consider is one in which players mutually acknowledge that some computers are more likely to be infected than others. We can exemplify this scenario by using a distribution $D_{\varepsilon,i}$ that discriminates among risk parameters for different players. $D_{\varepsilon,i}$ assigns a probability to the matrix $p_{ij} \in (0,1)^{n \times n}$ according to the following procedure. First draw each p_{ii} independently from the uniform distribution on $(0, \varepsilon_i)$. Then draw each p_{ij} from the uniform distribution on $(0, p_{ii})$. The distribution $D_{\varepsilon,i}$ reflects the same uncertain sentiment regarding risk as D_ε, yet it also accommodates a notion – certainly realized in practice – that some assets bear higher risk level than others.

Under the distribution $D_{\varepsilon,i}$, the computations involved in determining each player's strategic response to the behavior of others are analogous to those computations under the distribution D_ε. Again the individual variables in the relevant expressions are drawn independently so that linearity of expectation can be applied. For example, when all other players are failing to protect, player i will also fail to protect if and only if $c > \frac{\varepsilon_i}{2} \prod_{j \neq i}(1 - \frac{\varepsilon_j}{4})$. Unfortunately, determining all possible Bayesian Nash equilibria requires addressing a number of caveats, because players have different incentives due to the homogeneity in their beliefs about their respective risks. We defer a thorough analysis of this scenario to future work.

5 Discussion and Conclusions

Interdependent models of information security in corporate networks seem especially well-motivated, but it is difficult to utilize the sharpness of these models due to uncertainty regarding real world risk factors. Our approach has been to make these models smoother, by incorporating players' uncertainty about various risk parameters.

Our objective has been to develop a mechanism for dealing with risk uncertainty in a security context. We focused on a single IDS model involving a computer network, and we adapted the model to capture a notion that players have only a very rough idea of security threats and underlying structural ramifications. We formally resolved this uncertainty by means of a probability distribution on risk parameters, one that was common knowledge to all players. We postulated a reasonable such distribution, computed Bayesian Nash equilibria and tipping conditions for the resulting model, and compared those to the same conditions for the original model.

Crucially from a practical standpoint, we incorporated this new probabilistic machinery while actually assuming less – indeed our adapted model using the example distribution D_ε reduced the number of free parameters. Nonetheless, we found that the adapted model maintains characteristic equilibrium properties and asymptotic behaviors when information assumptions are relaxed. There are still only the two extreme equilibria. There is still a range of cost and risk distribution parameters for which the equilibrium can be tipped the other way by encouraging some players to switch strategies. Even the boundary conditions for equilibrium conditions and tipping effects are similar to those obtained from the original model, and we would expect such similarities to extend to other well-motivated probability distributions in other contexts.

There were some mild differences compared to the full knowledge model using the distribution's expected values of model parameters. In our homogeneous model incorporating uncertainty, a generally low contamination risk facilitated the possibility of slightly more defections, while a generally moderate to high contamination risk facilitated fewer defections. An application of this phenomenon is that when risks are small, it may be better from a social planner's standpoint to communicate such risks by using expected values of parameters, while if risks are large it may be better to present them in a manner that incorporates uncertainty using a distribution.

As a general rule, when we apply a security model to a real world situation, we expect that some real world data will be substituted for the parameters in the model. Unfortunately this is oftentimes difficult or impossible to do, especially for risk parameters. Without knowing the risks, we are left with the problem of how to use the model for anything at all. Our approach addresses this situation in a reasonable way for a very simple model. The approach itself is quite general and we expect to find additional applications in future work.

References

1. Acquisti, A., Grossklags, J.: What can behavioral economics teach us about privacy? In: Acquisti, A., Gritzalis, S., Di Vimercati, S., Lambrinoudakis, C. (eds.) Digital Privacy: Theory, Technologies, and Practices, pp. 363–380. Auerbach Publications, Boca Raton (2007)
2. Alpcan, T., Basar, T.: An intrusion detection game with limited observations. In: Proceedings of the 12th International Symposium on Dynamic Games and Applications, Sophia Antipolis, France (July 2006)
3. Bashir, M., Christin, N.: Three case studies in quantitative information risk analysis. In: Proceedings of the CERT/SEI Making the Business Case for Software Assurance Workshop, Pittsburgh, PA, pp. 77–86 (September 2008)
4. Campbell, K., Gordon, L., Loeb, M., Zhou, L.: The economic cost of publicly announced information security breaches: Empirical evidence from the stock market. Journal of Computer Security 11(3), 431–448 (2003)
5. Franklin, J., Paxson, V., Perrig, A., Savage, S.: An inquiry into the nature and causes of the wealth of internet miscreants. In: Proceedings of 14th ACM Conference on Computer and Communications Security (CCS), Alexandria, VA, pp. 375–388 (October 2007)
6. Freudiger, J., Manshaei, M., Hubaux, J.-P., Parkes, D.: On non-cooperative location privacy: A game-theoretic analysis. In: Proceedings of the 16th ACM Conference on Computer and Communications Security (CCS 2009), Chicago, IL, pp. 324–337 (November 2009)
7. Fultz, N., Grossklags, J.: Blue versus red: Towards a model of distributed security attacks. In: Dingledine, R., Golle, P. (eds.) FC 2009. LNCS, vol. 5628, pp. 167–183. Springer, Heidelberg (2009)

8. Gal-Or, E., Ghose, A.: The economic incentives for sharing security information. Information Systems Research 16(2), 186–208 (2005)
9. Grossklags, J., Christin, N., Chuang, J.: Secure or insure? A game-theoretic analysis of information security games. In: Proceedings of the 2008 World Wide Web Conference (WWW 2008), Beijing, China, pp. 209–218 (April 2008)
10. Grossklags, J., Christin, N., Chuang, J.: Security and insurance management in networks with heterogeneous agents. In: Proceedings of the 9th ACM Conference on Electronic Commerce (EC 2008), Chicago, IL, pp. 160–169 (July 2008)
11. Grossklags, J., Johnson, B.: Uncertainty in the weakest-link security game. In: Proceedings of the International Conference on Game Theory for Networks (GameNets 2009), Istanbul, Turkey, pp. 673–682 (May 2009)
12. Grossklags, J., Johnson, B., Christin, N.: The price of uncertainty in security games. In: Proceedings (online) of the Eighth Workshop on the Economics of Information Security (WEIS), London, UK (June 2009)
13. Grossklags, J., Johnson, B., Christin, N.: When information improves information security. In: Proceedings of the 2010 Financial Cryptography Conference (FC 2010), Tenerife, Spain, pp. 416–423 (January 2010)
14. Heal, G., Kunreuther, H.: IDS models of airline security. Journal of Conflict Resolution 49(2), 201–217 (2005)
15. Heal, G., Kunreuther, H.: The vaccination game. Technical report, Columbia Business School & The Wharton School (January 2005)
16. Hess, R., Holt, C., Smith, A.: Coordination of strategic responses to security threats: Laboratory evidence. Experimental Economics 10(3), 235–250 (2007)
17. Kearns, M., Ortiz, L.: Algorithms for interdependent security games. In: Thrun, S., Saul, L., Schölkopf, B. (eds.) Advances in Neural Information Processing Systems, vol. 16, pp. 561–568. MIT Press, Cambridge (2004)
18. Kunreuther, H., Heal, G.: Interdependent security. Journal of Risk and Uncertainty 26(2-3), 231–249 (2003)
19. Le, F., Lee, S., Wong, T., Kim, H., Newcomb, D.: Detecting network-wide and router-specific misconfigurations through data mining. IEEE/ACM Trans. Netw. 17(1), 66–79 (2009)
20. Liu, Y., Comaniciu, C., Man, H.: A Bayesian game approach for intrusion detection in wireless ad hoc networks. In: Proceedings of the Workshop on Game Theory for Communications and Networks, Pisa, Italy (October 2006)
21. Miura-Ko, A., Yolken, B., Mitchell, J., Bambos, N.: Security decision-making among interdependent organizations. In: Proceedings of the 21st IEEE Computer Security Foundations Symposium (CSF 2008), Pittsburgh, PA, pp. 66–80 (June 2008)
22. Moore, T., Clayton, R., Anderson, R.: The economics of online crime. Journal of Economic Perspectives 23(3), 3–20 (2009)
23. Paruchuri, P., Pearce, J., Marecki, J., Tambe, M., Ordonez, F., Kraus, S.: Playing games for security: An efficient exact algorithm for solving Bayesian Stackelberg games. In: Proceedings of the 7th Int. Conf. on Autonomous Agents and Multiagent Systems (AAMAS 2008), Estoril, Portugal, pp. 895–902 (May 2008)
24. Telang, R., Wattal, S.: An empirical analysis of the impact of software vulnerability announcements on firm stock price. IEEE Transactions on Software Engineering 33(8), 544–557 (2007)
25. Varian, H.R.: System reliability and free riding. In: Camp, L.J., Lewis, S. (eds.) Economics of Information Security. Advances in Information Security, vol. 12, pp. 1–15. Kluwer Academic Publishers, Dordrecht (2004)

Attack–Defense Trees and Two-Player Binary Zero-Sum Extensive Form Games Are Equivalent

Barbara Kordy*, Sjouke Mauw,
Matthijs Melissen**, and Patrick Schweitzer***

University of Luxembourg

Abstract. Attack–defense trees are used to describe security weaknesses of a system and possible countermeasures. In this paper, the connection between attack–defense trees and game theory is made explicit. We show that attack–defense trees and binary zero-sum two-player extensive form games have equivalent expressive power when considering satisfiability, in the sense that they can be converted into each other while preserving their outcome and their internal structure.

1 Introduction

Attack trees [1], as popularized by Bruce Schneier at the end of the 1990s, form an informal but powerful method to describe possible security weaknesses of a system. An attack tree basically consists of a description of an attacker's goal and its refinement into sub-goals. In case of a *conjunctive* refinement, all sub-goals have to be satisfied to satisfy the overall goal, while for a *disjunctive* refinement satisfying any of the sub-goals is sufficient to satisfy the overall goal. The non-refined nodes (i.e., the leaves of the tree) are basic attack actions from which complex attacks are composed.

Due to their intuitive nature, attack trees prove to be very useful in understanding a system's weaknesses in an informal and interdisciplinary context. The development of an attack tree for a specific system may start by building a small tree that is obviously incomplete and describes the attacks at a high level of abstraction, while allowing to refine these attacks and to add new attacks later as to make a more complete description. Over the last few years, attack trees have developed into an even more versatile tool. This is due to two developments. The first development consists of the formalization of the attack trees method [2] which provides an attack tree with a precise meaning. As a consequence, formal analysis techniques were designed [3,4] and computer tools were made commercially available [5,6].

* B. Kordy was supported by the grant No. C08/IS/26 from FNR Luxembourg.
** M. Melissen was supported by the grant No. PHD–09–082 from FNR Luxembourg.
*** P. Schweitzer was supported by the grant No. PHD–09–167 from FNR Luxembourg.

T. Alpcan, L. Buttyan, and J. Baras (Eds.): GameSec 2010, LNCS 6442, pp. 245–256, 2010.
© Springer-Verlag Berlin Heidelberg 2010

The second development comes from the insight that a more complete description can be achieved by modeling the activities of a system's defender in addition to those of the attacker. Consequently, one can analyze which set of defenses is optimal from the perspective of, for instance, cost effectiveness. Several notions of protection trees or defense nodes have already been proposed in the literature [7,8]. They mostly consist of adding one layer of defenses to the attack tree, thus ignoring the fact that in a dynamic system new attacks are mounted against these defenses and that, consequently, yet more defenses are brought into place. Such an alternating nature of attacks and defenses is captured in the notion of attack–defense trees [9]. In this recently developed extension of attack trees, the iterative structure of attacks and defenses can be visualized and evolutionary aspects can be modeled.

These two developments, the formalization of attack trees and the introduction of defenses, imply that an attack–defense tree can be formally considered as a description of a game. The purpose of this paper is to make the connection between attack–defense trees and game theory explicit. We expect that the link between the relatively new field of attack modeling and the well-developed field of game theory can be exploited by making game theoretic analysis methods available to the attack modeling community. As a first step, we study the relation between attack–defense trees and games in terms of expressiveness. Rather than studying the graphical attack–defense tree language, we consider an algebraic representation of such trees, called *attack–defense terms* (ADTerms) [9], which allows for easier formal manipulation.

The main contribution of this paper is to show that ADTerms with a satisfiability attribute are equivalent to two-player binary zero-sum extensive form games. Whenever we talk about games, we refer to a game in this class. We show equivalence by defining two mappings: one from games to ADTerms and one from ADTerms to games. Then, we interpret a *strategy* in the game as a *basic assignment* for the corresponding ADTerm and vice versa. Such a basic assignment expresses which attacks and defenses are in place. Equivalence then roughly means that for every winning strategy, there exists a basic assignment that yields a satisfiable term, and vice versa. Although the two formalisms have much in common, their equivalence is not immediate. Two notions in the domain of ADTerms have no direct correspondence in the world of games: conjunctive nodes and refinements. The mapping from ADTerms into games will have to solve this in a semantically correct way.

This paper is structured as follows. We introduce attack–defense terms and two-player binary zero-sum extensive form games in Section 2. In Section 3 we define a mapping from games to attack–defense terms and prove that a player can win the game if and only if he is successful in the corresponding ADTerm. A reverse mapping is defined in Section 4.

Proofs of theorems are not included due to space restrictions, and can be found in a technical report [10].

2 Preliminaries

2.1 Attack–Defense Trees

A limitation of attack trees is that they cannot capture the interaction between attacks carried out on a system and defenses put in place to fend off the attacks. To mitigate this problem and in order to be able to analyze an attack–defense scenario, attack–defense trees are introduced in [9]. Attack–defense trees may have two types of nodes: attack nodes and defense nodes, representing actions of two opposing players. The attacker and defender are modeled in a purely symmetric way. To avoid differentiating between attack–defense scenarios with an attack node as a root and a defense node as a root, the notions of *proponent* (denoted by p) and *opponent* (denoted by o) are introduced. The root of an attack–defense tree represents the main goal of the proponent. To be more precise, when the root is an attack node, the proponent is an attacker and the opponent is a defender, and vice versa.

To formalize attack–defense trees we use attack–defense terms. Given a set S, we write S^* for the set of all strings over S and ε for the empty string.

Definition 1. *Attack–defense terms (ADTerms) are typed ground terms over a signature $\Sigma = (S, F)$, where*

- *$S = \{p, o\}$ is a set of types (we denote $-p = o$ and $-o = p$),*
- *$F = \{(\vee_k^p)_{k \in \mathbb{N}}, (\wedge_k^p)_{k \in \mathbb{N}}, (\vee_k^o)_{k \in \mathbb{N}}, (\wedge_k^o)_{k \in \mathbb{N}}, c^p, c^o\} \cup \mathbb{B}^p \cup \mathbb{B}^o$ is a set of functions equipped with a mapping $\mathrm{type} \colon F \to S^* \times S$, which expresses the type of each function as follows. For $k \in \mathbb{N}$,*

$$\mathrm{type}(\vee_k^p) = (p^k, p) \qquad\qquad \mathrm{type}(\vee_k^o) = (o^k, o)$$
$$\mathrm{type}(\wedge_k^p) = (p^k, p) \qquad\qquad \mathrm{type}(\wedge_k^o) = (o^k, o)$$
$$\mathrm{type}(c^p) = (po, p) \qquad\qquad\quad \mathrm{type}(c^o) = (op, o)$$
$$\mathrm{type}(b) = (\varepsilon, p), \ \textit{for } b \in \mathbb{B}^p \qquad \mathrm{type}(b) = (\varepsilon, o), \ \textit{for } b \in \mathbb{B}^o.$$

The elements of $\mathbb{B}^p$ and $\mathbb{B}^o$ are typed constants, which represent basic actions of the proponent and the opponent, respectively. The functions $\vee_k^p, \wedge_k^p, \vee_k^o, \wedge_k^o$ represent disjunctive ($\vee$) and conjunctive ($\wedge$) refinement operators of arity k, for a proponent (p) and an opponent (o), respectively. Whenever it is clear from the context, we omit the subscript k. The binary function c^s ('counter'), where $s \in S$, connects a term of the type s with a countermeasure. By T_Σ we denote the set of all ADTerms. We partition T_Σ into T_Σ^p (the set of terms of the proponent's type) and T_Σ^o (the set of terms of the opponent's type). To denote the type of a term, we define a function $\tau \colon T_\Sigma \to S$ by $\tau(t) = s$ if $t \in T_\Sigma^s$.

Example 1. The ADTerm $t = c^p(\wedge^p(E, F), \vee^o(G)) \in T_\Sigma^p$ is graphically displayed in Fig. 1 (left). For this ADTerm, we have $\tau(t) = p$. Subterms E and F are basic actions of the proponent's type, and G is a basic action of the opponent's type.

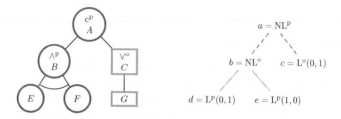

Fig. 1. An example of an ADTerm (left) and a two-player binary zero-sum extensive form game (right)

Assuming the proponent is the attacker, this means that the system can be attacked by combining the basic attack actions E and F. However the defender has the option to defend if he implements the basic defense action G.

In order to check whether an attack–defense scenario is feasible, we introduce the notion of satisfiability of an ADTerm by defining a satisfiability attribute sat. First, for player $s \in \{p, o\}$ we define a *basic assignment* for s as a function $\beta^s \colon \mathbb{B}^s \to \{true, false\}$. We gather the basic assignments for both players in a *basic assignment profile* $\beta = (\beta^p, \beta^o)$. Second, the function sat$\colon T_\Sigma \to \{true, false\}$ is used in order to calculate the satisfiability value of an ADTerm. It is defined recursively as follows

$$
\mathrm{sat}(t) = \begin{cases}
\beta^s(t^s), & \text{if } t = t^s \in \mathbb{B}^s, \\
\vee(\mathrm{sat}(t_1), \ldots, \mathrm{sat}(t_k)), & \text{if } t = \vee^s(t_1, \ldots, t_k), \\
\wedge(\mathrm{sat}(t_1), \ldots, \mathrm{sat}(t_k)), & \text{if } t = \wedge^s(t_1, \ldots, t_k), \\
\mathrm{sat}(t_1) \wedge \neg \mathrm{sat}(t_2), & \text{if } t = c^s(t_1, t_2).
\end{cases}
$$

For instance, consider the term t from Example 1 and the basic assignment profile $\beta = (\beta^p, \beta^o)$, where $\beta^p(E) = true$, $\beta^p(F) = true$, $\beta^o(G) = false$. We get sat$(t) = true$. Assuming the proponent is the attacker, this means that the basic defense action G is absent and the system is attacked by combining the basic attack actions E and F.

The next definition formalizes the notion of a satisfiable ADTerm for a player.

Definition 2. *For every player s, strategy β^s and strategy profile β, we define the sets of ADTerms* Sat$_\beta^s$, Sat$_{\beta^s}^s$, Sat$^s \subseteq T_\Sigma$ *in the following way. Let $t \in T_\Sigma$.*

- *$t \in$ Sat$_\beta^s$ if either $\tau(t) = s$ and sat$(t) = true$, or $\tau(t) = -s$ and sat$(t) = false$. In this case we say that s is successful in t under β.*
- *$t \in$ Sat$_{\beta^s}^s$ if $t \in$ Sat$_{(\beta^p, \beta^o)}^s$ for every basic assignment β^{-s}. In this case we say that s is successful in t under β^s.*
- *$t \in$ Sat^s if there exists a basic assignment β^s for player s such that $t \in$ Sat$_{\beta^s}^s$. In this case we say that t is satisfiable for s.*

Theorem 1. *For every ADTerm t, we have that every basic assignment profile β partitions T_Σ into Sat^p_β and Sat^o_β.*

Proof. This follows immediately from the first item in Definition 2.

2.2 Two-Player Binary Zero-Sum Extensive Form Games

We consider *two-player binary zero-sum extensive form games*, in which a proponent p and an opponent o play against each other. In those games, we allow only for the outcomes $(1,0)$ and $(0,1)$, where $(1,0)$ means that the proponent succeeds in his goal (breaking the system if he is the attacker, keeping the system secure if he is the defender), and $(0,1)$ means that the opponent succeeds. Note that the proponent is not necessarily the player who plays first in the game. Finally, we restrict ourselves to extensive form games, i.e., games in tree format. Our presentation of games differs from the usual one, because we present games as terms. This eases the transformation of games into ADTerms. We formalize games in the next definition, where L stands for a leaf and NL for a non-leaf of the term.

Definition 3. *Let $S = \{p, o\}$ denote the set of players and $\mathrm{Out} = \{(1,0), (0,1)\}$ the set of possible outcomes. A two-player binary zero-sum extensive form game is a term $t ::= \psi^p \mid \psi^o$, where*

$$\psi^p ::= \mathrm{NL}^p(\psi^o, \ldots, \psi^o) \mid \mathrm{L}^p(1,0) \mid \mathrm{L}^p(0,1)$$
$$\psi^o ::= \mathrm{NL}^o(\psi^p, \ldots, \psi^p) \mid \mathrm{L}^o(1,0) \mid \mathrm{L}^o(0,1).$$

We denote the set of all two-player binary zero-sum extensive form games by $\mathcal{G}$. We define the first player *of a game ψ^s as the function $\tau \colon \mathcal{G} \to S$ such that $\tau(\psi^s) = s$.*

Example 2. An example of a two-player binary zero-sum extensive form game is the expression $\mathrm{NL}^p(\mathrm{NL}^o(\mathrm{L}^p(0,1), \mathrm{L}^p(1,0)), \mathrm{L}^o(0,1))$. This game is displayed in Fig. 1 (right). When displaying extensive form games, we use dashed edges for choices made by the proponent, and solid edges for those made by the opponent. In this game, first the proponent can pick from two options; if he chooses the first option, the opponent can choose between outcomes $(0,1)$ and $(1,0)$. If the proponent chooses the second option, the game will end with outcome $(0,1)$.

Definition 4. *A function σ^s is a* strategy *for a game $g \in \mathcal{G}$ for player $s \in S$ if it assigns to every non-leaf of player s in g $\mathrm{NL}^s(\psi_1^{-s}, \ldots, \psi_n^{-s})$ a term ψ_k^{-s} for some $k \in \{1, \ldots, n\}$.*

A strategy profile *for a game $g \in \mathcal{G}$ is a pair $\sigma = (\sigma^p, \sigma^o)$, where σ^p is a strategy of g for p, and σ^o a strategy of g for o.*

If $g = \mathrm{NL}^s(\psi_1^{-s}, \ldots, \psi_n^{-s})$ and $\sigma = (\sigma^{\mathrm{P}}, \sigma^{\mathrm{o}})$, sometimes we abuse notation and write $\sigma(g) = \psi_k^{-s}$ where $\psi_k^{-s} = \sigma^s(g)$.

Now we define the outcome of a game in three steps.

Definition 5. *We say that* $(0, 1) \leq^{\mathrm{P}} (1, 0)$ *and* $(1, 0) \leq^{\mathrm{o}} (0, 1)$, *so that* $(\mathrm{Out}, \leq^{\mathrm{P}})$ *and* $(\mathrm{Out}, \leq^{\mathrm{o}})$ *are totally ordered sets. Let* $(r^{\mathrm{P}}, r^{\mathrm{o}})$ *be an element of* Out, *and* $\psi_1^{-s}, \ldots, \psi_n^{-s}$ *be games with player* $-s$ *as the first player.*

1. *The* outcome $\mathrm{out}_{(\sigma^{\mathrm{P}}, \sigma^{\mathrm{o}})} \colon \mathcal{G} \to \mathrm{Out}$ *of a game* g *under strategy profile* $\sigma = (\sigma^{\mathrm{P}}, \sigma^{\mathrm{o}})$ *is defined by:*

$$\mathrm{out}_{(\sigma^{\mathrm{P}}, \sigma^{\mathrm{o}})}(\mathrm{L}^s(r^{\mathrm{P}}, r^{\mathrm{o}})) = (r^{\mathrm{P}}, r^{\mathrm{o}})$$

$$\mathrm{out}_{(\sigma^{\mathrm{P}}, \sigma^{\mathrm{o}})}(\mathrm{NL}^s(\psi_1^{-s}, \ldots, \psi_n^{-s})) = \mathrm{out}_{(\sigma^{\mathrm{P}}, \sigma^{\mathrm{o}})}(\sigma^s(\mathrm{NL}^s(\psi_1^{-s}, \ldots, \psi_n^{-s})))$$

2. *The* outcome $\mathrm{out}_{\sigma^s} \colon \mathcal{G} \to \mathrm{Out}$ *of a game* g *under strategy* σ^s *is defined by:*

$$\mathrm{out}_{\sigma^s}(\mathrm{L}^s(r^{\mathrm{P}}, r^{\mathrm{o}})) = (r^{\mathrm{P}}, r^{\mathrm{o}})$$

$$\mathrm{out}_{\sigma^s}(\mathrm{NL}^s(\psi_1^{-s}, \ldots, \psi_n^{-s})) = \mathrm{out}_{\sigma^s}(\sigma^s(\mathrm{NL}^s(\psi_1^{-s}, \ldots, \psi_n^{-s})))$$

$$\mathrm{out}_{\sigma^s}(\mathrm{NL}^{-s}(\psi_1^{-s}, \ldots, \psi_n^{-s})) = \max_{1 \leq i \leq n} {}^{\leq^{-s}} \{\mathrm{out}_{\sigma^s}(\psi_i^{-s})\}$$

3. *The* outcome $\mathrm{out} \colon \mathcal{G} \to \mathrm{Out}$ *of a game* g *is defined by:*

$$\mathrm{out}(\mathrm{L}^s(r^{\mathrm{P}}, r^{\mathrm{o}})) = (r^{\mathrm{P}}, r^{\mathrm{o}})$$

$$\mathrm{out}(\mathrm{NL}^s(\psi_1^{-s}, \ldots, \psi_n^{-s})) = \max_{1 \leq i \leq n} {}^{\leq^s} \{\mathrm{out}_{\sigma^s}(\psi_i^{-s})\}$$

$$\mathrm{out}(\mathrm{NL}^{-s}(\psi_1^{-s}, \ldots, \psi_n^{-s})) = \max_{1 \leq i \leq n} {}^{\leq^{-s}} \{\mathrm{out}_{\sigma^s}(\psi_i^{-s})\}$$

Here $\mathrm{out}_{(\sigma^{\mathrm{P}}, \sigma^{\mathrm{o}})}$ denotes the outcome of the game when p and o play according to strategy σ^{P} and σ^{o}, respectively. Furthermore out_{σ^s} denotes the outcome if player s plays strategy σ^s, and player $-s$ tries to achieve the best possible outcome for himself. Finally, out denotes the outcome of the game if both players try to maximize their own outcome.

3   From Games to ADTerms

In this section, we show how to transform binary zero-sum two-player extensive form games into ADTerms. We define a function that transforms games into ADTerms, and a function that transforms a strategy for a game into a basic assignment for the corresponding ADTerm. First we show that the player who wins the game is also the player for whom the corresponding ADTerm is satisfiable, if both players play the basic assignment corresponding to their strategy in the game. Then we show that if a player has a strategy in a game which guarantees him to win, he is successful in the corresponding ADTerm under the corresponding basic assignment. For this purpose, we first define a function $[\cdot]_{\mathrm{AD}}$ that maps games into ADTerms.

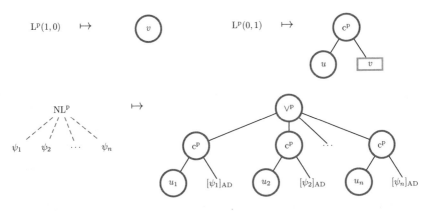

Fig. 2. Transformation of a game in extensive form into an ADTerm by function $[\cdot]_{AD}$

Definition 6. *Let v^s, u^s, and $u_1^s, \ldots, u_n^s$, for $s \in S$, represent fresh basic actions from $\mathbb{B}^s$. The function $[\cdot]_{AD} \colon \mathcal{G} \to T_\Sigma$ is defined in the following way.*

$$\mathrm{L}^\mathrm{P}(1,0) \mapsto v^\mathrm{P} \tag{1a}$$

$$\mathrm{L}^\mathrm{O}(1,0) \mapsto \mathrm{c}^\mathrm{o}(u^\mathrm{o}, v^\mathrm{P}) \tag{1b}$$

$$\mathrm{L}^\mathrm{P}(0,1) \mapsto \mathrm{c}^\mathrm{P}(u^\mathrm{P}, v^\mathrm{o}) \tag{1c}$$

$$\mathrm{L}^\mathrm{O}(0,1) \mapsto v^\mathrm{o} \tag{1d}$$

$$\mathrm{NL}^\mathrm{P}(\psi_1, \ldots, \psi_n) \mapsto \vee^\mathrm{P}(\mathrm{c}^\mathrm{P}(u_1^\mathrm{P}, [\psi_1]_{AD}), \ldots, \mathrm{c}^\mathrm{P}(u_n^\mathrm{P}, [\psi_n]_{AD})) \tag{1e}$$

$$\mathrm{NL}^\mathrm{O}(\psi_1, \ldots, \psi_n) \mapsto \vee^\mathrm{o}(\mathrm{c}^\mathrm{o}(u_1^\mathrm{o}, [\psi_1]_{AD}), \ldots, \mathrm{c}^\mathrm{o}(u_n^\mathrm{o}, [\psi_n]_{AD})). \tag{1f}$$

The rules for player p are visualized in Fig. 2 (the rules for player o are symmetric). The rules specify that a winning leaf for a player in the game is transformed into a satisfiable ADTerm for this player, i.e., an ADTerm consisting of only a leaf belonging to this player (Rule (1a)–(1d)), and that non-leaves in the game are transformed into disjunctive ADTerms of the same player (Rule (1e)–(1f)). These disjunctions have children of the form $\mathrm{c}^s(u_k^\mathrm{P}, [\psi_k]_{AD})$ for some k. The intended meaning here is that player s selects u_k^P exactly when his strategy selects ψ_k in the game. An example of a transformation of a game into an ADTerm is depicted in Fig. 3.

The resulting ADTerm is thus conjunction-free. Note that because terms in games alternate between p and o, this procedure results in valid ADTerms (i.e., in terms of the form $\mathrm{c}^s(u^{s_1}, v^{s_2})$, $s_1 = s$ and $s_2 = -s$, and disjunctive terms for player s have children for player s as well).

Now we define how to transform a strategy profile for a game into a basic assignment profile for an ADTerm. First we define a transformation $[\![\cdot]\!]_{AD}$ from a strategy σ^s ($s \in \{\mathrm{p}, \mathrm{o}\}$) for game g into a basic assignment $\beta^s = [\![\sigma^s]\!]_{AD}$ for ADTerm $[g]_{AD}$. Intuitively, if a player's strategy for the game selects a certain branch, the basic assignment for the ADTerm assigns *true* to the node u_k in the corresponding branch, and *false* to the nodes u_k in the other branches. Furthermore, ADTerms resulting from leaves in the game are always selected.

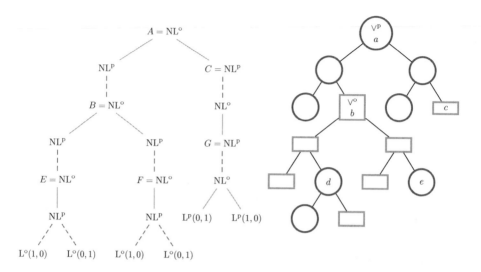

Fig. 3. The result of the transformation of the ADTerm from Fig. 1 into a game (left), and the game from Fig. 1 into an ADTerm (right)

Definition 7. *Let s be a player, g be a game and σ^s be a strategy of player s for g. The function $\beta^s = [\![\sigma^s]\!]_{AD}$ is defined as follows. For all ADTerms $c^s(u^s, v^{-s})$ and v^s resulting from the first four cases in Definition 6, we set $\beta^s(u^s) = \beta^s(v^s) = true$. For ADTerms obtained from game g by one of the last two cases in Definition 6, if $\sigma^s(g) = \psi_k$, we set $\beta^s(u_k^s) = true$ and $\beta^s(u_i^s) = false$ for $1 \leq i \leq n$, $i \neq k$.*

The strategy profile (β^P, β^o) can be transformed into a basic assignment profile by $[\![(\beta^P, \beta^o)]\!]_{AD} = ([\![\beta^P]\!]_{AD}, [\![\beta^o]\!]_{AD})$.

The next theorem states that a player is the winner in a game under a certain strategy profile if and only if he is successful in the corresponding ADTerm under the basic assignment profile corresponding to the strategy profile.

Theorem 2. *Let g be a game and σ a strategy profile for g. Then $\text{out}_\sigma(g) = (1,0)$ if and only if $[g]_{AD} \in \text{Sat}^P_{[\sigma]_{AD}}$.*

The following theorem states that a strategy in a game guarantees player s to win if and only if s is successful in the corresponding ADTerm under the corresponding basic assignment. Surprisingly, this is not a consequence of Theorem 2: there might be a basic assignment β^s for the ADTerm, for which there exists no strategy σ^s such that $\beta^s = [\![\sigma^s]\!]_{AD}$ (i.e, the function $[\![\cdot]\!]_{AD}$ is not surjective). Therefore it is not immediately clear that if a player has a strategy σ^s that wins from the other player independent of his strategy, a player with a basic assignment $[\![\sigma^s]\!]_{AD}$ wins from the other player independent of his basic assignment.

Theorem 3. *Let g be a game and σ^{p} be a strategy for p on g. Then* $\mathrm{out}_{\sigma^{\mathrm{p}}}(g) = (1,0)$ *if and only if* $[g]_{\mathrm{AD}} \in \mathrm{Sat}^{\mathrm{p}}_{[\sigma^{\mathrm{p}}]_{\mathrm{AD}}}$.

From this theorem, we immediately get the following corollary.

Corollary 1. *Whenever g is a game, $\mathrm{out}(g) = (1,0)$ if and only if $[g]_{\mathrm{AD}} \in \mathrm{Sat}^{\mathrm{p}}$.*

4   From ADTerms to Games

We proceed with the transformation in the other direction. We define two transformations, namely from ADTerms into games, and from basic assignment profiles into strategy profiles. Then we show that if a player has a basic assignment
for an ADTerm with which he is successful, the corresponding strategy in the
corresponding game guarantees him to win.

Definition 8. *We define a function $[\cdot]_{\mathrm{G}}$ from ADTerms to games as follows:*

$$v^{\mathrm{p}} \mapsto \mathrm{NL}^{\circ}(\mathrm{NL}^{\mathrm{p}}(\mathrm{L}^{\circ}(0,1), \mathrm{L}^{\circ}(1,0))) \tag{2a}$$

$$v^{\circ} \mapsto \mathrm{NL}^{\mathrm{p}}(\mathrm{NL}^{\circ}(\mathrm{L}^{\mathrm{p}}(1,0), \mathrm{L}^{\mathrm{p}}(0,1))) \tag{2b}$$

$$\vee^{\mathrm{p}}(\psi_1, \ldots, \psi_n) \mapsto \mathrm{NL}^{\circ}(\mathrm{NL}^{\mathrm{p}}([\psi_1]_{\mathrm{G}}, \ldots, [\psi_n]_{\mathrm{G}})) \tag{2c}$$

$$\vee^{\circ}(\psi_1, \ldots, \psi_n) \mapsto \mathrm{NL}^{\mathrm{p}}(\mathrm{NL}^{\circ}([\psi_1]_{\mathrm{G}}, \ldots, [\psi_n]_{\mathrm{G}})) \tag{2d}$$

$$\wedge^{\mathrm{p}}(\psi_1, \ldots, \psi_n) \mapsto \mathrm{NL}^{\circ}(\mathrm{NL}^{\mathrm{p}}([\psi_1]_{\mathrm{G}}), \ldots, \mathrm{NL}^{\mathrm{p}}([\psi_n]_{\mathrm{G}})) \tag{2e}$$

$$\wedge^{\circ}(\psi_1, \ldots, \psi_n) \mapsto \mathrm{NL}^{\mathrm{p}}(\mathrm{NL}^{\circ}([\psi_1]_{\mathrm{G}}), \ldots, \mathrm{NL}^{\circ}([\psi_n]_{\mathrm{G}})) \tag{2f}$$

$$\mathrm{c}^{\mathrm{p}}(\psi_1, \psi_2) \mapsto \mathrm{NL}^{\circ}(\mathrm{NL}^{\mathrm{p}}([\psi_1]_{\mathrm{G}}), [\psi_2]_{\mathrm{G}}) \tag{2g}$$

$$\mathrm{c}^{\circ}(\psi_1, \psi_2) \mapsto \mathrm{NL}^{\mathrm{p}}(\mathrm{NL}^{\circ}([\psi_1]_{\mathrm{G}}), [\psi_2]_{\mathrm{G}}) \tag{2h}$$

A graphical representation of the rules for player p is displayed in Fig. 4 (the
rules for player o are symmetric). It can easily be checked that this construction
guarantees valid games (in which p-moves and o-moves alternate). According
to these rules, we transform leaves for player s into two options for player s, a
losing and a winning one (Rules (2a) and (2b)). These choices correspond to not
choosing and choosing the leaf in the ADTerm, respectively. Disjunctive terms
for player s are transformed into choices for player s in the game (Rules (2c)
and (2d)). There is no direct way of representing conjunctions in games. We can
still handle conjunctive terms though, by transforming them into choices for the
other player (Rules (2e) and (2f)). This reflects the fact that a player can succeed
in all his options exactly when there is no way for the other player to pick an
option which allows him to succeed. Finally, countermeasures against player s
are transformed into a choice for player $-s$ (Rules (2g) and (2h)). Here, the first
option corresponds to player $-s$ not choosing the countermeasure, so that it is
up to player s whether he succeeds or not, while the second option corresponds
to player $-s$ choosing the countermeasure. The transformation of a game into
an ADTerm is illustrated in Fig. 3.

We proceed by defining a transformation $[\cdot]_{\mathrm{G}}$ from a basic assignment for an
ADTerm into a strategy for the corresponding game. We only give the definition
for $s = \mathrm{p}$; the definition for $s = \mathrm{o}$ is symmetric.

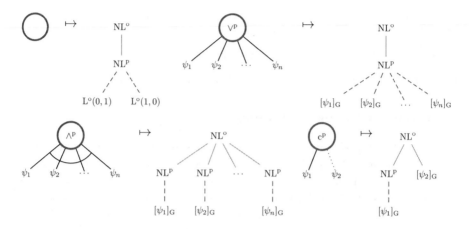

Fig. 4. Transformation of an ADTerm into a game by means of function $[\cdot]_G$

Definition 9. *Function* $[\![\cdot]\!]_G$ *is a transformation from a basic assignment* β^P *for ADTerm* t *into a strategy* $\sigma^P = [\![\beta^P]\!]_G$ *for the game* $[t]_G$. *If a (sub)term from* $[t]_G$ *is obtained by rule* (2n) *in Definition 8, then* σ^P *of that (sub)term is defined by rule* (3n) *in this definition.*

$$\sigma^P(NL^P(L^o(0,1), L^o(1,0))) \qquad = L^o(1,0) \quad if \ \beta^s(v) = true. \quad (3a)$$
$$= L^o(0,1) \quad otherwise.$$
$$\sigma^P(NL^P([\psi_1]_G, \ldots, [\psi_n]_G)) \qquad = [\psi_k]_G \qquad\qquad\qquad (3c)$$
$$where \ k \ is \ the \ smallest \ number \ such \ that \ \psi_k \in Sat^P_{\beta^P}.$$
$$= [\psi_1]_G$$
$$if \ there \ exists \ no \ such \ number.$$
$$\sigma^P(NL^P(NL^o([\psi_1]_G), \ldots, NL^o([\psi_n]_G))) = NL^o([\psi_k]_G) \qquad (3f)$$
$$where \ k \ is \ the \ smallest \ number \ such \ that \ \psi_k \in Sat^P_{\beta^P}.$$
$$= NL^o([\psi_1]_G)$$
$$if \ there \ exists \ no \ such \ number.$$
$$\sigma^P(NL^P(NL^o([\psi_1]_G), [\psi_2]_G)) \qquad = NL^o([\psi_1]_G) \ if \ \psi_2 \notin Sat^P_{\beta^P}. \quad (3h)$$
$$= [\psi_2]_G \qquad otherwise.$$

For Rules (3b), (3d), (3e) and (3g), σ^P *is trivially defined as there is only one refinement.*

Note that some of the rules, namely Rules (3c), (3f) and (3h), are non-local in the sense that we need to evaluate all subterms of the ADTerm before we can decide what to play in the game.

Theorem 4. *Let* t *be an ADTerm and* β^P *a basic assignment for* t. *Then* $t \in$ $Sat^P_{\beta^P}$ *if and only if* $out_{[\![\beta^P]\!]_G}([t]_G) = (1, 0)$.

Now we obtain immediately the following corollary by definition of out and Sat^P.

Corollary 2. *Whenever* t *is an ADTerm,* $t \in Sat^P$ *if and only if* $out([t]_G) = (1, 0)$.

5 Conclusion

We showed that attack–defense terms and binary zero-sum two-player extensive form games have equivalent expressive power when considering satisfiability, in the sense that they can be converted into each other while preserving their outcome. Moreover, the transformations preserved internal structure, in the sense that there exists injections between subterms in the game and subterms in the ADTerm such that if a player wins in the subterm of the game, the corresponding subterm in the ADTerm is satisfiable for this player, and vice versa. Therefore attack–defense trees with a satisfiability attribute and binary zero-sum two-player extensive form games can be seen as two different representations of the same concept. Both representations have their advantages. On the one hand, attack–defense trees are more intuitive, because conjunctions and refinements can be explicitly modeled. On the other hand, the game theory representation profits from the well-studied theoretical properties of games.

We saw that two notions in the domain of ADTerms had no direct correspondence to notions in the world of games: conjunctive nodes and refinements. The first problem has been solved by transforming conjunctive nodes for one player to disjunctive nodes for the other player. This also shows that, when considering the satisfiability attribute, the class of conjunction-free ADTerms has equal expressive power to the full class of ADTerms (note that the transformation from ADTerms into games and vice versa are not each other's inverse, i.e., $[[t]_{AD}]_G \neq t$ and $[[t]_G]_{AD} \neq t$). The second problem has been solved by adding extra dummy moves with only one option for the other player in between refining and refined nodes.

In the future, we plan to consider attack–defense trees accompanied with more sophisticated attributes, such that a larger class of games can be converted. An example of these are non-zero-sum games, where $(1, 1)$ can be interpreted as an outcome where both the attacker and the defender profit (for example, if the attacker buys his goal from the defender), and $(0, 0)$ as an outcome where both parties are damaged (when the attacker fails in his goal, but his efforts damage the defender in some way). Also the binary requirement can be lifted, so that the outcome of a player represents for instance the cost or gain of his actions. Furthermore, it would be interesting to look for a correspondence of incomplete and imperfect information in attack–defense trees.

Acknowledgments. The authors would like to thank Leon van der Torre and Wojciech Jamroga for valuable discussions on the topic of this paper.

References

1. Schneier, B.: Secrets and lies. Wiley, Indianapolis (2004)
2. Mauw, S., Oostdijk, M.: Foundations of Attack Trees. In: Won, D.H., Kim, S. (eds.) ICISC 2005. LNCS, vol. 3935, pp. 186–198. Springer, Heidelberg (2005)
3. Willemson, J., Jürgenson, A.: Serial Model for Attack Tree Computations. In: Lee, D., Hong, S. (eds.) ICISC 2009. LNCS, vol. 5984, pp. 118–128. Springer, Heidelberg (2010)

4. Rehák, M., Staab, E., Fusenig, V., Pěchouček, M., Grill, M., Stiborek, J., Bartoš, K., Engel, T.: Runtime Monitoring and Dynamic Reconfiguration for Intrusion Detection Systems. In: Kirda, E., Jha, S., Balzarotti, D. (eds.) RAID 2009. LNCS, vol. 5758, pp. 61–80. Springer, Heidelberg (2009)
5. Amenaza: SecurITree, http://www.amenaza.com/
6. Isograph: AttackTree+, http://www.isograph-software.com/atpover.htm
7. Edge, K.S., Dalton II, G.C., Raines, R.A., Mills, R.F.: Using Attack and Protection Trees to Analyze Threats and Defenses to Homeland Security. In: MILCOM, pp. 1–7. IEEE, Los Alamitos (2006)
8. Bistarelli, S., Dall'Aglio, M., Peretti, P.: Strategic Games on Defense Trees. In: Dimitrakos, T., Martinelli, F., Ryan, P.Y.A., Schneider, S. (eds.) FAST 2006. LNCS, vol. 4691, pp. 1–15. Springer, Heidelberg (2006)
9. Kordy, B., Mauw, S., Radomirović, S., Schweitzer, P.: Foundations of Attack–Defense Trees. In: FAST. LNCS. Springer, Heidelberg (2010), http://satoss.uni.lu/members/barbara/papers/adt.pdf
10. Kordy, B., Mauw, S., Melissen, M., Schweitzer, P.: Attack–defense trees and two-player binary zero-sum extensive form games are equivalent – technical report with proofs, http://arxiv.org/abs/1006.2732

Methods and Algorithms for Infinite Bayesian Stackelberg Security Games

(Extended Abstract)

Christopher Kiekintveld[1], Janusz Marecki[2], and Milind Tambe[3]

[1] University of Texas at El Paso, El Paso, TX
Department of Computer Science
cdkiekintveld@utep.edu
[2] IBM T.J. Watson Research Center, Yorktown Heights, NY
marecki@us.ibm.com
[3] University of Southern California, Los Angeles, CA
Department of Computer Science
tambe@usc.edu

Abstract. Recently there has been significant interest in applications of game-theoretic analysis to analyze security resource allocation decisions. Two examples of deployed systems based on this line of research are the ARMOR system in use at the Los Angeles International Airport [20], and the IRIS system used by the Federal Air Marshals Service [25]. Game analysis always begins by developing a model of the domain, often based on inputs from domain experts or historical data. These models inevitably contain significant uncertainty—especially in security domains where intelligence about adversary capabilities and preferences is very difficult to gather. In this work we focus on developing new models and algorithms that capture this uncertainty using continuous payoff distributions. These models are richer and more powerful than previous approaches that are limited to small finite Bayesian game models. We present the first algorithms for approximating equilibrium solutions in these games, and study these algorithms empirically. Our results show dramatic improvements over existing techniques, even in cases where there is very limited uncertainty about an adversaries' payoffs.

Keywords: Bayesian Stackelberg games, security games, approximate algorithms, sampling techniques.

1 Introduction

Game theory offers a powerful framework for modeling security decisions, both for protecting critical infrastructure [22,6] and computer networks [3,16,18]. In two recent real-world applications game theory is used to make security resource allocation decisions at the Los Angeles International Airport (LAX) [20] and for the Federal Air Marshals Service (FAMS) [25]. The game models at the heart of these systems capture the capabilities of both the police and the adversaries, as well as information about the potential consequences of different outcomes. An important consideration in these domains is that the security policy should be unpredictable, which comes naturally from the game analysis assuming a rational and adaptive adversary.

T. Alpcan, L. Buttyan, and J. Baras (Eds.): GameSec 2010, LNCS 6442, pp. 257–265, 2010.

Specifying an accurate game model to represent the domain is a crucial first step in any application of game theory. These models are typically based on input from domain experts, including the ones developed for the LAX and FAMS applications. Even though the models are based on the best available information they are *inherently uncertain*. In security games in particular it is very problematic to provide precise and accurate information about the preferences and capabilities of possible attackers. Our goal in this work is to develop new models and algorithms that explicitly reason about this uncertainty in the context of security games. Most game-theoretic models and solution algorithms make strong assumptions about perfect information and common knowledge. Bayesian games [11] are commonly used to represent uncertainty in games, but unfortunately the available algorithms for solving Bayesian games are limited. The best available algorithm for Bayesian security games is DOBSS [19] which applies to games with a finite number of attacker types. Unfortunately, this method does not scale well with the number of types and in practice can only solve relatively small games with few types.

We define a model for Bayesian Stackelberg Security Games with continuous payoff distributions for the attacker, leading to a infinite Bayesian game. For example, in this model we can represent the uncertainty that security forces have about the attacker's payoffs using normal distributions, or uniform distributions over an interval. Solving these game to find equilibrium solutions presents significant challenges; even for the finite case the problem is NP-hard [9], no exact method is known for the infinite case. We explore methods for approximating equilibrium solutions for these problems, and test these approaches experimentally. Specifically, we describe two different methods for computing the defender's optimal strategy. The first applies DOBSS to a find optimal solutions for games with a finite number of sample attacker types. The second uses replicator dynamics to approximate an optimal strategy using the Monte-Carlo sampling for approximating the attacker response. In this shortened version of the paper we present only an abbreviated summary of the main results; in a longer version we will present a full detailed experimental evaluation.

2   Related Work

Recent interest in applying game theory to security decision includes fielded applications at the Los Angeles International Airport [20] and the Federal Air Marshals Service [25], work on patrolling strategies for robots and unmanned vehicles [10,2,5], policy recommendations for protecting critical infrastructure [22,6], and applications in computer networks [3,16,18]. Bayesian games [11] are the dominant paradigm for modeling uncertainty in game theory, and there are many examples of specific games that have been solved analytically, including many types of auctions [14]. Unfortunately, algorithms for finding equilibria of Bayesian games are quite limited, and no general algorithms exist for infinite Bayesian games. Recent research efforts have focused primarily on developing approximation techniques [21,4,8]. Monte-Carlo sampling approaches similar to those we consider in our work have been applied to some kinds of auctions [7]. In addition, the literature on stochastic choice [15,17] studies problems that are simplified versions of the choices that attackers face in our model. Finally, the literature on robust optimization has also inspired distribution-free alternatives to Bayes-Nash equilibrium [1].

3 Bayesian Security Games

We define a new class of Bayesian Security Games, extending the model in Kiekintveld et. al. [12] to include uncertainty about the attacker's payoffs. The key difference between our model and existing approaches (such as in Paruchuri et. al [19]) is that we allow the defender to have a continuous distribution over the possible payoffs of the attacker. Previous models have restricted this uncertainty to a small, finite number of possible attacker types, limiting the kinds of uncertainty that can be modeled.

A security game has two players, a *defender*, Θ, and an *attacker*, Ψ, a set of *targets* $T = \{t_1, \ldots, t_n\}$ that the defender wants to protect (the attacker wants to attack) and a set of *resources* $R = \{r_1, \ldots, r_m\}$ (e.g., police officers) that the defender may deploy to protect the targets. Resources are identical in that any resource can be deployed to protect any target, and any resource provides equivalent protection. A defender's pure strategy, denoted σ_Θ, is a subset of targets from T with size less than or equal to m. An attacker's pure strategy, σ_Ψ, is exactly one target from T. Σ_Θ denotes the set of all defender's pure strategies and Σ_Ψ is the set of all attacker's pure strategies.

We model the game as a Stackelberg game [23] which unfolds as follows: (1) the defender commits to a mixed strategy δ_Θ that is a probability distribution over the pure strategies from Σ_Θ, (2) nature chooses a random attacker type $\omega \in \Omega$ with probability $Pb(\omega)$, (3) the attacker observes the defender's mixed strategy δ_Θ, and (4) the attacker responds to δ_Θ with a best-response strategy from Σ_Ψ that provides the attacker (of type ω) with the highest *expected* payoff given δ_Θ.

The payoffs for the defender depend on which target is attacked and whether the target is protected (covered) or not. Specifically, for an attack on target t, the defender receives a payoff $U_\Theta^u(t)$ if the target is uncovered, and $U_\Theta^c(t)$ if the target is covered. The payoffs for an attacker of type $\omega \in \Omega$ is $U_\Psi^u(t, \omega)$ for an attack on an uncovered target, and $U_\Psi^c(t, \omega)$ for an attack on a covered target. We assume that both the defender and the attacker know the above payoff structure exactly. However, the defender is uncertain about the attacker's type, and can only estimate the expected payoffs for the attacker. We choose not to model uncertainty that the attacker may have over the defender's payoffs because the attacker already observes the defender's strategy perfectly.

3.1 Bayesian Stackelberg Equilibrium

A Bayesian Stackelberg Equilibrium (BSE) for a security game consists of a strategy profile where every attacker type is playing a best-response to the defender strategy, and the defender is playing a best-response to the distribution of actions chosen by the attacker types. We first define the equilibrium condition for the attacker and then the equilibrium condition for the defender. We conveniently represent the defender's mixed strategy δ_Θ by the compact *coverage vector* $C = (c_t)_{t \in T}$ that gives the probabilities c_t that each target $t \in T$ is covered by at least one resource. Note that $\sum_{t \in T} c_t \leq m$

because there are m resources at the defender's disposal. In equilibrium each attacker type ω must best respond to the coverage C with a pure strategy $\sigma_\Psi^*(C, \omega)$ given by:

$$\sigma_\Psi^*(C, \omega) = \arg \max_{t \in T} (c_t \cdot U_\Psi^c(t, \omega) + (1 - c_t) \cdot U_\Psi^u(t, \omega)) \tag{1}$$

To define the equilibrium condition for the defender we first define the *attacker response function* $A(C) = (a_t(C))_{t \in T}$ that returns the probabilities $a_t(C)$ that each target $t \in T$ will be attacked, given the distribution of attacker types and a coverage vector C. Specifically:

$$a_t(C) = \int_{\omega \in \Omega} Pb(\omega) \mathbf{1}_t(\sigma_\Psi^*(C, \omega)) d\omega \tag{2}$$

where $\mathbf{1}_t(\sigma_\Psi^*(C, \omega))$ is the indicator function that returns 0 if $t = \sigma_\Psi^*(C, \omega)$ and 0 otherwise. Given the attacker response function $A(\cdot)$ and a set of all possible defender coverage vectors $\mathcal{C}$, the equilibrium condition for the defender is to execute its best-response mixed strategy $\delta_\Theta^* \equiv C^*$ given by:

$$\delta_\Theta^* = \arg \max_C \sum_{t \in T} a_t(C)(c_t \cdot U_\Theta^c(t) + (1 - c_t) \cdot U_\Theta^u(t)). \tag{3}$$

When the set of attacker types is infinite, calculating the attacker response function from Equation (2) is impractical. For this case we instead replace each payoff in the original model with a continuous distribution over possible payoffs. Formally, for each target $t \in T$ we replace values $U_\Psi^c(t, \omega)$, $U_\Psi^c(t, \omega)$ over all $\omega \in \Omega$ with two continuous probability density functions:

$$f_\Psi^c(t, r) = \int_{\omega \in \Omega} Pb(\omega) U_\Psi^c(t, \omega) d\omega \tag{4}$$

$$f_\Psi^u(t, r) = \int_{\omega \in \Omega} Pb(\omega) U_\Psi^u(t, \omega) d\omega \tag{5}$$

that represent the defender's *beliefs* about the attacker payoffs. For example, the defender expects with probability $f_\Psi^c(t, r)$ that the attacker receives payoff r for attacking target t when it is covered. This provides a convenient and general way for domain experts to express uncertainty about payoffs in the game model, whether due to their own beliefs or based on uncertain evidence from intelligence reports.

4   Solution Methods

To solve the model described in the previous section we need to find a Bayesian Stackelberg equilibrium, describing an optimal coverage strategy for the defender and the optimal response for every attacker type. If the space of possible attacker types is finite, an optimal defender strategy can be found using DOBSS [19]. Unfortunately, there are no known methods for finding exact equilibrium solutions for infinite Bayesian Stackelberg games, and DOBSS only scales to small numbers of types. Here we focus on methods for approximating solutions for infinite Bayesian Stackelberg games. The problem can be broken down into two parts:

1. Computing or estimating the attacker response function (Equation 2)
2. Optimizing over the space of defender strategies, given the attacker response function

Similarly to the previous work [13] we compute the attacker response function using Monte-Carlo sampling from the space of possible attacker types. In addition, we consider both optimal and approximate methods for optimizing the defender's strategy given the attacker response calculations. We now briefly describe these two methods for solving infinite Bayesian Stackelberg security games.

4.1 Sampled Bayesian ERASER

The first method we describe combines Monte-Carlo sampling from the space of attacker types with an exact optimization over the space of defender strategies. This approach is based on the DOBSS solver [19] for finite Bayesian Stackelberg games. However, we also incorporate several improvements from the ERASER solver [12] that offer faster solutions for the restricted class of security games. The resulting method can be encoded as a mixed-integer linear program (MIP), which we call *Bayesian ERASER* (not presented here due to space constraints).

To use Bayesian ERASER to approximate a solution for an infinite game we can draw a finite number of sample attacker types from the type distribution, assuming that each occurs with equal probability. The payoffs for each type are determined by drawing from the payoff distributions specified in Equations 4 and 5. This results in a constrained, finite version of the infinite game that can be solved using the Bayesian ERASER MIP. We refer to this method as *Sampled Bayesian ERASER* (SBE) and use SBE-x to denote this methods with x sample attacker types. Armantier et al. [4] develop an approach for approximated general infinite Bayesian games that relies on solving constrained versions of the original game. Given certain technical conditions, a sequence of equilibria of constrained games will converge to the equilibrium of the original game. Here, increasing the number of sample types corresponds to such a sequence of constrained games, so in the limit as the number of samples goes to infinity the equilibrium of SBE-∞ will converge to the true Bayes-Nash equilibrium.

4.2 Sampled Replicator Dynamics

The second approach we consider uses a local search method (replicator dynamics) to approximate the defender's optimal strategy, given the attacker response function. Given that we are already using sampling techniques to estimate the attacker response, it makes a great deal of sense to explore approximation methods for the defender optimization as well. This allows us to trade off whether additional computational resources should be devoted to improving the attacker response estimate, or improving the defender strategy. In our experimental results we show that this is key to scaling to large problem instances.

We implemented an approximation algorithm based on replicator dynamics [24], which we call *Sampled Replicator Dynamics* (SRD). Since this method is a form of local search, all we require is a black-box method to estimate the attacker response function, such as Monte-Carlo sampling. As above, we use SRD-x to denote the Monte-Carlo version of SRD with x sample attacker types. SRD proceeds in a sequence of iterations. At each step the current coverage strategy $C^n = (c_t^n)_{t \in T}$ is used to estimate the attacker response function, which in turn is used to estimate the expected payoffs for both players. A new coverage strategy $C^{n+1} = (c_t^{n+1})_{t \in T}$ is computed according to the replicator equation:

$$c_t^{n+1} \propto c_t^n \cdot (E_t(C) - U_\Theta^{min}), \tag{6}$$

where U_Θ^{min} represents the minimum possible payoff for the defender, and $E_t(C)$ is the expected payoff the defender gets for covering target t with probability 1 and all other targets with probability 0, given the estimated attacker response to C^n. The search runs for a fixed number of iterations, and returns the coverage vector with the highest expected payoff. We introduce a learning rate parameter α that interpolates between C^n and C^{n+1}, with C^{n+1} receiving weight α in the next population and C^n having weight $1 - \alpha$. Finally, we introduce random restarts to avoid becoming stuck in local optima. After initial experiments, we settled on a learning rate of $\alpha = 0.8$ and random restarts every 15 iterations, which generally yielded good results (though the solution quality was not highly sensitive to these settings).

5   Evaluation

We omit the majority of our evaluation due to space constraints, but present one result demonstrating the importance of modeling uncertainty rather than using a perfect-information approximation. We generate 500 random game instances with 5 targets and 1 defender resource. The defender's payoffs for a covered target are drawn from $U[0, 100]$, and the uncovered payoffs from $U[-100, 0]$. The attacker's payoffs are represented by Gaussian distributions, with mean values drawn from $U[-100, 0]$ for covered targets and $U[0, 100]$ for uncovered targets; we vary the standard deviation. A sample attacker type is defined by drawing one value from each of these distributions (two values for each target).

The baseline algorithm uses a single point to estimate each payoff, rather than a distribution. This is motivated by the standard methodology for eliciting game models from domain experts, where no information about the uncertainty of the parameters is included in the model. We model this with a perfect-information model where the attacker has only one type, corresponding to the mean value for each payoff distribution. This can be solved exactly using the SBE algorithm with a single attacker type, which we refer to as "SBE-Mean."

Figure 1 presents results for the solution quality for SBE-Mean, SBE, and SRD. We vary payoff uncertainty along the x-axis, measured by the standard deviation of the Gaussian distributions for the attacker payoffs (in the same units as the payoffs). We run each algorithm to generate a coverage strategy for the defender, and evaluate this coverage strategy against the true distribution of attacker types. Since we do not have a closed-from solution to compute this exactly, we rely on a very close approximation generated by sampling 10000 attacker types to evaluate the payoffs for each algorithm.

The expected payoffs are shown on the vertical axis. We run SBE with up to 7 sample types and SRD with up to 1000 due to large differences in the computational scalability of the algorithms. With only 7 types, SBE takes roughly 2 seconds to run, while SRD runs in less than half a second with 1000 types and 5000 search iterations.

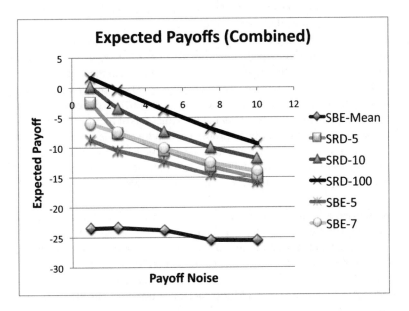

Fig. 1. Expected payoffs for SBE-Mean, SBE, and SRD with varying numbers of sample attacker types

In Figure 1 we see that the solution quality for both SBE and SRD is dramatically higher than the SBE-Mean baseline when there is payoff uncertainty, even if the uncertainty is relatively small. SBE and SRD show improvements over the baseline even with very small numbers of sample attacker types, with diminishing returns as the number of types increases. This is a strong indication that the perfect-information approach is not a good approximation for security games with uncertainty about the attacker's payoffs. SBE and SRD represent the first steps towards more robust methods that give high-quality solutions even when there is payoff uncertainty.

6   Conclusion

Stackelberg games are increasingly important in the analysis of a broad range of security domains, including deployed applications. The existing method to model uncertainty in these games are restricted to simple games with small, finite numbers of attacker types. We develop a class of infinite Bayesian Stackelberg security games in which attacker payoffs are provided as distributions, rather than point estimates of the payoffs. These games are able to more accurately capture the real payoff uncertainties in security domains, but present new computational challenges. We develop methods for approximating equilibrium solutions for these games using sampling and local search techniques. The SBE method exploits existing techniques for finite Bayesian Stackelberg games to solve constrained version of the infinite games. The SRD method combines replicator dynamics for searching the space of defender strategies with Monte-Carlo sampling techniques to estimate the attacker response function.

Our first important finding is that the baseline method that ignores uncertainty yields very poor results. *Modeling payoff uncertainty is critical in security games.* Both SBE and SRD give solutions with dramatically higher quality than the mean-approximation benchmark, even with just a few sample attacker types. The second major finding is that approximating the defender strategy enables scaling to much larger games and improves overall solution quality by enabling better approximations of the attacker response function. SRB is able to scale to very large problems (hundreds of targets) while using many more sample types than SBE. These results have immediate implications for the use of game theory in security domains, and open an exciting new research area in developing better approximation methods for Bayesian security games.

Acknowledgments. This research was supported by the United States Department of Homeland Security through the Center for Risk and Economic Analysis of Terrorism Events (CREATE) under grant number 2007- ST-061-000001. However, any opinions, conclusions or recommendations in this document are those of the authors and do not necessarily reflect views of the Department of Homeland Security. Janusz Marecki was supported in part by the DARPA GALE project, Contract No. HR0011-08-C-0110.

References

1. Aghassi, M., Bertsimas, D.: Robust game theory. Mathematical Programming: Series A and B 107(1), 231–273 (2006)
2. Agmon, N., Kraus, S., Kaminka, G.A., Sadov, V.: Adversarial uncertainty in multi-robot patrol. In: IJCAI (2009)
3. Alpcan, T., Basar, T.: A game theoretic approach to decision and analysis in network intrusion detection. In: Proc. of the 42nd IEEE Conference on Decision and Control, pp. 2595–2600 (2003)
4. Armantier, O., Florens, J.P., Richard, J.F.: Approximation of Bayesian Bash equilibrium. Journal of Applied Econometrics 23(7), 965–981 (2008)
5. Basiloco, N., Gatti, N., Amigoni, F.: Leader-follower strategies for robotic patrolling in environments with arbitrary topologies. In: AAMAS (2009)
6. Bier, V.M.: Choosing what to protect. Risk Analysis 27(3), 607–620 (2007)
7. Cai, G., Wurman, P.R.: Monte Carlo approximation in incomplete information, sequential auction games. Decision Support Systems 39(2), 153–168 (2005)
8. Ceppi, S., Gatti, N., Basilico, N.: Computing Bayes-Nash equilibria through support enumeration methods in Bayesian two-player strategic-form games. In: Proceedings of the ACM/IEEE International Conference on Intelligent Agent Technology (IAT), Milan, Italy, pp. 541–548 (September 15-18, 2009)
9. Conitzer, V., Sandholm, T.: Computing the optimal strategy to commit to. In: ACM EC, pp. 82–90 (2006)

10. Gatti, N.: Game theoretical insights in strategic patrolling: Model and algorithm in normal-form. In: ECAI, pp. 403–407 (2008)
11. Harsanyi, J.C.: Games with incomplete information played by Bayesian players (parts i–iii). Management Science 14 (1967-1968)
12. Kiekintveld, C., Jain, M., Tsai, J., Pita, J., Ordóñez, F., Tambe, M.: Computing optimal randomized resource allocations for massive security games. In: AAMAS (2009)
13. Kiekintveld, C., Marecki, J., Tambe, M.: Robust Bayesian methods for Stackelberg security games. In: Proceedings of the Ninth International Joint Conference on Autonomous Agents and Multi-agent systems (2010)
14. Krishna, V.: Auction Theory. Academic Press, London (2002)
15. Luce, R.D., Raiffa, H.: Games and Decisions. John Wiley and Sons, New York (1957); Dover republication (1989)
16. wei Lye, K., Wing, J.M.: Game strategies in network security. International Journal of Information Security 4(1-2), 71–86 (2005)
17. McFadden, D.: Quantal choice analysis: A survey. Annals of Economic and Social Measurement 5(4), 363–390 (1976)
18. Nguyen, K.C., Basar, T.A.T.: Security games with incomplete information. In: Proc. of IEEE Intl. Conf. on Communications, ICC 2009 (2009)
19. Paruchuri, P., Pearce, J.P., Marecki, J., Tambe, M., Ordonez, F., Kraus, S.: Playing games with security: An efficient exact algorithm for Bayesian Stackelberg games. In: AAMAS, pp. 895–902 (2008)
20. Pita, J., Jain, M., Western, C., Portway, C., Tambe, M., Ordonez, F., Kraus, S., Paruchuri, P.: Deployed ARMOR protection: The application of a game-theoretic model for security at the Los Angeles International Airport. In: AAMAS (Industry Track) (2008)
21. Reeves, D.M., Wellman, M.P.: Computing best-response strategies in infinite games of incomplete information. In: UAI (2004)
22. Sandler, T., Daniel, G., Arce, M.: Terrorism and game theory. Simulation and Gaming 34(3), 319–337 (2003)
23. von Stackelberg, H.: Marktform und Gleichgewicht. Springer, Vienna (1934)
24. Taylor, P., Jonker, L.: Evolutionary stable strategies and game dynamics. Mathematical Biosciences 16, 76–83 (1978)
25. Tsai, J., Rathi, S., Kiekintveld, C., Ordóñez, F., Tambe, M.: IRIS - A tools for strategic security allocation in transportation networks. In: AAMAS (Industry Track) (2009)

Author Index

Printing: Mercedes-Druck, Berlin
Binding: Stein+Lehmann, Berlin